# ENGINEERING YOUR FUTURE

## *A Short Course*

**Authors:**

**William C. Oakes, PhD** ..........................................*Purdue University*

**Les L. Leone, PhD** ................................... *Michigan State University*

**Craig J. Gunn, MS**.................................... *Michigan State University*

Contributors:

John B. Dilworth, PhD ...............................................................*Western Michigan University*
Heidi A. Diefes, PhD ...............................................................................*Purdue University*
Ralph E. Flori, PhD ....................................................................*University of Missouri-Rolla*

**GREAT LAKES PRESS, INC.**
**Okemos, MI     St. Louis, MO**
P.O. Box 550, Wildwood, MO 63040
(636) 273-6016
custserv@glpbooks.com   www.glpbooks.com

International Standard Book Number:  10: 1-881018-51-2
13: 978-1-881018-34-6

*All comments and inquiries should be addressed to:*

Great Lakes Press, Inc.
c/o John Gruender
PO Box 550
Wildwood, MO 63040-0550

Phone (636) 273-6016
Fax (636) 273-6086

www.glpbooks.com
custserv@glpbooks.com

Library of Congress Card Number:  2003102024

Printed in the USA by Sheridan Books, Inc. of Ann Arbor, Michigan.

10 9 8 7 6 5 4 3 2

# Important Information

**This book belongs to:** _____

_____

_____

_____

**phone:** _____

# Table of Contents

# Preface

You can't make an educated decision about what career to pursue without adequate information. *Engineering Your Future* endeavors to give you an introduction to the study and practice of engineering. In addition to presenting vital information, we've tried to make it interesting and easy to read as well.

You might find Chapter 3, *Profiles of Engineers*, to be of particular interest to you. The chapter includes information from real people—engineers practicing in the field. They discuss their jobs, their lives, and the things they wish they had known going into the profession. Chapter 2, *Engineering Majors*, also should be a tremendous help to you in determining what areas of engineering are most appealing to you at this stage of your education.

The rest of the book presents such things as a historical perspective of engineering; some thoughts about the future of the profession; some tips on how best to succeed in the classroom; advice on how to gain actual, hands-on experience; exposure to computer-aided design; and a nice introduction to several areas essential to the study and practice of engineering.

As you contemplate engineering, you should consider the dramatic impact engineers have had on our world. Note the eloquent words of American Association of Engineering Societies Chair Martha Sloan, a professor of electrical engineering at Michigan Technological University:

> *"In an age when technology helps turn fantasy and fiction into reality, engineers have played a pivotal role in developing the technologies that maintain our nation's economic, environmental and national security. They revolutionized medicine with pacemakers and MRI scanners. They changed the world with the development of television and the transistor, computers and the Internet. They introduced new concepts in transportation, power, satellite communications, earthquake-resistant buildings, and strain-resistant crops by applying scientific discoveries to human needs.*
>
> *"Engineering is sometimes thought of as applied science, but engineering is far more. The essence of engineering is design and making things happen for the benefit of humanity."*

Joseph Bordogna, President of IEEE, adds:

> *"Engineering will be one of the most significant forces in designing continued economic development and success for humankind in a manner that will sustain both the planet and its growing population. Engineers will develop the new processes and products. They will create and manage new systems for civil infrastructure, manufacturing, communications, health care delivery, information management, environmental conservation and monitoring, and everything else that makes modern society function."*

We hope that you, too, will find the field of engineering to be attractive, meaningful, and exciting—one that promises to be both challenging and rewarding, and one that matches well with your skills and interests.

You may be interested in knowing who authored each chapter. Dr. Oakes wrote the chapters on Majors, Classroom Success, and Problem Solving. Dr. Leone wrote Employment Statistics, Design, and Work Experience. Mr. Gunn wrote History and Communication. Dr. Diefes wrote Computer Tools. Dr. Flori wrote Teamwork. Dr. Dilworth wrote Ethics. A huge thanks is due Mr. John Gruender, executive editor of Great Lakes Press. His efforts contributed significantly to the final content and format of this book.

If you have comments or suggestions for us, please contact our editor, John Gruender, at jg@glpbooks.com or call (800) 837-0201. We would greatly appreciate your input.

*—The Authors*

# 1. The History of Engineering

## 1.1 Introduction

Engineering is a continuum, where every new innovation stands on the shoulders of those who have gone before—perhaps including yours some day. How do you get the necessary insight to participate? Look to the stories of history.

As we begin, you will notice that there were few engineering innovations in the early years. As time passed, innovations occurred more quickly. Today, engineering discoveries are made almost daily. The speed with which things now change indicates the urgency of understanding the process of innovation. But if we don't maintain a big-picture awareness regarding our projects, we may be surprised by unintended consequences. History provides a great opportunity to observe the context of before, during and after of some of the greatest engineering problems ever faced—and of those we face today.

*The study of history is the best way to know where we have been, where we are, and where we are going.*

A main thing to realize is that history is not about memorizing names and dates. History is about people. The insights you'll gain from stories about how engineers developed everything from kitchen appliances to high-tech industrial equipment can be very motivating. And you'll really be able to relate to the history of bridges, for instance, after you've been assigned to build a model bridge in class yourself!

Engineers are professionals. Professionals are leaders. To lead you need to understand the origins of your profession. The stories of history give you the foundation you need, as you learn from the great innovators and see how they handled all aspects of problem-solving. Such broad learning can greatly help your career and aid your development as a leader in your profession.

Many types of professionals are required to master the history of their trade as part of their degrees. The best professionals keep studying throughout their careers. Reading papers, magazines, and journals is simply contemporary "historic" study. This habit builds on their foundational knowledge of history. Such a background gives professionals their best chance to know what it takes to move forward in their field—an amazing challenge, as the stories to come will show!

The study of history, of course, not only helps us create new futures, but it also helps us understand what good qualities from the past are worth emulation. Craftsmanship, integrity, and dedication are clearly evident in our forebears' engineering artifacts. And history is full of interesting, educational stories, characters, and ingenious development. You can learn what it means to do quality work in a quality way, no matter what is the level of your contribution to the profession.

*"What news! How much more important to know what that is which was never old!"*
*—Henry David Thoreau*

## Definition of Engineering

Even if you already have a general knowledge of what engineering involves, a look at the definition of the profession may give you more insight. ABET—The Accreditation Board for Engineering and Technology—defines engineering as:

> *The profession in which knowledge of the mathematical and natural sciences, gained by study, experience, and practice, is applied with judgment to develop ways to use, economically, the materials and forces of nature for the benefit of mankind.*

In simple terms, engineering is about using natural materials and forces for the good of mankind—a noble endeavor. This definition places three responsibilities on an engineer: (1) to develop judgment so that you can (2) help mankind in (3) thrifty ways. Looking at case *histories* and *historic* overviews might help us achieve the insight needed to fulfill those responsibilities.

An engineering professor once said that the purpose of an engineer is *"to interpret the development and activity of man."* Technical coursework teaches us skills, but history can teach us how to *interpret* scenarios, and to sort out the pros and cons of various options. History helps us forge bonds of fellowship that connect us to the past and inspire us to be our best for tomorrow. A solid knowledge of history transforms our schooling from training to true education.

## 1.2 Getting Started

As we proceed, you should note that the engineer has always had a monumental impact on the human race at every stage of societal development. The few items mentioned here are only the tip of the iceberg when it comes to the contributions that engineers have made to the progress of humanity.

## Prehistoric Culture

If you look back at the definition of engineering given by ABET, you will notice an important statement: "The profession in which knowledge of the mathematical and natural sciences...is applied...." Individuals involved during prehistoric times in activities which we recognize today as engineering—problem-solving, tool-making, etc.—did not have a grasp of mathematical principles nor knowledge of natural science *as we know it today*. They designed and built needed items by trial and error and intuition. They built some spears that worked and some that failed, but in the end they perfected weapons that allowed them to bring down game animals and feed their families. Since written communication and transportation did not exist at that time, little information or innovation was exchanged with people from faraway places. Each group inched ahead on its own.

However, the innovators of yore would have made fine engineers today. Even in light of their limited skill, their carefully cultivated knowledge of their surroundings was more extensive than we typically can comprehend. Their skill in craftsmanship was often marvelous in its effectiveness, integrity, and intricacy. They passed on knowledge of all aspects of life, which they typically treated as a whole entity, with utmost seriousness to the next generation. This information was carefully memorized and kept accurate, evolving, and alive. Early man even tried to pass on vital information by way of coded cave paintings and etchings as an extra safeguard. Breakthroughs in transportation and exploration are being located ever earlier as we continue to make discoveries about various peoples traveling long before we thought they did— influencing others and bringing back knowledge. We can still see something of the prehistoric approach in some of today's cultures, such as Native Americans, aborigines, and others.

However, despite the strong qualities of prehistoric man—the importance placed on respect for life and a sense of the sacred—their physical existence was harshly limited. The work of engineers can be seen in this light as a quest to expand outer capacity without sacrificing inner integrity. The physical limitations of prehistoric man can perhaps be highlighted as follows. (How does engineering impact each of the areas listed?)

Physical limitations of prehistoric cultures:
- They had no written language.
- Their verbal language was very limited.
- They had no means of transportation.
- They had no separate concept of education or specialized methodology to discover new things.
- They lived by gathering food and trying to bring down game with primitive weapons.
- Improvement of the material aspects of life came about very slowly, with early, primitive engineering.

**Our Computer Age**

You, on the other hand, live in the information age. With such tools as the Internet, the answers to millions of questions are at your fingertips. We have traveled to the moon and our robots have crawled on Mars. Our satellites are exploring the ends of our known universe. Change is no longer questioned; it is expected. We excitedly await the next model in a series, knowing that as the current model is being sold it has already become obsolete. Speed, furious activity, and the compulsion to never sit still are part of our everyday lives. Let's look at the times when constant change was not the norm. As we progress from those primitive times into the 20th century, you will observe that fury of engineering activity.

We will present a panoramic view of engineering by briefly stating some of the more interesting happenings during specific time periods. Notice the kinds of innovations that were introduced. Take a careful look at the relationships that many inventions had with each other. Think about the present, and the connectivity between all areas of engineering and the critical importance of the computer. Innovations do not happen in a vacuum; they are interrelated with the needs and circumstances of the world at the time.

## Activity 1.1

Prepare a report that focuses on engineering in one of the following eras. Analyze the events that you consider to be engineering highlights and explain their importance to the progress of man.

a) Prehistoric man

b) Egypt and Mesopotamia

c) Greece and Rome

d) Europe in the Middle Ages

e) Europe in the Industrial Revolution

f) The 20th century

The assignment above might seem overwhelming—covering an enormous amount of time and information. Actually, the key developments might be simpler to identify than you think. In the pages that follow, however, we will only set the stage for your investigation of your profession. It will be your job to fill in the details for your particular discipline.

## The Speed of History

The rate of innovation brings up some interesting points. As we move through the past 6000 years, you should realize that the rate at which we currently introduce innovations is far more rapid than in the past. It used to take years to accomplish tasks that today we perform in a very short time—tasks that we simply take for granted. Think about the last time that your computer was processing slower than you thought it should. Your words may be echoing now: "Come on! Come on! I don't have all day!" In the past, there were often decades without noticeable technological progress. Think of the amount of time that it takes to construct a building today with the equipment that has been developed by engineers. It is not uncommon to see a complete house-frame constructed in a single day. Look in your history books and read about the time it took to create some of the edifices in Europe. You can visit churches today that took as long as 200 years to construct. Would we ever stand for that today?

*The ancients accomplished fantastic feats with only a rudimentary knowledge.*

Now, shift gears and evaluate the *purpose* behind the monumental efforts of the "slow" past. Was their only goal "to get the job done"? Ask them! Look into their stories. There you'll find coherent, colorful explanations for the case of keeping all aspects of culture connected to the main goals of life. The proper connection to God, truth, justice, fate, reality, life, and ancestry was the goal of early science and of many cultures. Not much was allowed to interfere. Even so, the ancients accomplished fantastic feats with only a rudimentary knowledge of the principles you learned as a child. But do we know what else they knew? Perhaps their lack of physical speed was partly voluntary! Perhaps it was surpassed by strengths and insights in other areas. Perhaps today's prowess comes at the loss of other qualities. We need to study history so that we can avoid making Faustian bargains. (Hey! Faust is a character from literary history!) Archimedes, for instance, refused to release information that could be used to make more effective weapons; he knew it would be used for evil and not the pursuit of wisdom; only when his home city of Syracuse was no longer able to hold off Roman attackers did he release his inventions to the military.

The study of history confronts us with dilemmas. Neither side of a true dilemma ever goes away. The story of history is never over. Speed is relative, after all!

Let's begin with a quick review of the history of engineering—six thousand years in a single paragraph...

Our technological roots can be traced back to the seed gatherer/hunter. Prehistoric man survived by collecting seeds and killing what animals he could chase down. He endured a very lean physical existence. As he gradually improved his security through innovation, humanity increased in numbers, and it was important to find ways to feed and control the growing population. To support larger populations reliably, the methods and implements of farming and security were improved. Much later, after many smaller innovations, specialized industrial man stepped onto the scene, ready to bring the world productivity and material wealth. The pursuit of science prospered due to its usefulness and profitability. Much abstract research was done with surplus funding. The microchip was developed. And today, the resulting Technological/Post-Industrial/Computer-Information Man is **you**—ready to use the vast information available in the present day to build the world of the future.

Most innovation would not have come to pass were it not for the work of engineers. The sections that follow will present a brief look at some of the highlights of those 6000 years. Spend some time poring over a few classic history texts to get the inside stories on the innovations that interest you the most. Though you might have interest in one particular field of engineering, you might find stories of innovation and discovery in other disciplines to be equally inspiring.

## 1.3 The Beginnings of Engineering

The foundations of engineering were laid with our ancestors' effort to survive and to improve their quality of life. From the beginning they looked around their environment and saw areas where life could be made easier and more stable. They found improved ways to hunt and fish. They discovered better methods for providing shelter for their families. Their main physical concern was day-to-day survival. As life became more complicated and small collections of families became larger communities, the need grew to look into new areas of concern: power struggles, acquisition of neighboring tribes' lands, religious observances. All of these involved work with tools. Engineering innovations were needed to further these interests. Of course, in those days projects weren't thought of as separate from the rest of life. In fact, individuals weren't generally thought of as being separate from their community. They didn't look at life from the point of view of specialties and individual interests. Every person was an engineer to an extent.

Modern aborigines still live today much as their ancestors did in prehistoric times. However, frequently even they take advantage of modern engineering in the form of tools, motors and medicine. The Amish can also fall into this category of being a roots-oriented culture. Such cultures tend to use tools only for physical necessities so that the significance of objects doesn't pollute their way of life. This struggle, as we know from contemporary "historic" media reports, has only been partly successful, and has caused conflict and misunderstanding on occasion.

## Egypt and Mesopotamia

As cities grew and the need for addressing the demands of the new fledgling societies increased, a significant change took place. People who showed special aptitude in certain areas were identified and assigned to ever more specialized tasks. This labeling and grouping was a scientific breakthrough. It gave toolmakers the time and resources to dedicate themselves to building and innovation. This new social function created the first real engineers, and for the first time innovation flourished rapidly.

Between 4000 and 2000 B.C., Egypt and Mesopotamia were the focal points for engineering activity. Stone tools were developed to help man in his quest for food. Copper and bronze axes were perfected through smelting. These developments were not only aimed at hunting. The development of the plow was allowing man to become a farmer so that he could reside in one place and leave the nomadic life. Mesopotamia also made its mark on engineering by giving birth to the wheel, the sailing boat, and methods of writing. Engineering skills that were applied to the development of everyday items immediately improved life as they knew it. We will never be able to understand completely the vast importance of the Greeks, Romans, and Egyptians in the life of the engineer.

*We will never be able to understand completely the vast importance of the Greeks, Romans, and Egyptians in the life of the engineer.*

Figure 1.1  The stepped Pyramid of Sakkara.

During the construction of the Pyramids (c.2700 – 2500 B.C.) the number of engineers required was immense. They had to make sure that everything fit correctly, that stones were properly transported long distances, and that the tombs would be secure against robbery. Imhotep (chief engineer to King Zoser) was building the stepped pyramid at Sakkara (pictured in Fig. 1.1) in Egypt about 2700 B.C. The more elaborate Great Pyramid of Khufu (pictured in Fig. 1.2) would come about 200 years later. The story of the construction of the pyramids is one that any engineer would appreciate, so consider doing some research on the Pyramids on your own. Or perhaps some day you'll have a chance to take a trip to Egypt to see them for yourself. By investigating the construction of the Pyramids, you will receive a clear and fascinating education about the need for designing, building, and testing with any engineering project. These early engineers, using simple tools, performed with great acuity, insight, and technical rigor, tasks that even today give us a sense of pride in their achievements.

The Great Pyramid of Khufu (pictured in Fig. 1.2) is the largest masonry structure ever built. Its base measures 756 feet on each side. The 480-foot structure was constructed of over 2.3 million limestone blocks with a total weight of over 58,000,000 tons. Casing blocks of fine limestone were attached to all four sides. These casing stones, some weighing as much as 15 tons, have been removed over the centuries for a wide variety of other uses. It is hard for us to imagine the engineering expertise needed to quarry and move these base and casing stones, and then piece them together so that they would form the pyramid and its covering.

Figure 1.2  The Great Pyramid of Khufu.

Here are additional details about this pyramid given by Roland Turner and Steven Goulden in *Great Engineers and Pioneers in Technology, Volume 1: From Antiquity through the Industrial Revolution:*

Buried within the pyramid are passageways leading to a number of funeral chambers, only one of which was actually used to house Khufu's remains. The granite-lined King's Chamber, measuring 17 by 34 feet, is roofed with nine slabs of granite which weigh 50 tons each. To relieve the weight on this roof, located 300 feet below the apex of the pyramid, the builder stacked five hollow chambers at short intervals above it. Four of the "relieving chambers" are roofed with granite lintels, while the topmost has a corbelled roof. Although somewhat rough and ready in design and execution, the system effectively distributes the massive overlying weight to the sturdy walls of the King's Chamber.

Sheer precision marks every other aspect of the pyramid's construction. The four sides of the base are practically identical in length—the error is a matter of inches—and the angles are equally accurate. Direct measurement from corner to corner must have been difficult, since the pyramid was built on the site of a rocky knoll (now completely enclosed in the structure). Moreover, it is an open question how the builder managed to align the pyramid almost exactly north-south. Still, many of the techniques used for raising the pyramid can be deduced.

After the base and every successive course was in place, it was leveled by flooding the surface with Nile water, no doubt retained by mud banks, and then marking reference points of equal depth to guide the final dressing. Complications were caused by the use of blocks of different heights in the same course.

The above excerpt mentions a few of the fascinating details of the monumental job undertaken to construct a pyramid with primitive tools and only human labor.

# 1.4 The Overview Approach

The following sections will give you a feel for what was going on from 2000 B.C. to the present. In this section we will review one specific engineering feat of each of the cultures of ancient Greece, Rome, and China. These overviews are meant to demonstrate the effort that went into engineering activities of the past. They represent only a small portion of the many developments of the time. It is important for you to look closely at what was being accomplished and the impact it had upon the people of the time. We will mention a few of those accomplishments, but it is important that you consider on your own what good came from the activities of those early engineers. How did their innovations set the groundwork for what was to come?

Figure 1.3  The Parthenon in Athens.

**Engineering the Temples of Greece**

The Parthenon, shown in Fig. 1.3, was constructed by Iktinos in Athens in 447 B.C. and was completed by 438 B.C. The temple as we know it was to be built on the foundation of a previous temple. The materials that were used came from the salvaged remains of the previous temple. The Parthenon was designed to house a statue of Athena, which was to be carved by Phidias and stand almost forty feet tall. The temple was to make the statue seem proportioned relative to the space within which it was to be housed. Iktinos performed the task that he was assigned, and the temple exists today as a monument to engineering capability.

Structural work on the Parthenon enlarged the existing limestone platform of the old temple to a width of 160 feet and a length of 360 feet. The building itself, constructed entirely of marble, measured 101 feet by 228 feet; it was the largest such temple on the Greek mainland. Around the body of the building Iktinos built a colonnade, customary in Greek temple architecture. The bases of the columns were 6 feet in diameter and were spaced 14 feet apart. Subtle harmonies were thus established, for these distances were all in the ratio of 4:9. Moreover, the combined height of the columns and entablatures (lintels) bore the same ratio to the width of the building.

Remember that this was the year 438 B.C. If we asked a contractor today about the skill needed to build this structure with tools that were available 2500 years ago, we'd get quite a reply.

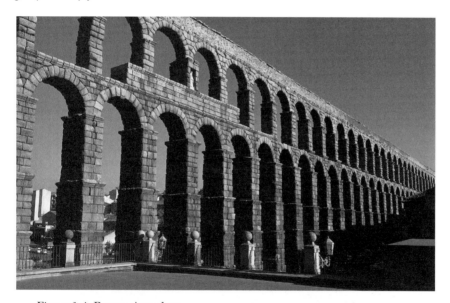

Figure 1.4  Roman Aqueduct.

## The Roman Roads and Aqueducts

Construction of the first great Roman Road, the Appian Way, began around 312 B.C. It connected Rome and Capua, a distance of 142 miles. The Appian Way eventually stretched to Brundisium at the very southern-most point in Italy, and covered 360 miles. With this start, the Roman engineers continued building roads until almost A.D. 200. Twenty-nine major roads eventually connected Rome to the rest of the empire. By A.D. 200 construction ceased, and repair and maintenance were the only work done on these roads. Aqueducts were part of this construction. One such aqueduct is shown in Fig. 1.4.

For those interested in civil engineering, the Roman roads followed elaborate principles of construction. A bedding of sand, 4 to 6 inches thick, or sometimes mortar one inch thick, was spread upon the foundation. The first course of large flat stones cemented together with lime mortar were placed upon this bedding of sand. If lime was not available, the stones (none smaller than a man's hand) were cemented together with clay. The largest were placed along the edge to form a retaining wall. This course varied from 10 inches thick on good ground to 24 inches on bad ground. A

*After the fall of Rome, road building was no longer practiced by any society in the world.*

layer of concrete about 9 inches deep was placed on top of this, followed by a layer of rich gravel or sand concrete. The roadway would generally be 12 inches thick at the sides of the road and 18 inches in the middle, thus creating a crown which caused runoff. While this third course was still wet, the fourth or final course was laid. This was made of carefully cut hard stones. Upon completion these roads would be from 2 feet to 5 feet thick, quite a feat for hand labor.

It is interesting to note that after the fall of Rome, road building was no longer practiced by anyone in the world. It would be many hundreds of years before those who specialized in road building again took on the monumental task of linking the peoples of the world.

Figure 1.5  The Great Wall of China.

## The Great Wall of China

*The Great Wall of China and the Alaskan Pipeline are two of the few man-made features that are visible from outer space.*

In 220 B.C of the Ch'in Dynasty, Meng T'ien, a military general, led his troops along the borders of China. His primary role was that of a commander of troops charged with the task of repelling the nomadic hordes of Mongolians who occasionally surged across the Chinese border. The Ch'in emperor, Shih Huang Ti, commissioned him to begin the building of what would become known as the Great Wall of China; see Fig. 1.5.

The emperor himself conceived the idea to link all the fortresses that guarded the northern borders of China. The general and the emperor functioned as engineers, even though this was not their profession. They solved a particular problem by applying the knowledge they possessed in order to make life better for their people. The ancient wall is estimated to have been 3,080 miles in length, while the modern wall runs about 1,700 miles. The original wall is believed to have passed Ninghsia, continuing north of a river and then running east through the southern steppes of Mongolia at a line north of the present Great Wall. It is believed to have reached the sea near the Shan-hal-huan

River. After serving as a buffer against the nomadic hordes for six centuries, the wall was allowed to deteriorate until the sixth and seventh centuries A.D., when it underwent major reconstruction under the Wei, Ch'i and Sui dynasties. Although the vast structure had lost military significance by the time of China's last dynasty, the Ch'ing, it never lost its significance as a "wonder of the world" and as a feat of massive engineering undertaking.

# 1.5 Traveling Through the Ages

At the same time that the previous monuments were being built, there were a number of other engineering feats under way. Let's look at a few of those other undertakings.

As you read through the following, we encourage you to investigate on your own any of the engineering accomplishments which grab your attention from this list. You should get an overall sense of both the pace and the focus of development over the centuries from these lists.

Note that as we enter the modern age, the scope of invention appears to narrow, with much of the activity relating to computers. What impact does this have on the notion of "ever-faster historic development"?

What is the role of the inventor in history? Sometimes the name is important, other times not. At times during certain eras, innovations were being made simultaneously by a number of people. So no individual really stands out. Perhaps the developments were more collective during such times. Perhaps the players involved were often racing each other to the patent office. At other times, with certain inventions, a single person made a significant breakthrough on his own.

When is it the person and when is it the times? Here are a few clues. When it's the person, his peers might think he's a nut, his work might even be outlawed or ignored. When it's the times, there are frequently many innovators doing similar work in close proximity or even in cooperation. Sometimes even when a name stands out, you still get the impression that his effort was more communal than singular. These are concepts to consider as you watch time and innovations flow past in these lists and in your studies.

**1200 B.C. – A.D. 1**

- The quality of wrought iron is improved.
- Swords are mass produced.
- Siege towers are perfected.
- The Greeks develop manufacturing.
- Archimedes introduces mathematics in Greece.
- Concrete is used for the arched bridges, roads and aqueducts in Rome.

## Activities 1.2

a)   Investigate the nature of manufacturing in these early times.

b)   Investigate warfare as it was first waged. What was the engineer's role in the design of war-related equipment? What did these engineers build during peacetime?

c)   Concrete was being used in Rome before A.D. 1. Trace the history of concrete from its early use to its prominence in building today.

d)   Metals have always been an important part of the history of engineering. Investigate the progression in the use of metals from copper and bronze to iron and steel. What effect did the use of these metals have on the societies in which they were used?

**A.D. 1 – 1000**

- The Chinese further develop the study of mathematics.
- Gunpowder is perfected.
- Cotton and silk are manufactured.
- There is growth in the silk and glass industries.
- Leonardo Fibonacci (1170-1240), medieval mathematician, writes the first Western text on algebra.

**1000 – 1400**

## Activity 1.3

What influence does the introduction of the first Western algebra text have on the society of the time? Why would engineers be interested in this text?

**1400 – 1700**

- Georgius Agricola's *De re metallica*, a treatise on mining and metallurgy, is published posthumously.
- Federigo Giambelli constructs the first time bomb for use against Spanish forces besieging Antwerp, Belgium.
- The first water closet (toilet) is invented in England.
- Galileo begins constructing a series of telescopes, with which he observes the rotation of the sun and other phenomena supporting the Copernican heliocentric theory.
- Using dikes and windmills, Jan Adriaasz Leeghwater completes drainage of the Beemstermeer, the largest project of its kind in Holland (17,000 acres).
- Otto von Guericke, mayor of Magdeburg, first demonstrates the existence of a vacuum.
- Christian Huygens begins work on the design of a pendulum-driven clock.
- Robert Hooke develops the balance spring to power watches.
- Charles II charters the Royal Society, England's first organization devoted to experimental science.
- Isaac Newton constructs the first reflecting telescope.
- Work is completed on the Languedoc Canal, the largest engineering project of its kind in Europe.
- Thomas Savery patents his "miner's friend," the first practical steam pump.
- The agriculture, mining, textile and glassmaking industries are expanded.
- The concept of the scientific method of invention and inquiry is originated.
- The humanities and science are first thought to be two distinctly separate entities.

- Robert Boyle finds that gas pressure varies inversely with volume (Boyle's Law).
- Leibniz makes a calculating machine to multiply and divide.

- The Leyden jar stores a large charge of electricity.
- The Industrial Revolution begins.
- James Watt makes the first rotary engine.
- The instrument-maker Benjamin Huntsman develops the crucible process for manufacturing steel, improving quality and sharply reducing cost.
- Louis XV of France establishes the Ecole des Ponts et Chausses, the world's first civil engineering school.
- John Smeaton completes construction of the Eddystone lighthouse.
- James Brindley completes construction of the Bridgewater Canal, beginning a canal boom in Britain.
- James Watt patents his first steam engine.
- The spinning jenny and water frame, the first successful spinning machines, are patented by James Hargreaves and Richard Arkwright, respectively.
- Jesse Ramsden invents the first screw-cutting lathe, permitting the mass production of standardized screws.
- The Society of Engineers, Britain's first professional engineering association, is formed in London.
- David Bushnell designs the first human-carrying submarine.
- John Wilkinson installs a steam engine to power machinery at his foundry in Shropshire, the first factory use of the steam engine.
- Abraham Darby III constructs the world's first cast iron bridge over the Severn River near Coalbrookdale.
- Claude Jouffroy d'Abbans powers a steamboat upstream for the first time.
- Joseph-Michel and Jacques-Etienne Montgolfier construct the first passenger-carrying hot air balloon.
- Henry Cort patents the puddling furnace for the production of wrought iron.
- Joseph Bramah designs his patent lock, which remains unpicked for 67 years.
- British civil engineer John Rennie completes the first building made entirely of cast iron.

**1700 – 1800**

## Activities 1.5

a) The Industrial Revolution changed the whole landscape of the world. How did the engineer fit into this revolution? What were some of the major contributions?

b) Research and compare the first rotary engine to the rotary engines of today.

**1800 – 1825**

- Automation is first used in France.
- The first railroad locomotive is unveiled.
- Jean Fourier, French mathematician, states that a complex wave is the sum of several simple waves.
- Robert Fulton begins the first regular steamboat service with the Clermont on the Hudson River in the U.S.
- Chemical symbols as they are used today are developed.
- The safety lamp for protecting miners from explosions is first used.
- The single wire telegraph line is developed.
- Photography is born.
- Electromagnetism is studied.
- The thermocouple is invented.
- Aluminum is prepared.
- Andre Ampere shows the effect of electric current in motors.
- Sadi Carnot finds that only a fraction of the heat produced by burning fuel in an engine is converted into motion. This forms the basis of modern thermodynamics.

**1825 – 1875**

- Rubber is vulcanized by Charles Goodyear in the United States.
- The first iron-hulled steamer powered by a screw propeller crosses the Atlantic.
- The rotary printing press comes into service.
- Reinforced concrete is used.
- Isaac Singer invents the sewing machine.
- George Boole develops symbolic logic.
- The first synthetic plastic material—celluloid—is created by Alexander Parkes.
- Henry Bessemer originates the process to mass-produce steel cheaply.
- The first oil well is drilled near Titusville, Pennsylvania.
- The typewriter is perfected.
- The Challenge Expedition (1871-1876) forms the basis for future oceanographic study.

**1875 – 1900**

- The telephone is patented in the United States by Alexander Graham Bell.
- The phonograph is invented by Thomas Edison.
- The incandescent light bulb also is invented by Edison.
- The steam turbine appears.
- The gasoline engine is invented by Gottlieb Daimler.
- The automobile is introduced by Karl Benz (see photo next page).

- The Wright brothers complete the first sustained flight.
- Detroit becomes the center of the auto industry.
- Stainless steel is introduced in Germany.
- Tractors with diesel engines are produced by Ford Motor Company.
- The first commercial airplane service between London and Paris commences.
- Diesel locomotives appear.
- Modern sound recordings are introduced.
- John Logie Baird invents a high-speed mechanical scanning system, which leads to the development of television.
- The Volkswagen Beetle goes into production.
- The first nuclear bombs are used.
- The transistor is invented.

**1900 – 1925**

**1925 – 1950**

- Computers first enter the commercial market.
- Computers are in common use by 1960.
- The first artificial satellite—Sputnik 1, USSR—goes into space.
- Explorer I, the first U.S. satellite, follows.
- The laser is introduced.
- Manned space flight begins.
- The first communication satellite—Telstar—goes into space.
- Integrated circuits are introduced.
- The first manned moon landing occurs.

**1950 – 1975**

- Supersonic transport from U.S. to Europe begins.
- Cosmonauts orbit the earth for a record 180 days.
- The Columbia space shuttle is reused for space travel.
- The first artificial human heart is implanted.

**1975 – 1990**

**1990 – Today**

- Robots walk on Mars.
- Computer processor speed is dramatically improved.
- The Channel Tunnel (the "Chunnel") between England and France is completed.
- World's new tallest building opens in Kuala Lumpur, Malaysia (1, 483 feet).
- Global Positioning Satellite (GPS) technology is declassified, resulting in hundreds of safety, weather and consumer applications.

You are the potential innovator of tomorrow. Research the progress made over the past decade in engineering on your own. Consider possible engineering innovations to which you might one day contribute.

## 1.6 A Case Study of Two Historic Engineers

The important thing to realize about history is that it's about people. Much of it is based on biography, in fact. To fully understand the story of an invention, you need to investigate the people involved. An invention and its inventor are inextricably woven together.

As you study the details of history, you'll likely find yourself most interested as you get to know more about the people involved and the challenges they were up against. Hopefully, the two studies that follow will inspire you to further investigation.

**Leonardo da Vinci**

Leonardo da Vinci, shown in Fig. 1.6, had an uncanny ability to envision mechanized innovations that would one day see common usage—especially in the field of weaponry (helicopters, tanks, artillery, see Figs. 1.7-1.11). He seemed to know almost everything that was knowable in his time. He was a true Renaissance man and harnessed his immense genius to make improvements to most every aspect of the lives of his contemporaries, and to greatly impact the lives of future generations. Da Vinci's handicap, however, was his inability to read Latin, which prevented him from learning from the common scientific writings of the day, which focused on the works of Aristotle and other Greeks and their relation to the Bible. Instead, da Vinci was forced to make his assessments solely from his observations of the world around him. He was not interested in the thoughts of the ancients; he simply wanted to use his engineering skills to improve his environment.

Figure 1.6 Leonardo da Vinci (born April 15, 1452, Vinci, Italy, died May 2, 1519).

He did indeed bring his genius to bear on an enormously wide range of subjects, but little of his work had any relationship to that of his contemporaries. He designed, he built, and he tested. But various essential aspects of development were always lacking from his innovations. He could do many great things, but he couldn't do it all. He lacked a community of engineering peers with whom he could integrate his efforts. Being so far ahead of his time relative to the development of the field of engineering in his day, it was centuries before many of his innovations came to fruition.

Da Vinci's life as an engineer was one of total immersion in the intellectual activities of his time. He created frescoes, he tried his hand at sculpture, he was an architect, and he engineered hundreds of useful devices and many others that never came to life. He was a respected member of the community and was consistently in the employment of members of the aristocracy.

Figure 1.7 Weapon design.

Figure 1.8  A flying machine.

To obtain his first commission as an engineer he wrote a letter claiming that he could: construct movable bridges; remove water from the moats of fortresses under siege; destroy any fortification not built of stone; make mortars, dart-throwers, flame-throwers and cannon capable of firing stones and making smoke; design ships and weapons for war at sea; dig tunnels without making any noise; make armored wagons to break up the enemy in advance of the infantry; design buildings; sculpt in any medium; and paint.

With that resume, he was hired with the title of Painter and Engineer to the Duke. Did he do all the things that he said he could? We don't know. But we do know that, among other things, he worked diligently perfecting the canal system around Milan.

DaVinci's chief court function during the period he was employed by the Duke was to produce spectacular shows for the entertainment of the aristocracy who came to court. He produced musical events, designed floats, and delighted the court with processions that dazzled the eyes and flying devices that wowed the audience. The interesting thing is that during this time he wrote over 5000 pages of notes, detailing every conceivable kind of invention. This truly was a great engineer at work.

Figure 1.9  Cannon design.

FIG. 13. Leonardo's machine guns. (CA, 56v-a)

Figure 1.10 Machine guns.

There is so much you will encounter if you investigate Leonardo da Vinci. It is interesting to note that he seemingly examined every aspect of his world and left behind a significant record from which we can learn.

Figure 1.11 Stone throwers.

### Gutenberg and His Printing Press

Another important innovator of this era developed one of the most history-altering inventions of all time—the printing press. In 1455, Johannes Gutenberg, pictures in Fig. 1.12, printed the first book, a Bible. He lived before Leonardo da Vinci and wasn't a generalist like da Vinci, but his invention changed society forever. With the invention of the printing press, man was able to use, appreciate, and disseminate information as never before.

Figure 1.12 Johannes Gutenberg, creator of the first mass-producing printing press (born 1394-1398, Mainz, Germany, died Feb. 3, 1468).

The development of printing came at a time when there was a growing need for the ability to spread information. There had been a long phase of general societal introversion from which mankind was prepared to emerge. Gutenberg's printing press was the spark that ignited the flame of widespread communication.

Figure 1.13 The printing press.

There were already printing presses when Gutenberg introduced his. As early as the 11th century, the Chinese had developed a set of movable type from a baked mixture of clay and glue. The type pieces were stuck onto a plate where impressions could be taken by pressing paper onto the stationary type. Since the type was glued to the plate, the plate could be heated and the type removed and re-situated. The process was not used in any form of mass production, but it did form a basis for the future invention of the Gutenberg printing press. In Europe there were attempts at creating presses that would produce cheap playing cards and frivolous items, but it took Gutenberg to see the practical need for a press that could provide the common man with reading material. The time was right; the inventor was prepared. History is made by the engineer who delivers the major innovation society is waiting for—whether society has been eagerly awaiting the breakthrough or is caught pleasantly by surprise.

Johannes Gutenberg was born into a noble family of the city of Mainz, Germany. His early training is reputed to have been in goldsmithing. By 1428 he had moved to Strasbourg. It was there that he began to formulate the idea for his printing press. His first experiments with movable type stemmed from his notions of incorporating the techniques of metalworking—such as casting, punch-cutting, and stamping—for the mass production of books. Since all European books at this time were hand-written by scribes with elaborate script, Gutenberg decided to reproduce this writing style with a font of over 300 characters, far larger than the fonts of today. To make this possible, he invented the variable-width mold, and perfected a rugged blend of lead, antimony, and tin used by type foundries up to the present century.

*History is made by the engineer who delivers the major innovation society is waiting for.*

## Activities 1.4

a)   Study some of the inventions of Leonardo da Vinci. Draw conclusions on how these early works influenced later inventions or innovations.

b)   Research and build one of da Vinci's inventions, or a model of one, from his plans.

c)   Where does Boyle's Law fit in your future studies?

d)   Do you agree that the humanities and science should be separated? Why or why not?

e)   Explain why the printing press revolutionized the world and the life of the engineer.

# 1.7  The History of the Disciplines

As you investigate the many areas of engineering throughout this textbook, you will discover that there are a wide variety of career paths from which to choose. As a prelude to the in-depth look at engineering fields which we provide in Chapter 2, this section takes a brief look at the historic backgrounds of the following disciplines:

- Aerospace Engineering
- Agricultural Engineering
- Chemical Engineering
- Civil Engineering
- Computer Engineering
- Electrical Engineering
- Industrial Engineering
- Mechanical Engineering

From the core areas listed, many additional engineering specialties have evolved over time, for a total today of over 30 different fields in engineering.

The information here touches only lightly on the backgrounds of some of these fields. We recommend that you research in greater detail the history of disciplines of interest to you. An in-depth description of a number of these fields is found in the next chapter.

The individual disciplines of engineering were actually named only a short time ago relative to the history of the world. What we present here is a short history of the major disciplines dating from after the time that they were individually recognized.

## Aerospace Engineering

Aerospace engineering is concerned with engineering applications in the areas of *aeronautics* (the science of air flight) and *astronautics* (the science of space flight). Aerospace engineering deals with flight of every kind: balloon flight, sailplanes, propeller- and jet-powered aircraft, missiles, rockets, satellites, and advanced interplanetary concepts such as ion-propulsion rockets and solar-wind vehicles. It is the field of future interplanetary travel. The challenge is to produce vehicles that can traverse the long distances of space in ever-shorter periods of time. New propulsion systems will require that engineers venture into areas never before imagined. By reading histories and biographies, you get a feel for the changes that aerospace pioneers brought about, which will help you to understand the mindset needed to make great leaps forward yourself.

### Activity 1.7

Where do you think the next breakthrough will come for the aerospace engineer? How should this affect the approach we take in the education of the next generation of aerospace engineers? Do you think your generation will walk on Mars?

## Agricultural Engineering

At the turn of the 20th century, a large majority of the working population was engaged in agriculture. The mass entry into industrial employment would come a few decades later. This movement to the factories of America lowered the number of workers in agriculture to under five percent. This drastic change occurred for a variety of reasons, the principal reason being the integration of technology and engineering into agriculture, which allowed modern farmers to feed approximately ten times as many families per farmer as their ancestors did a hundred years prior.

As you investigate the many facets of agricultural engineering, you will discover that America leads the world in agricultural technology. The world depends upon the United States to feed those in areas of famine, to supply agricultural implements to countries that do not have the resources to perfect such implements, and to continue to perfect technologies to feed more and more of the world's growing population. The agricultural engineer has one of the largest responsibilities in the engineering community. The modern world cannot survive without the efficiencies of a mass-produced food supply.

Agricultural engineering focuses on the following areas:
- soil and water
- structures and environment

- electrical power and processing
- food engineering
- power and machinery

## Activities 1.6

1. Compare the growth of industry with the number of people leaving the farms of America. What innovations in farming helped to offset this exodus? How were we able to continue to feed the population with such a change?

2. Inspect an early piece of farm machinery. Find a current model of that machine. How are they different? What has not changed in their design?

3. Farm machinery can be highly dangerous if not used properly. Investigate the dangers posed by farm machinery and how the operators are protected from those dangers.

4. Follow a piece of farm machinery through its many changes from the 17th century to the present day.

5. Investigate where agriculture has gone in the late 20th century. What new aspects has it undertaken?

## Chemical Engineering

Chemical Engineering is one of the newer disciplines among engineering professions. A closely related field, Chemistry, has been studied for centuries in its many forms from alchemy to molecular structure, but chemical engineering began its rise to prominence shortly after World War I. The number of activities involving chemical processing forced engineers and chemists to combine their efforts, bonding the two disciplines together. The chemical engineer needed to understand the way in which chemicals moved and how they could be joined together. This necessity lead the chemical engineer to begin experimenting with the ways in which mechanisms were used to separate and combine chemicals. Experimenting with mass transfer, fluid flow, and heat transfer proved beneficial. The chemical engineering field is relatively new and is a critical engineering profession.

Chemical engineering applies chemistry to industrial processes which change the composition or properties of an original substance for useful purposes. Chemical engineering is involved in the manufacture of drugs, cements, paints, lubricants, pesticides, fertilizers, cosmetics, foods; in oil refining, combustion, extraction of metals from ores; and in the production of ceramics, brick, and glass. The petrochemical industry is a sizable market for the chemical engineer. Chemical engineers can apply their skills in food engineering, process dynamics and control, environmental control, electrochemical engineering, polymer science technology, unit operations, and plant design and economics.

## Activities 1.8

1. Pick a product that you believe has had to be engineered by a chemical engineer. Explain the process that started with the raw materials and ended with the product you identified. Now look into the recent history of each step of this process. Where have the latest improvements occurred? Try to identify the oldest technique used today in the process. Any book covering the background of this discipline and its products should include such information.

2. Look at the history of the study of chemistry and try to pinpoint those times when the application of chemistry involves the chemical engineer.

3. What inventions and innovations did the first chemical engineers develop?

4. Contact a chemical engineering society and collect information on their history—especially about their early members.

## Civil Engineering

The 17th and 18th centuries gave birth to most of the major modern engineering disciplines in existence today. But civil engineering is the oldest. Civil engineering traces its roots to early eighteenth century France, though the first man definitely to call himself a "civil engineer" was a keen Englishman, John Smeaton, in 1761, [Kirby 1956, p. xvi]. Smeaton was a builder of lighthouses who helped distinguish the profession of civil engineering from architecture and military engineering. The surveying of property, the building of roads and canals to move goods and people, and the building of bridges to allow safe passage over raging waters all fall within the parameters of what we now call civil engineering. Construction was the primary focus of the civil engineer while the military engineer focused on destruction. In America, construction of the earliest railroads began in 1827 and demonstrated the dedication of those early engineers in opening up the West to expansion.

Today, Civil engineering focuses primarily on structural issues such as rapid transit systems, bridges, highway systems, skyscrapers, recreational facilities, houses, industrial plants, dams, nuclear power plants, boats, shipping facilities, railroad lines, tunnels, harbors, offshore oil and gas facilities, pipelines, and canals. Civil engineers are heavily involved in improving the movement of populations through their many physical systems. Mass transit development allows the civil engineer to protect society and the environment, reducing traffic and toxic emissions from individual vehicles. Highway design provides motorists with safe motorways and access to recreation and work. The civil engineer is always at the center of discussion when it comes to transport and buildings.

## Activities 1.9

1. Trace the evolution of the first accounts of civil engineering in history. How has it changed? What new areas have entered the field and why did they become part of civil engineering?

2. Research a significant American civil engineering project and explain its importance in the history of the United States.

3. Investigate one of the following and describe in simple terms how it was constructed: a railroad, a bridge, a canal, or a sewer.

4. Research and discuss the connection between mining engineering and civil engineering.

The role that computers play in engineering design has reached a level where no major business can exist comfortably without access to computer technology. Large, medium, and small organizations alike all depend on computers to perform inventories, create billing, record sales, and order new stock. The storage of data alone makes the computer indispensable. The once-mammoth computers of the past have been downsized to the size of your fist, and computing speed today was unthinkable just years ago. Even homeowners seem to be vying with their neighbors to see who can purchase the fastest computer. The world has become almost completely computer-centered, and the computer's role should only increase.

**Computer Engineering**

## Activity 1.10

It's time to use your imagination: what will computer engineers of the future be required to do in order to both lead and to keep up with innovations? What new things will computers do?

The amount of historical perspective in this discipline is vast. "The records of magnetic effects date back to remotest antiquity...The whole of Electrical Engineering is based on magnetic and electrical phenomenon." [Dunsheath 1962, p. 21] In the late 1700s, experiments concerning electricity and its properties began. By the early 1800s, Volta began his studies of electric conduction through a liquid. Electric current and its movement became the topic of the day. As the need for electricity to power the many devices that were being fabricated by the mechanical engineers grew, the profession of electrical engineering prospered. Direct current, alternating current, the electric telegraph—all became areas to be investigated by the early electrical engineer. The profession started with magnetism and electrical phenomenon, but it has grown more recently in new directions as new requirements have been imposed by modern society.

**Electrical Engineering**

Electrical engineering is the largest branch of engineering, employing over 400,000 engineers. Among the major specialty areas in electrical engineering are electronics and solid-state circuitry, communication systems, computers and automatic control, instrumentation and measurements, power generation and transmission, and industrial applications.

## Activities 1.11

1. Investigate early electrical phenomenon and how it was perceived by early man.

2. Take one pioneer in electrical engineering and show how his/her experiments aided the growth of electrical engineering.

3. What are some of the highpoints in the history of electrical engineering?

## Industrial Engineering

Industrial engineering is a growing branch of the engineering family because of the awareness that the application of engineering principles and techniques can help create better working conditions. Industrial engineers must design, install, and improve systems which integrate people, materials, and equipment to provide efficient production of goods. They must coordinate their understanding of the physical and social sciences with the activities of workers to design areas in which the workers will produce the best results. The daily lives of the people with whom industrial engineers work are closely tied to the designs that they create.

## Activity 1.12

Observe an area where people work. How would you as an industrial engineer improve the working conditions and productivity of these workers?

## Mechanical Engineering

As coke replaced charcoal in the blast furnaces of England in the early 1700s, the beginnings of the modern Mechanical Engineering profession dawned. The introduction of coke allowed for larger blast furnaces and a greater ability to use iron. Higher quality wrought iron could also be produced. These improvements laid the foundation for the Industrial Revolution, during which the production of great quantities of steel was made possible. As these materials became more readily available, the mechanical engineer began to design improved lathes and milling and boring machines. The mechanical engineer realized that a wide variety of devices needed to be created to work with the quantities of iron and steel being produced. As the pace of manufacturing increased, the numbers of mechanical engineers also showed a marked increase. The profession would continue to grow in its size and importance. Steam power became vitally important during the industrial revolution. The mechanical engineer was needed to create the tools to harness the power of steam. From the steam engine to the automobile, from the automobile to the airplane, from the airplane to the space shuttle, the mechanical engineer has been and always will be in great demand for the development of new devices for the betterment of man.

Automobiles, engines, heating and air-conditioning systems, gas and steam turbines, air and space vehicles, trains, ships, servomechanisms, transmission mechanisms, radiators, mechatronics, and pumps are a few of the systems and devices requiring mechanical engineering knowledge. Mechanical engineering deals with power, its generation, and its application. Power affects the rate of change or "motion" of something. This can be change of temperature or change of motion due to outside stimulus. Mechanical engineering is the broadest-based discipline in engineering. The breadth of study required to be a mechanical engineer allows one to diversify into many of the other engineering areas. The major specialty areas of mechanical engineering are: applied mechanics; control; design; engines and power plants; energy; fluids; lubrication; heating, ventilation, and air-conditioning (HVAC); materials, pressure vessels and piping; and transportation and aerospace.

## Activities 1.13

1. Look around you. How many things can you find that have been influenced by the work of a mechanical engineer?

2. Investigate the early lathes and explain the improvements made in today's models.

3. Explore in greater depth the effect the iron industry had upon the Industrial Revolution.

4. Who are some of the important names in the early days of the Industrial Revolution and what did they do for the revolution?

5. Where would mechanical engineering be without the Industrial Revolution?

6. Pick a period in the time between 1700–1999. Look closely at the contributions made by mechanical engineers.

# References

*A History of Technology, Volume II*, Ed. Charles Singer, Oxford University Press, New York, 1956.

American Society of Civil Engineers, *The Civil Engineer: His Origins,* New York: ASCE, 1970

Burghardt, M. David, *Introduction to Engineering,* 2nd Ed., New York, HarperCollins, 1995.

Burstall, A., *A History of Mechanical Engineering*, London: Faber, 1963.

De Camp, L. Sprague, *The Ancient Engineers*, Cambridge, The MIT Press, 1970.

Dunsheath, P., *A History of Electrical Engineering,* London: Faber, 1962.

Gray, R. B., *The Agricultural Tractor 1855-1950*, St. Joseph, MI: American Society of Agricultural Engineers, 1975

*Great Engineers and Pioneers in Technology, Volume 1: From Antiquity through the Industrial Revolution,* Eds. Roland Turner and Steven L. Goulden, New York, St. Martin's Press, 1981.

Greaves, W. F., and J. H. Carpenter, *A Short History of Electrical Engineering*, London: Longmans, Green and Co Ltd, 1969.

Kirby, R., et al., *Engineering in History*, New York, McGraw-Hill, 1956.

Miller, J. A., *Master Builders of Sixty Centuries*, Freeport, New York, Books for Libraries Press, 1972.

Red, W. Edward, *Engineering—The Career and The Profession*, Monterey, California, Brooks/Cole Engineering Division, 1982.

Rosenberg, S. H., *Rural America A Century Ago*, St. Joseph, MI: American Society of Agricultural Engineers, 1976

## Exercises and Activities

1.1    The history of engineering is long and varied. It contains many interesting inventions and refinements. Select one of these inventions and discuss the details of its creation. For example, you might explain how the first printing presses came into being and what previous inventions were used to create the new device.

1.2    Build a simple model of one of the inventions mentioned in this chapter—a bridge, aqueduct or submarine, for instance. Explain the difficulties of building these devices during the time they were invented.

1.3    Explain how easy it would be to create some inventions of the past using our present-day knowledge and capability.

1.4    Engineering history is filled with great individuals who have advanced the study and practice of engineering. Investigate an area of engineering that is interesting to you and write a detailed report on an individual who made significant contributions in that area.

1.5    Explain what it would have been like to have been an engineer during any particular historical era.

1.6    Compare the lives of any two engineers from the past. Are there similarities in their experiences, projects and education?

1.7    What kind of education were engineers of old able to obtain?

1.8    What period of engineering history interests you most? Why? Explain why this period is so important in the history of engineering.

1.9    If you had to explain to a 7-year-old child why engineering is important to society, what information from the history of engineering would you relate? Why?

# 2. Engineering Majors

## 2.1 Introduction

Engineers produce things that impact us every day. They invent, design, develop, manufacture, test, sell, and service products and services that improve the lives of people. The Accreditation Board for Engineering and Technology (ABET), which is the national board that establishes accreditation standards for all engineering programs, defines engineering as follows [Landis]:

*Engineering is the profession in which a knowledge of the mathematical and natural sciences, gained by study, experience, and practice, is applied with judgment to develop ways to utilize, economically, the materials and forces of nature for the benefit of mankind.*

Frequently, students early in their educational careers find it difficult to understand exactly what engineers do, and often more to the point, where they fit best in the vast array of career opportunities available to engineers.

Common reasons for a student to be interested in engineering include:
1. Proficiency in math and science
2. Suggested by a high school counselor
3. Has a relative who is an engineer
4. Heard it's a field with tremendous job opportunity
5. Read that it has high starting salaries

While these can be valid reasons, they don't imply a firm understanding of engineering. What is really important is that a student embarking upon a degree program, and ultimately a career, understands what that career entails and the options it presents. We all have our own strengths and talents. Finding places to use those strengths and talents is the key to a rewarding career.

The purpose of this chapter is to provide information about some of the fields of engineering in order to help you decide if this is an area that you might enjoy. We'll explore the role of engineers, engineering job functions and the various engineering disciplines.

*Engineers invent, design, develop, manufacture, test, sell, and service pro ducts and services.*

## The Engineer and the Scientist

To better understand what engineers do, let's contrast the roles of engineers with those of the closely related field of the scientist. Many students approach both fields for similar reasons: they were good at math and science in high school. While this is a prerequisite for both fields, it is not a sufficient discriminator to determine which is the right career for a given individual.

The main difference between the engineer and the scientist is in the object of each one's work. The scientist searches for answers to technological questions to obtain a knowledge of why a phenomenon occurs. The engineer also searches for answers to technological questions, but always with an application in mind.

Theodore Von Karman, one of the pioneers of America's aerospace industry, said, "Scientists explore what is; engineers create what has not been." [Paul Wright, *Introduction to Engineering*].

*"Scientists explore what is; engineers create what has not been."*

In general, science is about discovering things or acquiring new knowledge. Scientists are always asking, "Why?" They are interested in advancing the knowledge base that we have in a specific area. The answers they seek may be of an abstract nature, such as understanding the beginning of the universe, or more practical, such as the reaction of a virus to a new drug.

The engineer also asks, "Why?" but it is because of a problem which is preventing a product or service from being produced. The engineer is always thinking about the application when asking why. The engineer becomes concerned with issues such as the demand for a product, the cost of producing the product, the impact on society and the environment of the product.

Scientists and engineers work in many of the same fields and industries but have different roles. Here are some examples:

- Scientists study the planets in our solar system to understand them; engineers study the planets so they can design a spacecraft to operate in the environment of that planet.
- Scientists study atomic structure to understand the nature of matter; engineers study the atomic structure in order to build smaller and faster microprocessors.
- Scientists study the human neurological system to understand the progression of neurological diseases; engineers study the human neurological system to design artificial limbs.
- Scientists create new chemical compounds in a laboratory; engineers create processes to mass-produce new chemical compounds for consumers.
- Scientists study the movement of tectonic plates to understand and predict earthquakes; engineers study the movement of tectonic plates to design safer buildings.

## The Engineer and the Engineering Technologist

Another profession closely related to engineering is engineering technology. Engineering technology and engineering have similarities, yet there are differences; they have different career opportunities. ABET, which accredits engineering technology programs as well as engineering programs, defines engineering technology as follows:

> *Engineering technology is that part of the technological field which requires the application of scientific and engineering knowledge and methods combined with technical skills in support of engineering activities; it lies in the occupational spectrum between the craftsman and engineering at the end of the spectrum closest to the engineer.*

Technologists work with existing technology to produce goods for society. Technology students spend time in their curricula working with actual machines and equipment that are used in the jobs they will accept after graduation. By doing this, technologists are equipped to be productive in their occupation from the first day of work.

Both engineers and technologists apply technology for the betterment of society. The main difference between the two fields is that the engineer is able to create new technology through research, design and development. Rather than being trained to use specific machines or processes, engineering students study additional mathematics and engineering science subjects. This equips engineers to use these tools to advance the state of the art in their field and move technology forward.

*The main difference between the two fields is that the engineer is able to create new technology through research, design and development.*

There are areas where engineers and engineering technologists perform very similar jobs. For example, in manufacturing settings, engineers and technologists are employed as supervisors of assembly line workers. Also, in technical service fields both are hired to work as technical support personnel supporting equipment purchased by customers. However, most opportunities are different for engineering and engineering technology graduates.

- The technologist identifies the computer networking equipment necessary for a business to meet its needs and oversees the installation of that equipment; the engineer designs new computer boards to transmit data faster.
- The technologist develops a procedure to manufacture a shaft for an aircraft engine using a newly developed welding technique; the engineer develops the new welding machine.
- The technologist analyzes a production line and identifies new robotic equipment to improve production; the engineer develops a computer simulation of the process to analyze the impact of the proposed equipment.
- The technologist identifies the equipment necessary to assemble a new CD player; the engineer designs the new CD player.
- The technologist identifies the proper building materials and oversees the construction of a new building; the engineer determines the proper support structures, taking into account the local soil, proposed usage, earthquake risks and other design requirements.

Engineering is an exciting field with a vast range of career opportunities. In trying to illustrate the wide range of possibilities, Professors Jane Daniels and Richard Grace from Purdue University constructed the cubic model of Figure 2.1. One edge of the cube represents the engineering disciplines that most students identify as the potential majors. A second edge of the cube represents the different job functions an engineer can have within a specific engineering discipline. The third edge of the cube represents industrial sectors where engineers work. A specific engineering position, such as a mechanical engineering design position in the transportation sector, is the intersection of these three axes. As one can see from the cube, there is a vast number of possible engineering positions.

## What Do Engineers Do?

*It is not until the junior or senior year that students take technical electives.*

In the following sections of this chapter, the engineering functions and majors are described. The remaining axis, the industrial sectors, are dependent on the companies and governmental agencies that employ engineers.

To obtain more information about the various industrial sectors:

- Explore your school's placement center
- Visit job fairs
- Attend seminars on campus sponsored by various companies
- Search the Internet (visit Web sites describing career opportunities)
- Talk to faculty familiar with a certain industry
- "Shadow" a practicing engineer
- Work as an intern or co-op engineer
- Take an engineering elective course

Most engineering curricula include common courses within an engineering major for the first two years of study. It is not until the junior or senior year that students take technical electives that can be industry specific. Students are encouraged to explore various career opportunities so they can make better decisions when required to select their junior and senior level electives.

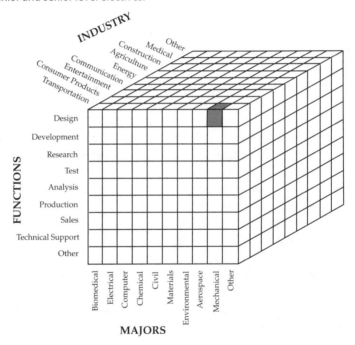

Figure 2.1 Engineering positions. [Grace and Daniels]

## Example 2.1

Let's consider a mechanical engineer (an "ME") who performs a design function to illustrate how a specific job in one industrial sector can vary from that in another sector.

- **Aerospace** – *Design of an aircraft engine fan blade*: Detailed computer analyses and on-engine testing are required for certification by the FAA. Reliability, efficiency, cost and weight are all design constraints. The design engineer must push current design barriers to optimize the design constraints, potentially making tradeoffs between efficiency, cost and weight.
- **Biomedical** – *Design of an artificial leg and foot prosthesis giving additional mobility and control to the patient*: Computer modeling is used to model the structure of the prosthesis and the natural movement of a leg and foot. Collaboration with medical

personnel to understand the needs of the patient is a critical part of the design process. Reliability, durability and functionality are the key design constraints.

- **Power** – *Design of a heat recovery system in a power plant, increasing the plant productivity*: Computer analyses are performed as part of the design process. Cost, efficiency and reliability are the main design constraints. Key mechanical components can be designed with large factors of safety since, weight is not a design concern.
- **Consumer products** – *Design of a pump for toothpaste for easier dispensing*: Much of the development work might be done with prototypes. Cost is a main design consideration. Consumer appeal is another consideration and necessitates extensive consumer testing as part of the development process.
- **Computer** – *Design of a new ink-jet nozzle with resolution approaching laser printer quality*: Computer analyses are performed to ensure that the ink application is properly modeled. Functionality, reliability and cost are key design concerns.

# 2.2 Engineering Functions

Within engineering there are basic classifications of jobs that are common across the various engineering disciplines. What follows are brief descriptions of these different engineering job functions. A few examples are provided for each function. It is important to realize that all the fields of engineering have roles in each of the main functions described here.

## Research

The role of the engineering researcher is the closest to that of a scientist of all the engineering functions. Research engineers explore fundamental principles of chemistry, physics, biology and mathematics in order to overcome barriers preventing advancement in their field. Engineering researchers differ from scientists in that they are interested in the application of a breakthrough, whereas scientists are concerned with the knowledge that accompanies a breakthrough.

*Research engineers explore fundamental principles of chemistry, physics, biology and mathematics.*

Research engineers conduct investigations to extend knowledge using various means. One of the means is conducting experiments. Research engineers may be involved in the design and implementation of experiments and the interpreting of the results. Typically, the research engineer does not perform the actual experiment. Technicians are usually called upon for the actual testing. Large-scale experiments may involve the coordination of additional supporting personnel including other engineers, scientists, technologists, technicians and craftspeople.

Research is also conducted using the computer. Computational techniques are developed to calculate solutions to complex problems without having to conduct costly and time-consuming experiments. Computational research requires the creation of mathematical models to simulate the natural occurring phenomena under study. Research engineers also might develop the computational techniques to perform the complex calculations in a timely and cost-effective fashion.

*An advanced degree is required, and often a Ph.D. is needed.*

Most research engineers work for some type of research center. A research center might be a university, a government laboratory such as NASA, or an industrial research center. In most research positions an advanced degree is required, and often a Ph.D. is needed. If research appeals to you, a great way to explore it is by doing an undergraduate research project with an engineering professor. This will allow you to observe the operation of a laboratory first-hand and find out how well you enjoy being part of a re-search team.

## Development

*Development engineers take the knowledge acquired by the researchers and apply it to a specific product.*

Development engineers bridge the gap between laboratory research and full-scale production. The development function is often coupled with research in so-called R&D (research and development) divisions. Development engineers take the knowledge acquired by the researchers and apply it to a specific product or application. The re-searcher may prove something is possible in a laboratory setting; the development engineer shows that it will work on a large, production-size scale and under actual conditions encountered in the field. This is done in pilot manufacturing plants or by using prototypes.

Development engineers are continuously looking for ways to incorporate the findings of researchers into prototypes to test their feasibility for use in tomorrow's products. Often, an idea proven in a laboratory needs to be significantly altered before it can be introduced on a mass production scale. It is the role of development engineers to identify these areas and work with the design engineers to correct them before full-scale production begins.

An example of a development process is the building of concept cars within the au-tomotive industry. These are unique cars that incorporate advanced design concepts and technology. The cars are then used as a test case to see if the design ideas and technology actually perform as predicted. The concept cars are put through exhaustive tests to determine how well the new ideas enhance a vehicle's performance. Each year, new technology is introduced into production automobiles that was first proven in development groups using concept vehicles.

## Testing

*Test engineers are responsible for designing and implementing tests to verify the integrity, reliability and quality of products.*

Test engineers are responsible for designing and implementing tests to verify the in-tegrity, reliability and quality of products before they are introduced to the public. The test engineer devises ways to simulate the conditions a product will be subjected to during its life. Test engineers work closely with development engineers in evaluating prototypes and pilot facilities. Data from these initial development tests are used to decide whether full production versions will be made or if significant changes are needed before a full-scale release. Test engineers work with design engineers to iden-tify the changes in the product to ensure its integrity.

A challenge that engineers face is simulating the conditions a product will face during its life span, and doing so in a timely, cost-effective manner. Often the condi-tions the product will face are difficult to simulate in a laboratory. A constant problem for the test engineer is simulating the aging of a product. An example of such a testing challenge is the testing of a pacemaker for regulating a patient's heart which is de-signed to last several decades. An effective test of this type cannot take 20 years or the product will be obsolete before it is introduced. The test engineer must also simulate conditions within the human body without exposing people to unnecessary risks.

Other challenges facing test engineers involve acquiring accurate and reliable data. The test engineer must produce data that show that the product is functioning properly or that identify areas of concern. Test engineers develop data acquisition and in-strumentation methods to achieve this. Techniques such as radio telemetry may be used to transmit data from the inside of a device being tested. The measurement techniques must not interfere with the operation of the device, presenting a tremen-dous challenge for small, compact products. Test engineers must cooperate with design engineers to determine how the device being tested can be fitted with instrumentation yet still meet its design intent.

Test engineers must have a wide range of technical and problem-solving skills. They must also be able to work in teams involving a wide range of people. They work with design and development engineers, technicians, and craftspeople, as well as management.

## Example 2.2

Test engineers must understand the important parameters of their tests. The development of a certain European high-speed train provides an example of the potential consequences which may result when test engineers fail to understand these parameters. A test was needed to show that the windshield on the locomotive could withstand the high-velocity impacts of birds or other objects it might encounter. This is also a common design constraint encountered in airplane design. The train test engineers borrowed a "chicken gun" from an aerospace firm for this test. A chicken gun is a mechanism used to propel birds at a target, simulating in-flight impact. With modern laws governing cruelty to animals, the birds are humanely killed and frozen until the test.

On the day of the test, the test engineers aimed the gun at the locomotive windshield, inserted the bird and fired. The bird not only shattered the windshield but put a hole through the engineer's seat.

The design engineers could not understand what went wrong. They double-checked their calculations and determined that the windshield should have held. The problem became clear after the test engineers reviewed their procedure with the engineers from the aerospace firm from whom they had borrowed the equipment. The aerospace test engineers asked how long they had let the bird thaw.

The response was, "Thaw the bird?"

There is a significant difference in impact force between a frozen eight-pound bird and a thawed bird. The test was successfully completed later with a properly thawed bird.

## Design

The design function is what many people think of when they think of engineering, and this is where the largest number of engineers are employed. The design engineer is responsible for providing the detailed specifications of the products society uses.

Rather than being responsible for an entire product, most design engineers are responsible for a component or part of the product. The individual parts are then assembled into a product such as a computer, automobile, or airplane. Design engineers produce detailed dimensions and specifications of the part to ensure that the component fits properly with adjoining pieces. They use modern computer design tools and are often supported by technicians trained in computer drafting software.

The form of the part is also a consideration for the design engineer. Design engineers use their knowledge of scientific and mathematical laws, coupled with experience, to generate a shape to meet the specifications of the part. Often, there is a wide range of possibilities and considerations. In some fields, these considerations are ones

*The design engineer is responsible for providing the detailed specifications of the products society uses.*

that can be calculated. In others, such as in consumer products, the reaction of a potential customer to a shape may be as important as how well the product works.

The design engineer also must verify that the part meets the reliability and safety standards established for the product. The design engineer verifies the integrity of the product. This often requires coordination with analysis engineers to simulate complex products and field conditions, and with test engineers to gather data on the integrity of the product. The design engineer is responsible for making corrections to the design based on the results of the tests performed.

In today's world of ever increasing competition, the design engineer must also involve manufacturing engineers in the design process. Cost is a critical factor in the design process and may be the difference between a successful product and one that fails. Communication with manufacturing engineers is therefore critical. Often, simple design changes can radically change a part's cost and affect the ease with which the part is made.

Design engineers also work with existing products. Their role includes redesigning parts to reduce manufacturing costs and time. They also work on redesigning products that have not lived up to expected lives or have suffered failure in the field. They also modify products for new uses. This usually requires additional analysis, minor redesigns and significant communication between departments.

## Analysis

Analysis is an engineering function performed in conjunction with design, development and research. Analysis engineers use mathematical models and computational tools to provide the necessary information to design, development or research engineers to help them perform their function.

*Analysis engineers use mathematical models and computational tools to provide the necessary information to design, development or research engineers.*

Analysis engineers typically are specialists in a technology area important to the products or services being produced. Technical areas might include heat transfer, fluid flow, vibrations, dynamics, system modeling, and acoustics, among others. They work with computer models of products to make these assessments. Analysis engineers often possess an advanced degree and are experienced in their area of expertise.

In order to produce the information required of them, they must validate their computer programs or mathematical models. To do so may require comparing test data to their predictions. This requires coordination with test engineers in order to design an appropriate test and to record the relevant data.

An example of the role of analysis is the prediction of temperatures in an aircraft engine. Material selection, component life estimates and design decisions are based in large part on the temperature the parts attain and the duration of those temperatures. Heat transfer analyses are used to determine these temperatures. Engine test results are used to validate the temperature predictions. The permissible time between engine overhauls can depend on these temperatures. The design engineers then use these results to ensure reliable aircraft propulsion systems.

## Systems

Systems engineers work with the overall design, development, manufacture and operation of a complete system or product. Design engineers are involved in the design of individual components, but systems engineers are responsible for the integration of the components and systems into a functioning product.

Systems engineers are responsible for ensuring that the components interface properly and work as a complete unit. Systems engineers are also responsible for identifying the overall design requirements. This may involve working with customers or marketing personnel to accurately determine market needs. From a technical stand-point, systems engineers are responsible for meeting the overall design requirements.

Systems engineering is a field that most engineers enter only after becoming proficient in an area important to the systems, such as component design or development. Some graduate work often is required prior to taking on these assignments. However, there are some schools where an undergraduate degree in systems engineering is offered.

*Systems engineers work with the overall design, development, manufacture and operation of a complete system.*

## Manufacturing and Construction

Manufacturing engineers turn the specifications of the design engineer into a tangible reality. They develop processes to make the products we use every day. They work with diverse teams of individuals, from technicians on the assembly lines to management, in order to maintain the integrity and efficiency of the manufacturing process.

It is the responsibility of manufacturing engineers to develop the processes for taking raw materials and changing them into the finished pieces that the design engineers detailed. They utilize state-of-the-art machines and processes to accomplish this. As technology advances, new processes often must be developed for manufacturing the products.

*It is the responsibility of manufacturing engineers to develop the processes for taking raw materials and changing them into the finished pieces.*

The repeatability or quality of manufacturing processes is an area of increasing concern to modern manufacturing engineers. These engineers use statistical methods to determine the precision of a process. This is important, since a lack of precision in the manufacturing process may result in inferior parts that cannot be used or that may not meet the customer's needs. Manufacturing engineers are very concerned about the quality of the products they produce. High quality manufacturing means lower costs since more parts are usable, resulting in less waste. Ensuring quality means having the right processes in place, understanding the processes and working with the people involved to make sure the processes are being maintained at peak efficiency.

Manufacturing engineers also keep track of the equipment in a plant. They schedule and track required maintenance to keep the production line moving. They also must track the inventories of raw materials, partially finished parts and completely finished parts. Excessive inventories tie up substantial amounts of cash that could be used in other parts of the company. The manufacturing engineer is also responsible for maintaining a safe and reliable workplace, including the safety of the workers at the facility and the environmental impact of the processes.

Manufacturing engineers must be able to work with diverse teams of people including design engineers, tradesmen and management. Current "just in time" manufacturing practices reduce needed inventories in factories but require manufacturing engineers at one facility to coordinate their operation with manufacturing engineers at other facilities. They also must coordinate the work of the line workers who operate the manufacturing equipment. It is imperative that manufacturing engineers maintain a constructive relationship with their company's trade workers.

Manufacturing engineers play a critical role in modern design practices. Since manufacturing costs are such an important component in the success of a product, the design process must take into account manufacturing concerns. Manufacturing engineers identify high cost or high risk operations in the design phase of a product. When problems are identified, they work with the design engineers to generate alternatives.

In the production of large items such as buildings, dams and roads, the production engineer is called a construction engineer rather than a manufacturing engineer.

However, the role of the construction engineer is very similar to that of the manufacturing engineer. The main difference is that the construction engineer's production facility is typically outdoors while the manufacturing engineer's is inside a factory. The functions of construction engineers are the same as mentioned above when "assembly line" and "factory" are replaced with terms like "job site," reflecting the construction of a building, dam or other large-scale project.

## Operations and Maintenance

After a new production facility is brought on-line, it must be maintained. The operations engineer oversees the ongoing performance of the facility. Operations engineers must have a wide range of expertise dealing with the mechanical and electrical issues involved with maintaining a production line. They must be able to interact with manufacturing engineers, line workers and technicians who service the equipment. They must coordinate the service schedule of the technicians to ensure efficient service of the machinery, minimizing its downtime impact on production.

Maintenance and operations engineers also work in non-manufacturing roles. Airlines have staffs of maintenance engineers who schedule and oversee safety inspections and repairs. These engineers must have expertise in sophisticated inspection techniques to identify all possible problems.

Large medical facilities and other service sector businesses require operations or maintenance engineers to oversee the operation of their equipment. It is obviously critical that emergency medical equipment be maintained in peak working order.

## Technical Support

A technical support engineer serves as the link between customer and product and assists with installation and setup. For large industrial purchases, technical support may be included in the purchase price. The engineer may visit the installation site and oversee a successful start-up. For example, a new power station for irrigation might require that the technical support engineer supervises the installation and helps the customer solve problems to get the product operational. To be effective, the engineer must have good interpersonal and problem-solving skills as well as solid technical training.

The technical support engineer may also trouble-shoot problems with a product. Serving on a computer company's help line is one example. Diagnosing design flaws found in the field once the product is in use is another example.

Technical support engineers do not have to have in-depth knowledge of each aspect of the product. However, they must know how to tap into such knowledge at their company.

Modern technical support is being used as an added service. Technical support engineers work with customers to operate and manage their own company's equipment as well as others. For example, a medical equipment manufacturer might sell its services to a hospital to manage and operate its highly sophisticated equipment. The manufacturer's engineers would not only maintain the equipment, but also help the hospital use its facilities in the most efficient way.

## Customer Support

Customer support functions are similar to those of technical support as a link between the manufacturer and the customer. However, customer support personnel also are involved in the business aspect of the customer relationship. Engineers are often used for this function because of their technical knowledge and problem solving ability. Typically, these positions require experience with the products and customers, and also require some business training.

The customer support person works with technical support engineers to ensure proper customer satisfaction. Customer support is also concerned with actual or perceived value for customers. Are they getting what they paid for? Is it cost-effective to continue current practices? Customer support persons are involved in warranty issues, contractual agreements and the value of trade-in or credits for existing equipment. They work very closely with the technical support engineers and also with management personnel.

## Sales

Engineers are valuable members of the sales force in numerous companies. These engineers must have interpersonal skills conducive to effective selling. Sales engineers bring many assets to their positions.

Engineers have the technical background to answer customer questions and concerns. They are trained to identify which products are right for the customer and how they can be applied. Sales engineers can also identify other applications or other products that might benefit the customer once they become familiar with their customer's needs.

In some sales forces, engineers are utilized because the customers are engineers themselves and have engineering-related questions. When airplane manufacturers market their aircraft to airlines, they send engineers. The airlines have engineers who have technical concerns overseeing the maintenance and operation of the aircraft. Sales engineers have the technical background to answer these questions.

As technology continues to advance, more and more products become technically sophisticated. This produces an ever-increasing demand for sales engineers.

## Consulting

Consulting engineers are either self-employed or they work for a firm that does not provide goods or services directly to consumers. Such firms provide technical expertise to organizations that do. Many large companies do not have technical experts on staff in all areas of operation. Instead, they use consultants to handle issues in those technical areas.

For example, a manufacturing facility in which a cooling tower is used in part of the operation might have engineers who are well versed in the manufacturing processes but not in cooling tower design. The manufacturer would hire a consulting firm to design the cooling tower and related systems. Such a consultant might oversee the installation of such a system or simply provide a report with recommendations. After the system is in place or the report is delivered, the consultant would move on to another project with another company.

Consulting engineers also might be asked to evaluate the effectiveness of an organization. In such a situation, a team of consultants might work with a customer and provide suggestions and guidelines for improving the company's processes. These might be design methods, manufacturing operations or even business practices. While some consulting firms provide only engineering-related expertise, other firms provide both engineering and business support and require the consulting engineers to work on business-related issues as well as technical issues.

Consulting engineers interact with a wide range of companies on a broad scope of projects, and come from all engineering disciplines. Often a consultant needs to be registered as a professional engineer in the state where he or she does business.

**Management**

In many instances, engineers work themselves into project management positions, and eventually into full-time management. National surveys show that more than half of all engineers will be involved in some type of management responsibilities—supervisory or administrative—before their career is over. Engineers are chosen for their technical ability, their problem solving ability and their leadership skills.

Engineers may manage other engineers or support personnel, or they may rise to oversee the business aspects of a corporation. Often, prior to being promoted to this level of management, engineers acquire some business or management training. Some companies provide this training or offer incentives for employees to take management courses in the evening on their own time.

**Other Fields**

Some engineering graduates enter fields other than engineering, such as law, education, medicine and business. Patent law is one area in which an engineering or science degree is almost essential. In patent law, lawyers research, write, and file patent applications. Patent lawyers must have the technical background to understand what an invention does so that they can properly describe the invention and legally protect it for the inventor.

Another area of law that has become popular for engineering graduates is corporate liability law. Corporations are faced with decisions every day over whether to introduce a new product, and must weigh potential risks. Lawyers with technical backgrounds have the ability to weigh technical as well as legal risks. Such lawyers are also used in litigation for liability issues. Understanding what an expert witness is saying in a suit can be a tremendous advantage in a courtroom and can enable a lawyer to effectively cross-examine the expert witness. Often, when a corporation is sued over product liability, the lawyers defending the corporation have technical backgrounds.

Engineers are involved in several aspects of education. The one students are most familiar with is engineering professors. College professors usually have their Ph.D.'s. Engineers with master's degrees can teach at community colleges and in some engineering technology programs. Engineering graduates also teach in high schools, middle schools, and elementary schools. To do this full-time usually requires additional training in educational methods, but the engineering background is a great start. Thousands of engineers are involved in part-time educational projects where they go to classes as guest speakers to show students what "real engineers" do. The beginning of this chapter described one such engineer. The American Society for Engineering Education (ASEE) is a great resource if you are interested in engineering education.

Engineers also find careers in medicine, business, on Wall Street and in many other professions. In modern society, with its rapid expansion of technology, the combination of problem solving ability and technical knowledge makes engineers very valuable and extremely versatile.

## Example 2.3

A dean of a New York engineering school passed along a story about her faculty, which was concerned that most of that year's graduates had not accepted traditional engineering positions. The largest employers were firms that hire stockbrokers. The school's engineering graduates were prized for their ability to look at data and make rational conclusions from the data. In other words, they were very effective problem solvers. They also had the ability to model the data mathematically. This combination made the engineering graduates a perfect fit for Wall Street. This is not an isolated story. Many engineering graduates find themselves in careers unrelated to their engineering education. All engineers are trained problem solvers, and in today's technically advanced society, such skills are highly regarded.

## 2.3 Engineering Majors

The following section is a partial listing of the various engineering disciplines in which engineering students can earn degrees. The list includes all of the most common majors. Some smaller institutions may offer programs that are not mentioned in this chapter. Be aware that some disciplines are referred to by different names at various institutions. Also, some programs may be offered as a subset of other disciplines. For instance, aeronautical and industrial engineering is sometimes combined with mechanical engineering. Environmental engineering might be offered as part of civil or chemical engineering.

These descriptions are not meant to be comprehensive. To describe a field completely, such as electrical engineering, would take an entire book by itself. These descriptions are meant to give you an overview of the fields and to answer some basic questions about the differences among the fields. It is meant to be a starting point. When selecting a major, a student might investigate several sources, including people actually working in the field.

An engineering student should keep in mind that the list of engineering fields is fluid. Areas such as aerospace, genetic, computer and nuclear engineering did not even exist 50 years ago. Yet men and women developed the technology to create these fields. In your lifetime, other new fields will be created. The objective is to gain the solid background and tools to handle future challenges. It is also important to find a field you enjoy.

### Example 2.4

**"Show me the money!"** As academic advisors, we see students regularly who are interested in engineering but not sure which discipline to pursue. Students are tempted to decide which field to enter by asking which offers the greatest salary. There are two issues a student should consider when looking at salaries of engineers.

First, all engineers make a good starting salary—well over the average American household income. However, a high salary would not make up for a job that is hated. The salary spread between the engineering disciplines is not large, varying only by about 10% from the average. The career you embark on after graduation from college will span some forty years—a long time to spend in a discipline you don't really enjoy.

Second, if you consider money a critical factor, you should consider earning potential, not starting salary. Earning potential in engineering is dependent, to a large extent, on job performance. Engineers who do well financially are those who excel professionally. Again, it is very rare for someone to excel in a field that he or she does not enjoy.

# Aerospace Engineering

**Aerospace engineering** was previously referred to as aeronautical and astronautical engineering. Technically, aerospace engineering involves flight within the Earth's atmosphere and had its birth when two bicycle repairmen from Dayton, Ohio, made the first flight at Kitty Hawk, North Carolina, in 1903. Since that time, aerospace engineers have designed and produced aircraft that have broken the sound barrier, achieved near-orbit altitudes, and become the standard mode of transportation for long journeys. Aerospace engineering involves flight in space which began on October 4, 1957, when the Soviet Union launched Sputnik into orbit. The United States achieved one of its proudest moments in the field of aerospace on July 20, 1969, when Neil Armstrong set foot on the moon's surface—the first person ever to do so. In order to describe what the broad field of aerospace engineering has become, it can be separated into the categories of aerodynamics, propulsion, structures, controls, orbital mechanics, and life sciences.

*Aerospace engineering can be separated into the categories of aerodynamics, propulsion, structures, controls, orbital mechanics, and life sciences.*

**Aerodynamics** is the study of the flow of air over a streamlined surface or body. The aerodynamicist is concerned with the lift that is produced by a body such as the wing of an aircraft. Similarly, there is a resistance or drag force when a body moves through a fluid. An engineer looks to optimize lift and minimize drag. While mechanical, civil and chemical engineers also study flow over bodies, the aerospace engineer is interested especially in high-speed air flows. When the speed of the aircraft approaches or exceeds the speed of sound, the modeling of the airflow becomes much more complex.

Air-breathing propulsion systems that power airplanes and helicopters include propellers, and gas turbine jets, as well as ramjets and scram-jets. Propulsion systems used to launch or operate spacecraft include liquid and solid rocket engines. **Propulsion engineers** are continually developing more efficient, quieter and cleaner-burning conventional engines as well as new engine technologies. New engine concepts include wave rotors, electric propulsion, and nuclear-powered craft.

The structural support of an aircraft or spacecraft is critical. It must be both lightweight and durable. These conditions are often mutually exclusive and provide challenges for structural design engineers. **Structural engineers** utilize new alloys, composites and other new materials to meet design requirements for more efficient and more maneuverable aircraft and spacecraft.

The control schemes for aircraft and spacecraft have evolved rapidly. The new commercial airplanes are completely digitally controlled. Aerospace engineers work with electrical engineers to design **control systems** and subsystems used to operate the craft. They must be able to understand the electrical and computational aspects of the control schemes as well as the physical systems that are being controlled.

Aerospace engineers are also interested in **orbital mechanics**. They calculate where to place a satellite to operate as a communication satellite or as a global positioning system (GPS). They might also determine how to use the gravity fields of the near-Earth planets to help propel a satellite to the outskirts of the solar system.

An aerospace engineer must be aware of human limitations and the effects of their craft on the human body. It would serve no useful purpose to design a fighter plane that was so maneuverable it incapacitated the pilot with a high G-force. Designing a plane and a system which keeps the pilot conscious is an obvious necessity. Understanding the physiological and psychological affects of lengthy exposure to weightlessness is important when designing a spacecraft or space station.

The American Institute of Aeronautics and Astronautics (AIAA) is one of the most prominent aerospace professional societies. It is composed of the following seven technical groups, which include 66 technical committees.

- Engineering and Technology Management
- Aircraft Technology Integration and Operations
- Propulsion and Energy
- Space and Missile Systems
- Aerospace Sciences
- Information and Logistics Systems
- Structures, Design and Testing

**Agricultural engineering** traces its roots back thousands of years to the time when people began to examine ways to produce food and food products more efficiently. Today, the role of the agricultural engineer is critical to our ability to feed the ever-expanding population of the world. The production and processing of agricultural products is the primary concern of agricultural engineers. Areas within agricultural engineering include power machinery, food processing, soils and water, structures, electrical technologies, and bioengineering.

## Agricultural Engineering

*Areas include power machinery, food processing, soils and water, structures, electrical technologies, and bioengineering.*

The mechanical equipment used on modern farms is highly sophisticated and specialized. **Harvesting equipment** not only removes the crop from the field but also begins to process it and provides information to the farmers on the quality of the harvest. This requires highly complicated mechanical and electrical systems that are designed and developed by the agricultural engineer.

Often, once the food is harvested it must be processed before it reaches the marketplace. Many different technologies are involved in the efficient and safe processing and delivery of food products. **Food process engineers** are concerned with providing healthier products to consumers who increasingly rely on processed food products. This often requires the agricultural engineer to develop new processes. Another modern-day concern is increased food safety. Agricultural engineers design and develop means by which food is produced free of contamination such as the irradiation techniques used to kill potentially harmful bacteria in food.

The effective management of **soil and water resources** is a key aspect of productive agriculture. Agricultural engineers design and develop means to address effective land use, proper drainage and erosion control. This includes such activities as designing and implementing an irrigation system for crop production and designing a terracing system to prevent erosion in a hilly region.

Agricultural **structures** are used to house harvested crops, livestock and their feed. Agricultural engineers design structures including barns, silos, dryers and processing centers. They look for ways to minimize waste or losses, optimize yields and protect the environment.

**Electrical and information technology** development is another area important to the agriculture community due to the fact that farms are typically located in isolated regions. Agricultural engineers design systems that meet the needs of the rural communities.

**Bioengineering** has rapidly evolved into a field with wide uses in health products as well as agricultural products. Agricultural engineers are working to harness these rapidly developing technologies to further improve the quality and quantity of the agricultural products necessary to continue to feed the world's population.

The American Society of Agricultural Engineers (ASAE) is one of the most prominent professional societies for agricultural engineers. It has eight technical divisions which address areas of agricultural engineering:

- Food Processing
- Information and Electrical Technologies
- Power and Machinery
- Structures and Environmental

- Soil and Water
- Forest
- Bioengineering
- Aqua culture

## Architectural Engineering

In ancient times, major building projects required a master builder. The master builder was responsible for the design and appearance of the structure, for selecting the appropriate materials, and for personally overseeing construction. With the advent of steel construction in the 19th century, it became necessary for master builders to consult with specialists in steel construction in order to complete their designs. As projects became more complex, other specialists were needed in such areas as mechanical and electrical systems. Modern architectural engineers facilitate the coordination between the creativity of the architect and the technical competencies of a variety of technology specialists.

Architectural engineers are well-grounded in the engineering fundamentals of structural, mechanical and electrical systems, and have at least a foundational understanding of aesthetic design. They know how to design a building and how the various technical systems are interwoven within that design. Their strengths are both creative and pragmatic.

The National Society of Architectural Engineers, which is now the Architectural Engineering Institute, defined architectural engineering as:

*". . . the profession in which a knowledge of mathematics and natural sciences, gained by study, experience, and practice, is applied with judgement to the development of ways to use, economically and safely, the materials and forces of nature in the engineering design and construction of buildings and their environmental systems."* [Belcher]

There are four main divisions within architectural engineering: structural; electrical and lighting; mechanical systems; and construction engineering and management.

**Structural** is primarily concerned with the integrity of the structure of buildings. Determining the integrity of a building's structure involves the analysis of loads and forces resulting from normal usage and operation as well as from earthquakes, wind forces, snow loads, or wave impacts. The structural engineer takes the information from the analysis, and designs the structural elements that will support the building while at the same time meeting the aesthetic and functional needs involved.

The area of **electrical and lighting** systems is concerned with the distribution of utilities and power throughout the building. This requires knowledge of electricity and power distribution as well as lighting concerns. Lighting restrictions may require energy-efficient lighting systems that are sufficient to meet the functional requirements related to use of the building. The architectural engineer also must insure that the lighting will complement the architectural designs of the building and the rooms within that structure.

**Mechanical systems** control the climate of a building, which includes cooling and heating the air in rooms as well as controlling humidity and air quality. Mechanical systems are also used to distribute water through plumbing systems. Other mechanical systems include transportation by way of elevators and escalators within a building. These systems must be integrated to complement the architectural features of the building.

The fourth area is **construction engineering and management**, which combine the technical requirements with the given financial and legal requirements to meet project deadlines. The architectural engineer is responsible for implementing the design in a way which assures the quality of the construction and meets the cost and schedule of the project. Modern computer tools and project management skills are implemented to manage such complex projects.

Other areas which are merging for architectural engineers include energy management, computerized controls, and new building materials including plastics and composites and acoustics.

Architectural engineers are employed by consulting firms, contractors, and government agencies. They may work on complex high-rise office buildings, factories, stadiums, research labs, or educational facilities. They also may work on renovating historic structures or developing affordable low-income housing. As the construction industry continues to grow and as projects continue to become more complex, the outlook for architectural engineers is bright.

There are less than 20 ABET accredited programs in architectural engineering in the United States. Some of these are four-year programs while others provide the combined engineering foundation and architectural insights in a five-year program.

The Architectural Engineering Institute (AEI) was formed in 1998 with the merger of the National Society of Architectural Engineers and the American Society of Civil Engineers Architectural Engineering Division. AEI was created to be the home for professionals in the building industry. It is organized into divisions, which include:

- Commercial Buildings
- Industrial Buildings
- Residential Buildings
- Institutional Buildings
- Military Facilities
- Program Management
- Mitigation of the Effects of Terrorism
- Building Systems
- Education
- Architectural Systems
- Fully Integrated and Automated Project Process
- Glass as an Engineered Material
- Sick/Healthy Buildings
- Designing for Facilities for the Aging

**Reference:** Belcher, M. C., Magill's Survey of Science: Applied Science, Salem Press, Pasadena, CA, 1993.

## Biomedical Engineering

**Biomedical engineering** is one of the newer fields of engineering, first recognized as a discipline in the 1940s. Its origins, however, date back to the first artificial limbs made of wood or other materials. Captain Ahab in "Moby Dick" probably didn't think of the person who made his wooden leg as a biomedical engineer, yet he was a predecessor for biomedical engineers who design modern prosthetics.

*Biomedical engineering encompasses the three basic categories of medical, clinical, and bioengineering.*

Biomedical engineering is a very broad field that overlaps with several other engineering disciplines. In some institutions, it may be a specialization within another discipline. Biomedical engineering applies the fundamentals of engineering to meet the needs of the medical community. Because of the wide range of skills needed in the fields of engineering and medicine, biomedical engineering often requires graduate work. The broad field of biomedical engineering encompasses the three basic categories of medical, clinical, and bioengineering.

**Bioengineering** is the application of engineering principles to biological systems. This can be seen in the production of food or in genetic manipulation to produce a disease-resistant strain of a plant or animal. Bioengineers work with geneticists to produce bioengineered products. New medical treatments are produced using genetically altered bacteria to produce human proteins needed to cure diseases. Bioengineers may work with geneticists in producing these new products in the mass quantities needed by consumers.

The medical aspect of biomedical engineering involves the design and development of devices to solve medical challenges. This includes designing mechanical devices such as a prosthesis which give individuals the desired mobility. It also involves the development of the chemical processes necessary to make an artificial kidney function and the electrical challenges in designing a new pacemaker.

**Medical engineers** develop instrumentation for medical uses including non-intrusive surgical instruments. Much of the trauma of surgery results from the incisions made to gain access to the area of concern. Procedures allowing a surgeon to make a small incision, such as orthoscopic surgery, have greatly reduced risk and recovery time for patients.

Rehabilitation is another area in which biomedical engineers work. It involves designing and developing devices for the physically impaired. Such devices can expand the capabilities of impaired individuals, thereby improving their quality of life or shortening recovery times.

**Clinical engineering** involves the development of systems to serve hospitals and clinics. Such systems exhaust anesthetic gases from an operating room without impairing the surgical team. Air lines in ventilating systems must be decontaminated to prevent micro-organisms from spreading throughout the hospital. Rehabilitation centers must be designed to meet the needs of patients and staff.

## Chemical Engineering

**Chemical engineering** differs from most of the other fields of engineering in its emphasis on chemistry and the chemical nature of products and processes. Chemical engineers take what chemists do in a laboratory and, applying fundamental engineering, chemistry and physics principles, design and develop processes to mass-produce products for use in our society. These products include detergents, paints, plastics, fertilizers, petroleum products, food products, pharmaceuticals, electronic circuit boards and many others.

The most common employment of chemical engineers is in the design, development and operation of large-scale chemical production facilities. In this area, the design function involves the **design of the processes** needed to safely and reliably produce the final product. This may involve controlling and using chemical reactions, separation processes, or heat and mass transfer. While the chemist might develop a new compound in the laboratory, the chemical engineer would develop a new process to make the compound in a pilot plant, which is a small-scale version of a full-size production facility. An example would be the design of a process to produce a lower-saturated-fat product with the same nutritional value yet still affordable. Another example would be the development of a process to produce a higher-strength plastic used for automobile bumpers.

*Chemical engineers design and develop processes to mass-produce detergents, paints, plastics, fertilizers, petroleum products, food products, pharmaceuticals, electronic circuit boards and many others.*

With respect to **energy sources,** chemical engineers are involved in the development of processes to extract and refine crude oil and natural gas. Petroleum engineering grew out of chemical engineering and is described in a separate section. Chemical engineers are also involved in alternative fuel development and production.

The processing of petroleum into plastics by chemical engineers has created a host of consumer products that are used every day. Chemical engineers are involved in adapting these products to meet new needs. GE Plastics, for instance, operates a research house in Massachusetts made entirely from plastic. It is used as a research facility to develop new building materials that are cheaper and more beneficial to the consumer.

Many of the chemicals and their byproducts used in industry can be dangerous to people and/or the environment. Chemical engineers must develop processes that minimize harmful waste. They work with both new and traditional processes to treat hazardous byproducts and reduce harmful emissions.

Chemical engineers are also very active in the bio-products arena. This includes the **pharmaceutical industry**, where chemical engineers design processes to manufacture new lines of affordable drugs. Geneticists have developed the means to artificially produce human proteins that can be used successfully to treat many diseases. They use genetically altered bacteria to produce these proteins. It is the job of the chemical engineer to take the process from the laboratory and apply it to a larger-scale production process to produce the quantities needed by society.

Chemical engineers are also involved in **bio-processes** such as dialysis, where chemical engineers work along with other biomedical engineers to develop new ways to treat people. Chemical engineers might be involved in the development of an artificial kidney. They could also be involved in new ways to deliver medicines, such as through skin implants or patches.

Chemical engineers have become very active in the production of **circuit boards**, such as those used in computers. To manufacture the very small circuits required in modern electronic devices, material must be removed and deposited very precisely along the path of the circuit. This is done using chemical techniques developed by chemical engineers. With the demand for smaller and faster electronics, challenges continue for chemical engineers to develop the processes to make them possible.

Chemical engineers are also involved in the **modeling of systems**. Computer models of manufacturing processes are made so that modifications can be examined without having to build expensive new facilities. Managing large processes involving facilities which could be spread around the globe requires sophisticated scheduling capabilities. Chemical engineers who developed these capabilities for chemical plants have found that other large industries with scheduling challenges, such as airlines, can utilize these same tools.

The American Institute of Chemical Engineering (AIChE) is one of the most prominent professional organizations for chemical engineers. It is organized into 13 technical divisions that represent the diverse areas of the chemical engineering field:

- Catalysis and Reaction Engineering
- Computing and Systems Technology
- Engineering and Construction Contracting
- Environmental
- Food, Pharmaceutical and Bioengineering
- Forest Products
- Fuels and Petrochemicals

- Heat Transfer and Energy Conversion
- Management
- Materials Engineering and Sciences
- Nuclear Engineering
- Safety and Health
- Separations

## Civil Engineering

*Civil engineering emphasizes structural, environmental, transportation, water, surveying, planning, and construction.*

Ancient examples of early **civil engineering** can be seen in the pyramids of Egypt, the Roman roads, bridges and aqueducts of Europe, and the Great Wall in China. These all were designed and built under the direction of the predecessors to today's civil engineers. Modern civil engineering continues to face the challenges of meeting the needs of society. The broad field of civil engineering includes these categories: structural, environmental, transportation, water resources, surveying, urban planning, and construction engineering.

A humorous aside perhaps illustrates one aspect of civil engineering.... An Air Force general with an engineering background was asked what the difference is between civil and aerospace engineers. "That's easy," he responded. "The aerospace engineers build modern, state-of-the-art weapon systems and the civil engineers build the targets."

**Structural engineers** are the most common type of civil engineer. They are primarily concerned with the integrity of the structure of buildings, bridges, dams and highways. Structural engineers evaluate the loads and forces to which a structure will be subjected. They analyze the structural design in regard to earthquakes, wind forces, snow loads or wave impacts, depending on the area in which the building will be constructed.

A related field of structural engineering is **architectural engineering**, which is concerned with the form, function and appearance of a structure. The architectural engineer works alongside the architect to ensure the structural integrity of a building. The architectural engineer combines analytical ability with the concerns of the architect.

Civil engineers in the **environmental** area may be concerned with the proper disposal of wastes—residential and industrial. They may design and adapt landfills and waste treatment facilities to meet community needs. Industrial waste often presents a greater challenge because it may contain heavy metals or other toxins which require special disposal procedures. Environmental engineering has come to encompass much more and is detailed in a later section of this chapter.

**Transportation engineers** are concerned with the design and construction of highways, railroads and mass transit systems. They are also involved in the optimization and operation of the systems. An example of this is **traffic engineering**. Civil engineers develop the tools to measure the need for traffic control devices such as signal lights and to optimize these devices to allow proper traffic flow. This can become very complex in cities where the road systems were designed and built long ago but the areas around those roads have changed significantly with time.

Civil engineers also work with water resources as they construct and maintain dams, aqueducts, canals and reservoirs. **Water resource engineers** are charged with providing safe and reliable supplies of water for communities. This includes the design and operation of purification systems and testing procedures. As communities continue to grow, so do the challenges for civil engineers to produce safe and reliable water supplies.

Before any large construction project can be started, the construction site and surrounding area must be mapped or surveyed. **Surveyors** locate property lines and establish alignment and proper placement of engineering projects. Modern surveyors use satellite technology as well as aerial and terrestrial photogrammetry. They also rely on computer processing of photographic data.

A city is much more than just a collection of buildings and roads. It is a home and working place for people. As such, the needs of the inhabitants must be taken into account in the design of the city's infrastructure. **Urban planning engineers** are involved in this process. They incorporate the components of a city (buildings, roads, schools, airports, etc.) to meet the overall needs of the population. These needs include adequate housing, efficient transportation and open spaces. The urban planning engineer is always looking toward the future to fix problems before they reach a critical stage.

Another civil engineering area is **construction engineering**. In some institutions, this may even be a separate program. Construction engineers are concerned with the management and operation of construction projects. They are also interested in the improvement of construction methods and materials. Construction engineers design and develop building techniques and building materials that are safer, more reliable, cost-effective and environmentally friendly. There are many technical challenges that construction engineers face and will continue to face in the coming decades.

An example of one such construction challenge is the rebuilding of a city's infrastructure, such as its sewers. While the construction of a sewer may not seem glamorous or state-of-the-art, consider the difficulty of rebuilding a sewer under an existing city. A real problem facing many cities is that their infrastructure was built decades or centuries ago, but it has a finite life. As these infrastructures near the end of their expected lives, the construction engineer must refurbish or reconstruct these systems without totally disrupting the city.

The American Society of Civil Engineers (ASCE) is one of the most prominent professional organizations for civil engineers. It is organized into the following 16 technical divisions covering the breadth of civil engineering:

- Aerospace
- Air Transport
- Architectural Engineering
- Construction Division
- Energy
- Engineering Mechanics
- Environmental Engineering
- Geomatics
- Highway
- Materials Engineering
- Pipeline
- Urban Planning and Development
- Urban Transportation
- Water Resources Engineering
- Water Resources Planning and Management
- Waterways, Ports, Coastal and Ocean Engineering

## Computer Engineering

*The computer engineer focuses primarily on computer hardware.*

Much of what computer and electrical engineers do overlaps. Many **computer engineering** programs are part of electrical engineering or computer science programs. However, computer technology and development have progressed so rapidly that specialization is required in this field, and thus computer engineering is treated separately. Given the wide range of computer applications and their continued growth, computer engineering has a very exciting future.

Computer engineering is similar to computer science, yet distinct. Both fields are extensively involved with the design and development of software. The main difference is that the computer scientist focuses primarily on the software and its optimization. The computer engineer, by contrast, focuses primarily on computer hardware—the machine itself. Software written by the computer engineer is often designed to control or to interface more efficiently with the hardware of the computer and its components.

Computer engineers are involved in the design and development of operating systems, compilers and other software that requires efficient interfacing with the components of the computer. They also work to improve computer performance by optimizing the software and hardware in applications such as computer graphics.

Computer engineers work on the design of computer architecture. Designing faster and more efficient computing systems is a tremendous challenge. As faster computers are developed, applications arise for even faster machines. The continual quest for faster and smaller microprocessors involves overcoming barriers introduced by the speed of light and by circuitry so small that the molecular properties of components become important.

Computer engineers develop and design electronics to interface with computers. These include modems, Ethernet connections and other means of data transmission. Computer engineers created the devices that made the Internet possible.

In addition to having computers communicate with each other, the computer engineer is also interested in having computers work together. This may involve increasing the communication speed of computer networks or using tightly coupled multiple processors to improve the computing speed.

Security is becoming a bigger concern as more and more information is transferred using computers. Computer engineers are developing new means of commercial and personal security to protect the integrity of electronic communications.

Artificial intelligence, voice recognition systems and touch screens are examples of other technologies with which computer engineers are involved. The computers of the future will look and operate much differently than they do now, and it is the computer engineer who will make this happen.

## Electrical Engineering

Considering the wide range of electronic devices people use every day, it is not surprising that **electrical engineering** has become the most populated of the engineering disciplines. Electrical engineers have a wide range of career opportunities in almost all of the industrial sectors. In order to provide a brief discussion of such a broad field, we will divide electrical engineering into eight areas: computers, communications, circuits and solid state devices, control, instrumentation, signal processing, bioengineering, and power.

Engineers specializing in **computer** technology are in such high demand that numerous institutions offer a separate major for computer engineering. Please refer back to the earlier section which described computer engineering separately.

Electrical engineers are responsible for the explosion in **communication** technologies. Satellites provide nearly instantaneous global communication. Global positioning systems (GPS) allow anyone with the required handheld unit to pinpoint precisely where they are located anywhere in the world. Fiber optics and lasers are rapidly improving the reliability and speed with which information can be exchanged. Wireless communication allows people to communicate anywhere, with anyone. Future breakthroughs in this field will have a tremendous impact on how we live in tomorrow's society.

Electrical engineers also design and develop electronic **circuits**. Circuit design has changed rapidly with the advent of microelectronics. As circuits continue to shrink, new barriers such as molecular size emerge. As the limits of current technology are approached, there will be incentives to develop new and faster ways to accomplish the same tasks.

Almost all modern machines and systems are digitally controlled. **Digital controls** allow for safer and more efficient operation. Electronic systems monitor processes and make corrections faster and more effectively than human operators. This improves reliability, efficiency and safety. The electrical engineer is involved in the design, development and operation of these control systems. The engineer must determine the kind of control required, what parameters to monitor, the speed of the correction, and many other factors. Control systems are used in chemical plants, power plants, automotive engines, airplanes and a variety of other applications.

For a control system to operate correctly, it must be able to measure the important parameters of whatever it is controlling. Doing so requires accurate **instrumentation**, another area for the electrical engineer. Electrical engineers who work in this area develop electrical devices to measure quantities such as pressure, temperature, flowrate, speed, heart rate and blood pressure. Often, the electrical devices convert the measured quantity to an electrical signal that can be read by a control system or a computer. Instrumentation engineers also design systems to transmit the measured information to a recording device, using telemetry. For example, such systems are needed for transmitting a satellite's measurements to the recording computers back on earth as it orbits a distant planet.

**Signal processing** is another area where electrical engineers are needed. In many instances the electrical signals coming from instrumentation or other sources must be conditioned before the information can be used. Signals may need to be electronically filtered, amplified or modified. An example is the active noise control system on a stethoscope which allows a paramedic to listen to a patient's heart while in a helicopter. The active noise control system can block out the sound of the helicopter so the paramedic can make a quick, accurate assessment of the patient. Signal processing also comes into play in areas such as voice recognition for computers.

Electrical engineers also work in biomedical, or **bioengineering,** applications, as described earlier. Electrical engineers work with medical personnel to design and develop devices used in the diagnosis and treatment of patients. Examples include nonintrusive techniques for detecting tumors through magnetic resonance imaging (MRI) or through computer automated tomography (CAT) scans. Other examples include pacemakers, cardiac monitors and controllable prosthetic devices.

*Electrical engineering includes computers, communications, circuits and solid state devices, control, instrumentation, signal processing, bioengineering, and power.*

The generation, transmission and distribution of electric **power** is perhaps the most traditional aspect of electrical engineering. Electrical engineers work closely with mechanical engineers in the production of electrical power. They also oversee the distribution of power through electrical networks and must ensure reliable supplies of electricity to our communities. With modern society's dependence on electricity, interruptions in the flow of electricity can be catastrophic. Many of today's power-related challenges revolve around reliability and cost of delivery. Electrical engineers also work with materials engineers to incorporate superconductivity and other technology in more efficient power transmission.

The Institute of Electrical and Electronics Engineers (IEEE) is the largest and most prominent professional organization for electrical engineers. It is organized into 37 technical divisions, which indicates the breadth of the field of electrical engineering. The technical divisions include:

- Aerospace and Electronic Systems
- Antennas and Propagation
- Broadcast Technology
- Circuits and Systems
- Communications
- Components Packaging and Manufacturing Technology
- Computer
- Consumer Electronics
- Control Systems
- Dielectrics and Electrical Insulation
- Education
- Electromagnetic Compatibility
- Electron Devices
- Engineering in Medicine and Biology
- Engineering Management
- Instrumentation and Measurement
- Lasers and Electro-Optics
- Magnetics
- Microwave Theory and Techniques
- Neural Networks
- Nuclear and Plasma Sciences
- Oceanic Engineering
- Power Electronics
- Power Engineering
- Professional Communication
- Reliability
- Robotics and Automation
- Signal Processing
- Social Implications of Technology
- Solid-State Circuits
- Geoscience and Remote Sensing
- Industrial Electronics
- Industrial Applications
- Information Theory
- Systems, Man and Cybernetics
- Ultrasonics, Ferroelectrics and Frequency Control
- Vehicular Technology

## Environmental Engineering

**Environmental engineering** is a field which has evolved to improve and protect the environment while maintaining the rapid pace of industrial activity. This challenging task has three parts to it: disposal, remediation, and prevention.

**Disposal** is similar to that covered under civil engineering. Environmental engineers are concerned with disposal and processing of both industrial and residential waste. Landfills and waste treatment facilities are designed for residential waste concerns. The heavy metals and other toxins found in industrial wastes require special disposal procedures. The environmental engineer develops the techniques to properly dispose of such waste.

**Remediation** involves the cleaning up of a contaminated site. Such a site may contain waste which was improperly disposed of, requiring the ground and/or water to be removed or decontaminated. The environmental engineer develops the means to remove the contamination and return the area to a usable form.

**Prevention** is an area that environmental engineers are becoming more involved in. Environmental engineers work with manufacturing engineers to design processes that reduce or eliminate harmful waste. One example is in the cleaning of machined parts. Coolant is sprayed on metal parts as they are cut to extend the life of the cutting tools. The oily fluid clings to the parts and has to be removed before the parts are assembled. An extremely toxic substance had been used in the past to clean the parts because there was not a suitable alternative. Environmental engineers discovered that oil from orange peels works just as well and is perfectly safe (even edible). Since this new degreasing fluid did not need to be disposed of in any special way, manufacturing costs were reduced, and the environment of the workers improved.

Environmental engineers must be well grounded in engineering fundamentals and current on environmental regulations. Within their companies, they are the experts on compliance with the ever-changing environmental laws. An environmental engineer must be able to understand the regulations and know how to apply them to the various processes they encounter.

**Industrial Engineering** is described by the Institute of Industrial Engineers as the design, improvement and installation of integrated systems of people, material and energy. Industrial engineering is an interdisciplinary field that involves the integration of technology, mathematical models and management practices. Traditional industrial engineering is done on a factory floor. However, the skills of an industrial engineer are transferable to a host of other applications. As a result, industrial engineers find themselves working within a wide variety of industries. Four of the main areas of emphasis for industrial engineers are production, manufacturing, human factors, and operations research.

## Industrial Engineering

*Areas of emphasis for industrial engineers are production, manufacturing, human factors, and operations research.*

The **production** area includes functions such as plant layout, material handling, scheduling, and quality and reliability control. An industrial engineer would examine the entire process involved in making a product, and optimize it by reducing cost and production time, and by increasing quality and reliability. In addition to factory layout, industrial engineers apply their expertise in other ways. For example, with an amusement park an industrial engineer would analyze the flow of people through the park to reduce bottlenecks and provide a pleasant experience for the patrons.

**Manufacturing** differs from production in that it addresses the components of the production process. While production concerns are on a global scale, manufacturing concerns address the individual production station. The actual material processing, such as machining, is optimized by the industrial engineer.

The **human factors area** involves the placement of people into the production system. An industrial engineer in this area studies the interfaces between people and machines in the system. The machines may include production machinery, computers or even office chairs and desks. The industrial engineer considers ergonomics in finding ways to improve the interfaces. He or she looks for ways to improve productivity while providing a safe environment for workers.

**Operations research** is concerned with the optimization of systems. This involves mathematically modeling systems to identify ways to improve them. Project management techniques such as critical path identification fall under operations research. Often, computer simulations are required to either model the system or to study the effects of changes to the system. These systems may be manufacturing systems or other organizations. The optimizing of sales territories for a pharmaceutical sales force provides a non-manufacturing example.

The Institute for Industrial Engineering (IIE) is one of the most prominent professional organizations for industrial engineers. It is organized into three societies, 10 technical divisions and 8 interest groups. The following technical divisions show the breadth of industrial engineering:

- Aerospace and Defense
- Energy, Environment and Plant Engineering
- Engineering Economy
- Facilities Planning and Design
- Financial Services
- Logistics Transportation and Distribution
- Manufacturing
- Operations Research
- Quality Control and Engineering Reliability
- Utilities

## Marine and Ocean Engineering

Nearly 80 percent of the Earth's surface is covered with water. Engineers concerned with the exploration of the oceans, the transportation of products over water, and the utilization of resources in the world's oceans, seas and lakes are involved in **marine and ocean engineering**.

Marine engineers focus on the design, development and operation of ships and boats. They work together with **naval architects** in this capacity. Naval architects are concerned with the overall design of the ship. They focus on the shape of the hull in order to provide the appropriate hydrodynamic characteristics. They are also concerned with the usefulness of the vessel for its intended purpose, and with the design of the ship's sub-systems including ventilation, water and sanitary systems to allow the crew to work efficiently.

The **marine engineer** is primarily concerned with the subsystems of the ship which allow the ship to serve its purpose. These include the propulsion, steering and navigation systems. The marine engineer might analyze the ship for vibrations or stability in the water. The ship's electrical power distribution and air-conditioning fall under the responsibility of marine engineers. They also might be involved in the analysis and design of the cargo handling systems of the ship.

The responsibilities of an **ocean engineer** involve the design, development and operation of vehicles and devices other than boats or ships. These include submersible vehicles used in the exploration of the oceans and in the obtaining of resources from the ocean depths. He or she might be involved in the design of underwater pipelines or cables, offshore drilling platforms and offshore harbor facilities.

Ocean engineers also are involved with the interaction of the oceans and things with which oceans come in contact. They study wave action on beaches, docks, buoys, moorings and harbors. Ocean engineers design ways to reduce erosion while protecting the marine environment. They study ways to protect and maintain marine areas which are critical to our food supply. Ocean engineers become involved with pollution control and treatment in the sea and alternative sources of energy from the ocean.

One of the professional societies to which these engineers may be involved is the Society of Naval Architects and Marine Engineers. The society is subdivided into the following nine technical and research committees:

- Hull Structure
- Hydrodynamics
- Ship's Machinery
- Ship Technical Operations
- Offshore

- Ship Production
- Ship Design
- Ship Repair and Conversion
- Small Craft

The origins of **materials engineering** can be traced to around 3000 B.C. when people began to produce bronze for use in creating superior hunting tools. Since that time, many new materials have been developed to meet the needs of society. Materials engineers develop these new materials and the processes to create them. The materials may be metals or non-metals: ceramics, plastics and composites. Materials engineers are generally concerned with four areas of materials: structure, properties, processes, and performance.

Materials engineers study the **structure** and composition of materials on a scale ranging from the microscopic to the macroscopic. They are interested in the molecular bonding and chemical composition of materials. The materials engineer is also concerned with the effect of grain size and structure on the material properties .

The **properties** in question might include strength, crack growth rates, hardness and durability. Numerous technological advances are impeded by a lack of materials possessing the properties required by the design engineers. Materials engineers seek to develop materials to meet these demands.

A given material may have very different properties depending on how the material is **processed.** Steel is a good example. Cooling can affect its properties drastically. Steel that is allowed to cool in air will have different properties than steel that is cooled through immersion in a liquid. The composition of a material also can affect its properties. Materials such as metallic alloys contain trace elements that must be evenly distributed throughout the alloy to achieve the desired properties. If the trace elements are not well distributed or form clumps in the metal, the material will have very different properties than the desired alloy. This could cause a part made with the alloy to fail prematurely. Materials engineers design processes and testing procedures to ensure that the material has the desired properties.

The materials engineer also works to ensure that a material meets the **performance** needs of its application by designing testing procedures to ensure that these requirements are met. Both destructive and nondestructive testing techniques are used to serve this process.

Materials engineers develop new materials, improve traditional materials and produce materials reliably and economically through synthesis and processing. Subspecialties of materials engineering, such as metallurgy and ceramics engineering, focus on classes of materials with similar properties.

**Metallurgy** involves the extraction of metals from naturally occurring ore for the development of alloys for engineering purposes. The metallurgical engineer is concerned with the composition, properties and performance of an alloy. Detailed investigation of a component failure often identifies design flaws in the system. The materials engineer can provide useful information regarding the condition of materials to the design engineer.

**Ceramics** is another area of materials engineering. In ceramic engineering, the naturally occurring materials of interest are clay and silicates, rather than an ore. These non-metallic minerals are employed in the production of materials that are used in a

## Materials Engineering

*Materials engineers are concerned with structure, properties, processes, and performance.*

wide range of applications, including the aerospace, computer and electronic industries.

Other areas of materials engineering focus on polymers, plastics and composites. **Composites** are composed of different kinds of materials which are synthesized to create a new material to meet some specific demands. Materials engineers are also involved in **biomedical** applications. Examples include the development of artificial tissue for skin grafts, or bone replacement materials for artificial joints.

One of the professional societies to which materials engineers may belong is the Minerals, Metals and Materials Society. It is organized into the following five technical divisions:

- Electronic, Magnetic and Photonic Materials
- Extraction and Processing
- Light Metals
- Materials Processing and Manufacturing
- Structural Materials

## Mechanical Engineering

*Mechanical engineers are involved in the design, development, production, control, operation and service of mechanical devices.*

**Mechanical engineering** is one of the largest and broadest of the engineering disciplines. It is second only to electrical engineering in the number of engineers employed in the field. Fundamentally, mechanical engineering is concerned with machines and mechanical devices. Mechanical engineers are involved in the design, development, production, control, operation and service of these devices. Mechanical engineering is composed of two main divisions: design and controls, and thermal sciences.

The **design** function is the most common function of mechanical engineering. It involves the detailed layout and assembly of the components of machines and devices. Mechanical engineers are concerned about the strength of parts and the stresses the parts will need to endure. They work closely with materials engineers to ensure that correct materials are chosen. Mechanical engineers must also ensure that the parts fit together by specifying detailed dimensions.

Another aspect of the design function is the design process itself. Mechanical engineers develop computational tools to aid the design engineer in optimizing a design. These tools speed the design process by automating time-intensive analyses.

Mechanical engineers are also interested in controlling the mechanical devices they design. **Control** of mechanical devices can involve mechanical or hydraulic controls. However, most modern control systems incorporate digital control schemes. The mechanical engineer models controls for the system and programs or designs the control algorithm.

The noise generated from mechanical devices is often a concern, so mechanical engineers are often involved in acoustics—the study of noise. The mechanical engineer works to minimize unwanted noise by identifying the source and designing ways to minimize it without sacrificing a machine's performance.

In the **thermal sciences**, mechanical engineers study the flow of fluids and the flow of energy between systems. Mechanical engineers deal with liquids, gases and two-phase flows, which are combinations of liquids and non-liquids. Mechanical engineers might be concerned about how much power is required to supply water through piping systems in buildings. They might also be concerned with aerodynamic drag on automobiles.

The flow of energy due to a temperature difference is called **heat transfer**, another thermal science area in which mechanical engineers are involved. They predict and study the temperature of components in environments of operation. Modern personal computers have microprocessors that require cooling. Mechanical engineers design the cooling devices to allow the electronics to function properly.

Mechanical engineers design and develop **engines**. An engine is a device that produces mechanical work. Examples include internal combustion engines used in automobiles and gas turbine engines used in airplanes. Mechanical engineers are involved in the design of the mechanical components of the engines as well as the overall cycles and efficiencies of these devices.

Performance and efficiency are also concerns for mechanical engineers involved in the production of power in large **power generation systems.** Steam turbines, boilers, water pumps and condensers are often used to generate electricity. Mechanical engineers design these mechanical components needed to produce the power that operates the generators. Mechanical engineers also are involved in alternative energy sources including solar and hydroelectric power, and alternative fuel engines and fuel cells.

Another area in the thermal sciences is **heating, ventilating and air-conditioning** (HVAC). Mechanical engineers are involved in the climate control of buildings, which includes cooling and heating the air in buildings as well as controlling humidity. In doing so, they work closely with civil engineers in designing buildings to optimize the efficiency of these systems.

Mechanical engineers are involved in the **manufacturing processes** of many different industries. They design and develop the machines used in these processes, and develop more efficient processes. Often, this involves automating time-consuming or expensive procedures within a manufacturing process. Mechanical engineers also are involved in the development and use of **robotics** and other automated processes.

In the area of **biomedical engineering**, mechanical engineers help develop artificial limbs and joints that provide mobility to physically impaired individuals. They also develop mechanical devices used to aid in the diagnosis and treatment of patients.

The American Society of Mechanical Engineering (ASME) is one of the most prominent professional societies for mechanical engineers. It is divided into 35 technical divisions, indicating the diversity of this field. These divisions are:

- Advanced Energy Systems
- Aerospace Engineering
- Applied Mechanics
- Basic Engineering Technical Group
- Bioengineering
- Design Engineering
- Dynamic Systems and Control
- Electrical and Electronic Packaging
- FACT
- Fluids Engineering
- Fluids Power Systems and Technology Systems
- Heat Transfer
- Information Storage/Processing
- Internal Combustion Engine
- Gas Turbine
- Manufacturing Engineering
- Materials
- Materials Handling Engineering
- Noise Control and Acoustics
- Non-destructive Evaluation Engineering
- Nuclear Engineering
- Ocean Engineering
- Offshore Mechanics / Arctic Engineering
- Petroleum
- Plant Engineering and Maintenance
- Power
- Pressure Vessels and Piping

- Process Industries
- Rail Transportation
- Safety Engineering and Risk Analysis
- Solar Energy
- Solid Waste Processing

- Technology and Society
- Textile Engineering
- Tribology

## Mining Engineering

Modern society requires a vast amount of products made from raw materials such as minerals. The continued production of these raw materials helps to keep society functioning. **Mining engineers** are responsible for maintaining the flow of these raw materials by discovering, removing and processing minerals into the products society requires.

**Discovering** the ore involves exploration in conjunction with geologists and geophysicists. The engineers combine the utilization of seismic, satellite and other technological data, utilizing a knowledge of rocks and soils. The exploration may focus on land areas, the ocean floor, or even below the ocean floor. In the future, mining engineers may also explore asteroids, which are rich in mineral deposits.

Once mineral deposits are identified, they may be **removed**. One way minerals are removed is by way of mining tunnels. The engineers design and maintain the tunnels and the required support systems including ventilation and drainage. Other times, minerals are removed from open pit mines. Again, the engineers analyze the removal site and design the procedure for removing the material. The engineer also develops a plan for returning the site to a natural state. Mining engineers use boring, tunneling and blasting techniques to create a mine. Regardless of the removal technique, the environmental impact of the mining operation is taken into account and minimized.

The mining engineer is also involved in the **processing** of the raw minerals into usable forms. Purifying and separating minerals involves chemical and mechanical processes. While mining engineers may not be involved in producing a finished product that consumers recognize, they must understand the form their customers can use and design processes to transform the raw materials into usable forms.

Mining engineers also become involved in the **design** of the specialized equipment required for use in the mining industry. The design of automated equipment capable of performing the most dangerous mining jobs helps to increase safety and productivity of a mining operation. Since mines typically are established in remote areas, mining engineers are involved in the **transportation** of minerals to the processing facility.

The expertise of mining engineers is not used exclusively by the mining industry. The same boring technology used in developing mines is used to create subway systems and railroad tunnels, such as the one under the English Channel.

**Nuclear engineers** are concerned primarily with the use and control of energy from nuclear sources. This involves electricity production, propulsion systems, waste disposal and radiation applications.

The production of **electricity** from nuclear energy is one of the most visible applications of nuclear engineering. Nuclear engineers focus on the design, development and operation of nuclear power facilities. This involves using current fission technology as well as the development of fusion, which would allow sea water to be used as fuel. Nuclear energy offers an environmentally friendly alternative to fossil fuels. A current barrier to production of nuclear facilities is the high cost of construction. This barrier provides a challenge for design engineers to overcome. Research is currently being performed on the viability of smaller, more efficient nuclear reactors.

Nuclear power is also used in **propulsion systems.** It provides a power source for ships and submarines, allowing them to go years without refueling. It is also used as a power source for satellites. Nuclear-powered engines are being examined as an alternative to conventional fossil-fueled engines, making interplanetary travel possible.

One of the main drawbacks to nuclear power is the production of **radioactive waste**. This also creates opportunities for nuclear engineers to develop safe and reliable means to dispose of spent fuel. Nuclear engineers develop ways to reprocess the waste into less hazardous forms.

Another area in which nuclear engineers are involved is the use of **radiation** for medical or agricultural purposes. Radiation therapy has proven effective in treating cancers, and radioactive isotopes are also used in diagnosing diseases. Irradiating foods can eliminate harmful bacteria and help ensure a safer food supply.

Due to the complex nature of nuclear reactions, nuclear engineers are at the forefront of advanced computing methods. High-performance computing techniques, such as parallel processing, constitute research areas vital to nuclear engineering.

The American Nuclear Society (ANS) is one of the professional societies to which nuclear engineers belong. It is divided into these 16 technical divisions:

# Nuclear Engineering

*Nuclear engineers are concerned with electricity production, propulsion systems, waste disposal and radiation applications.*

- Biology and Medicine
- Decommissioning, Decontamination and Reutilization
- Education and Training
- Environmental Sciences
- Fuel Cycle and Waste Management
- Fusion Energy
- Human Factors
- Isotopes and Radiation
- Materials Science and Technology
- Mathematics and Computations
- Nuclear Criticality Safety
- Nuclear Operations
- Nuclear Installations Safety
- Power
- Radiation Protection and Shielding
- Reactor Physics

## Petroleum Engineering

Petroleum and petroleum products are essential components in today's society. **Petroleum engineers** maintain the flow of petroleum in a safe and reliable manner. They are involved in the exploration for crude oil deposits, the removal of oil, and the transporting and refining of oil.

Petroleum engineers work with geologists and geophysicists to identify potential oil and gas reserves. They combine satellite information, seismic techniques and geological information to **locate deposits of gas or oil**. Once a deposit has been identified, it can be removed. The petroleum engineer designs, develops and operates the needed drilling equipment and facilities. Such facilities may be located on land or on offshore platforms. The engineer is interested in removing the oil or gas in a safe and reliable manner—safe for the people involved as well as for the environment. **Removal of oil** is done in stages, with the first stage being the easiest and using conventional means. Oil deposits are often located in sand. A significant amount of oil remains coating the sand after the initial oil removal. Recovery of this additional reserve requires the use of secondary and tertiary extraction techniques utilizing water, steam or chemical means.

*Petroleum engineers are involved in the exploration for crude oil deposits, the removal of oil, and the transporting and refining of oil.*

**Transporting** the oil or gas to a processing facility is another challenge for the petroleum engineer. At times this requires the design of a heated pipeline such as the one in Alaska to carry oil hundreds of miles over frozen tundra. In other instances this requires transporting oil in double-hulled tankers from an offshore platform near a wildlife refuge. Such situations necessitate extra precautions to ensure that the wildlife is not endangered.

Once the oil or gas arrives at the processing facility, it must be **refined** into usable products. The petroleum engineer designs, develops and operates the equipment to chemically process the gas or oil into such end products. Petroleum is made into various grades of gasoline, diesel fuel, aircraft fuel, home heating oil, motor oils, and a host of consumer products from lubricants to plastics.

## Other Fields

The most common engineering majors have been described in this chapter. However, there are other specialized engineering programs at some institutions. Here is a partial listing of some of these other programs:

- Automotive Engineering
- Acoustical Engineering
- Applied Mathematics
- Bioengineering
- Engineering Science
- Engineering Management
- Excavation Engineering
- Fire Engineering
- Forest Engineering
- General Engineering
- Genetic Engineering
- Geological Engineering
- Inventive Design
- Manufacturing Engineering
- Packaging Engineering
- Pharmaceutical Engineering
- Plastics Engineering
- Power Engineering
- Systems Engineering
- Theatre Engineering
- Transportation Engineering
- Welding Engineering

# 2.4 Emerging Fields

The fields of engineering have been, and will continue to be, dynamic. As new technologies emerge, new definitions are needed to classify disciplines. The boundaries will continue to shift and new areas will emerge. The explosion of technological advances means that there is a good chance you will work in a field that is not currently defined. Technological advances are also blurring the traditional delineations between fields. Areas not traditionally linked are coming together and providing cross-disciplinary opportunities for engineers. A good example of this is in the development of smart buildings which sense the onset of an earthquake, adapt their structures to survive the shaking, and then return to normal status afterward. Research is being conducted on these technologies that bridge computer engineering with structural (Civil) engineering to produce such adaptable buildings.

The incredible breakthroughs in biology have opened many new possibilities and will continue to impact most of the fields of engineering. Historically, biology has not been as integrated with engineering as has physics and chemistry, but that is rapidly changing and will have enormous impact on the future of engineering. The ability to modify genetic codes has implications in a wide range of engineering applications including the production of pharmaceuticals that are customized for individual patients, alternative energy sources, and environmental reclamation.

Nanotechnology is an area that is receiving a great deal of attention and resources, and is blurring the boundaries of the fields of engineering. Nanotechnology is an emerging field in which new materials and tiny structures are built atom-by-atom, or molecule-by-molecule, instead of the more conventional approach of sculpting parts from pre-existing materials. The possibilities for nano-applications include:

- the creation of entirely new materials with superior strength, electrical conductivity, resistance to heat, and other properties
- microscopic machines for a variety of uses, including probes that could aid diagnostics and repair
- a new class of ultra-small, super-powerful computers and other electronic devices, including spacecraft
- a technology in which biology and electronics are merged, creating "gene chips" that instantly detect food-borne contamination, dangerous substances in the blood, or chemical warfare agents in the air
- the development of "molecular electronics" and devices that "self assemble," similar to the growth of complex organic structures in living organisms

Nanotechnology requires specialized laboratory and production facilities that provide further challenges and opportunities for future engineers.

The explosion of information technology with the Internet and wireless communication has produced a melding of disciplines to form new fields in information science and technology. Information management and transfer is an important and emerging issue in all disciplines of engineering and has opened opportunities for engineers who want to bridge the gaps between the traditional fields and information technologies.

Undoubtedly, more new fields will open and be discovered as technology continues to advance. As the boundaries of the genetic code and molecular-level device are crossed, new frontiers will open. As an engineer, you will have the exciting opportunity to be part of the discovery and definition of these emerging fields which will have tremendous impact on society's future.

## 2.5 Closing Thoughts

The information presented in this chapter is meant to provide a starting point on the road to choosing a career. There may have been aspects of one or more of these engineering fields that appealed to you, and that's great. The goal, however, is not to persuade you that engineering is for everyone; it is not. The goal is to provide information to help you decide if engineering would be an enjoyable career for *you*.

As a student, it is important to choose a career path that will be both enjoyable and rewarding. For many, an engineering degree is a gateway to just such a career. Each person has a unique set of talents, abilities and gifts that are well matched for a particular career. In general, people find more rewarding careers in occupations where their gifts and talents are well used. Does engineering match your talents, abilities and interests?

Right now, choosing a career may seem overwhelming. But at this point, you don't have to. What you are embarking on is an education that will provide the base for such decisions. Think about where a degree in engineering could lead. One of the exciting aspects of an engineering education is that it opens up a wide range of jobs after leaving college. Most students will have several different careers before they retire, and the important objective in college is obtaining a solid background that will allow you to move into areas that you enjoy later in life.

As you try to make the right choice for you, seek out additional information from faculty, career centers, professional societies, placement services and industrial representatives. Ask a lot of questions. Consider what you would enjoy studying for four years in college. What kind of entry level job would you be able to get with a specific degree? What doors would such a degree open for you later in life? Remember, your decision is unique to you. No one can make it for you.

# 2.6 Engineering and Technical Organizations

The following is a list of many of the engineering technical societies available to engineers. They can be a tremendous source of information for you. Many have student branches which allow you to meet both engineering students and practicing engineers in the same field. Join one or more of these organizations during your freshman or sophomore year.

**American Association for the Advancement of Science**
1200 New York Avenue, NW
Washington, DC 20005
(202) 326-6400
www.aaas.org

**American Association of Engineering Societies**
1111 19th Street, NW
Suite 403
Washington, DC 20036
(202) 296-2237
www.aaes.org

**American Ceramic Society**
735 Ceramic Place
Westerville, OH 43081-8720
(614) 890-4700
www.acers.org

**American Chemical Society**
1155 16th Street, NW
Room 1209
Washington, DC 20036-1807
(202) 872-4600
www.acs.org

**American Concrete Institute**
38800 Country Club Drive
Farmington Hills, MI 48331
(248) 848-3700
www.aci-int.org

**American Congress on Surveying and Mapping**
5410 Grosvenor Lane
Suite 100
Bethesda, MD 20814-2122
(301) 493-0200

**American Consulting Engineers Council**
1015 15th St., NW, Suite 802
Washington, DC 20005
(202) 347-7474
www.acec.org

**American Gas Association**
1515 Wilson Blvd.
Arlington, VA 22209
(703) 841-8400
www.aga.com

**American Indian Science and Engineering Society**
5661 Airport Blvd.
Boulder, CO 80301-2339
(303) 939-0023
www.colorado.edu/AISES

**American Institute of Aeronautics and Astronautics**
1801 Alexander Bell Drive, Suite 500
Reston, VA 20191-4344
(800) NEW-AIAA or (703) 264-7500
www.aiaa.org

**American Institute of Chemical Engineers**
345 East 47th Street
New York, NY 10017-2395
(212) 705-7000 or (800) 242 4363
www.aiche.org

**American Institute of Mining, Metallurgical and Petroleum Engineers**
345 East 47th Street
New York, NY 10017
(212) 705-7695

**American Nuclear Society**
555 North Kensington Avenue
La Grange Park, IL 60526
(708) 352-6611
www.ans.org

**American Oil Chemists' Society**
1608 Broadmoor Drive
Champaign, IL 61821-5930
(217) 359-2344
www.aocs.org

**American Railway Engineering Association**
8201 Corporate Drive
Landover, MD
(301) 459-3200

**American Society of Agricultural Engineers**
2950 Niles Road
St. Joseph, MI 49085-9659
(616) 429-0300
www.asae.org

**American Society of Civil Engineers**
1801 Alexander Bell Drive
Reston, VA 20191-4400
(800) 548-2723 or (703) 295-6000
www.asce.org

**American Society of Naval Engineers**
1452 Duke Street
Alexandria, VA 22314
(703) 836-6727
www.jhuapl.edu/ASNE

**American Society for Engineering Education**
1818 N St., NW, Suite 600
Washington, DC 20036-2479
(202) 331-3500
www.asee.org

**American Society for Heating, Refrigeration and Air Conditioning Engineers**
1791 Tulie Circle NE
Atlanta, GA 30329
(404) 636-8400
www.ashrae.org

**American Society of Mechanical Engineers**
345 East 47th Street
New York, NY 10017-2392
(212) 705-7722
www.asme.org

**American Society of Plumbing Engineers**
3617 Thousand Oaks Blvd.
Suite 210
Westlake Village, CA 91362-3649
(805) 495-7120
www.aspe.org

**American Society of Nondestructive Testing**
1711 Arlingate Lane
P.O. Box 28518
Columbus, OH 43228-0518
(614) 274-6003
www.asnt.org

**American Society for Quality**
611 East Wisconsin Avenue
Milwaukee, WI 53202
(414) 272-8575
www.asq.org

**American Water Works Association**
6666 West Quincy Avenue
Denver, CO 80235
(303) 443-9353
www.aws.org

**The Architectural Engineering**
   **Institute**
1801 Alexander Bell Drive, 1$^{st}$ Floor
Reston, VA 20191-4400
(703) 295-6370   Fax (703) 295-6132
http://www.aeinstitute.org

**Association for the Advancement of**
   **Cost Engineering International**
209 Prairie Avenue
Suite 100
Morgantown, WV  26505
(304) 296-8444
www.aacei.org

**Board of Certified Safety Professionals**
208 Burwash Avenue
Savoy, IL 61874-9571
(217) 359-9263
www.bcsp.com

**Construction Specifications Institute**
601 Madison Street
Alexandria, VA 22314-1791
(703) 684-0300
www.csinet.org

**Information Technology Association of**
   **America**
1616 N. Fort Myer Drive, Suite 1300
Arlington, VA 22209
(703) 522-5055
www.itaa.org

**Institute of Electrical and Electronics**
   **Engineers**
1828 L Street NW, Suite 1202
Washington, DC 20036
(202) 785-0017
www.ieee.org

**Institute of Industrial Engineers**
25 Technology Park
Norcross, GA 30092
(770) 449-0461
www.iienet.org

**Iron and Steel Society**
410 Commonwealth Drive
Warrendale, PA 15086-7512
(412) 776-1535
www.issource.org

**Laser Institute of America**
12424 Research Parkway, Suite 125
Orlando, FL 32826
(407) 380-1553
www.laserinstitute.org

**Mathematical Association of**
   **America**
1529 18th Street, NW
Washington, DC 20036
(202) 387-5200
www.maa.org

**The Minerals, Metals and Materials**
   **Society**
420 Commonwealth Drive
Warrendale, PA 15086
(412) 776-9000
www.tms.org

**NACE International**
1440 South Creek Drive
Houston, TX 77084-4906
(281) 492-0535
www.nace.org

**National Academy of Engineering**
2101 Constitution Avenue, NW
Washington, DC 20418
(202) 334-3200

**National Action Council for Minorities in Engineering, Inc.**
The Empire State Building
350 Fifth Avenue, Suite 2212
New York, NY 10118-2299
(212) 279-2626
www.naof cme.org

**The National Association of Minority Engineering Program Administrators, Inc.**
1133 West Morse Blvd., Suite 201
Winter Park, FL 32789
(407) 647-8839
(407) 629-2502 Fax

**National Association of Power Engineers**
1 Springfield Street
Chicopee, MA 01013
(413) 592-6273
www.powerengineers.com

**National Institute of Standards and Technology**
Publications and Programs Inquiries
Public and Business Affairs
Gaithersburg, MD 20899
(301) 975-3058
www.nist.gov

**National Science Foundation**
4201 Wilson Blvd.
Arlington, VA 22230
(703) 306-1234
www.nsf.gov

**National Society of Black Engineers**
1454 Duke Street
Alexandria, VA 22314
(703) 549-2207
www.nsbe.org

**National Society of Professional Engineers**
1420 King Street
Alexandria, VA 22314
(888) 285-6773
www.nspe.org

**Society of Allied Weight Engineers**
5530 Aztec Drive
La Mesa, CA 91942
(619) 465-1367

**Society of American Military Engineers**
607 Prince Street
Alexandria, VA 22314
(703) 549-3800 or (800) 336-3097
www.same.org

**Society of Automotive Engineers**
400 Commonwealth Drive
Warrendale, PA 15096
(412) 776-4841
www.sae.org

**Society of Fire Protection Engineers**
7315 Wisconsin Avenue
Suite 1225W
Bethesda, MD 20814
(301) 718-2910

**Society of Hispanic Professional Engineers**
5400 East Olympic Blvd.
Suite 210
Los Angeles, CA 90022
(213) 725-3970
www.engr.umd.edu/organiza-
tions/shpe

**Society of Manufacturing Engineers**
One SME Drive
P.O. Box 930
Dearborn, MI 48121-0930
(313) 271-1500
www.sme.org

**Society for Mining, Metallurgy**
  **Exploration, Inc.**
8307 Shaffer Parkway
Littleton, CO 80127
(303) 973-9550
www.smenet.org

**Society of Naval Architects and**
  **Marine Engineers**
601 Pavonia Avenue
Jersey City, NJ  07306
(800) 798-2188
www.sname.org

**Society of Petroleum Engineers**
P.O. Box 833836
Richardson, TX 75083-3836

**Society of Plastics Engineers**
14 Fairfield Drive
Brookfield, CT 06804-0403
(203) 775-0471
www.4spe.org

**Society of Women Engineers**
120 Wall Street
11th Floor
New York, NY 10005
(212) 509-9577
www.swe.org

**SPIE–International Society for**
  **Optical Engineering**
P.O. Box 10
Bellingham, WA 98227-0010
(360) 676-3290
www.spie.org

**Tau Beta Pi**
508 Dougherty Engineering Hall
P.O. Box 2697
Knoxville, TN 37901-2697
(423) 546-4578
www.tbp.org

**Women in Engineering Initiative**
University of Washington
101 Wilson Annex
P.O. Box 352135
Seattle, WA 98195-2135
(206) 543-4810
www.engr.washington.edu/~wieweb

# References

Burghardt, M.D., *Introduction to the Engineering Profession, 2nd Edition,* Harper Collins College Publishers, 1995.

Garcia, J., *Majoring {Engineering},* The Noonday Press, New York, New York, 1995.

Grace, R., and J. Daniels, *Guide to 150 Popular College Majors,* College Entrance Examination Board, New York, New York, 1992, pp. 175-178.

Kemper, J.D., *Engineers and Their Profession,* 4th Edition, Oxford University Press, New York, 1990.

Irwin, J.D., *On becoming An Engineer,* IEEE Press,  New York, 1997.

Landis, R., *Studying Engineering, A Road Map to a Rewarding Career*, Discovery Press, Burbank, California, 1995.

Smith, R.J., B.R. Butler, W.K. LeBold, *Engineering as a Career,* 4th Edition, McGraw-Hill Book Company, New York, 1983.

Wright, P.H., *Introduction to Engineering,* 2nd Edition, John Wiley and Sons, Inc., New York, 1994.

# Exercises and Activities

2.1     Contact a practicing engineer in a field of engineering that interests you. Write a brief report on his or her activities and compare them to the description of that field of engineering as described in this chapter.

2.2     For a field of engineering that interests you, make a list of potential employers that hire graduates from your campus, and list the cities in which they are located.

2.3     Visit a job fair on your campus and briefly interview a company representative. Prepare a brief report on what that company does, what the engineer you spoke to does, and the type of engineers they are looking to hire.

2.4     Make a list of companies that hire co-op and/or intern students from your campus. Write a brief report on what these companies are looking for and their locations.

2.5     Select a company that employs engineers in a discipline that interests you, and visit their web page. (You can search for them using Yahoo or Alta Vista, or you can call them and ask for their Web address.) Prepare a brief report on what the company does, hiring prospects, engineering jobs in that organization, and where the company is located.

2.6     Contact a person with an engineering degree who is not currently employed in a traditional engineering capacity. Write a one-page paper on how that person uses his or her engineering background in their job.

2.7     Write a one-page paper on an engineering field that will likely emerge during your lifetime (a field that does not currently exist). Consider what background an engineering student should obtain in preparation for this emerging field.

2.8     Draft a sample letter requesting a co-op or intern position.

2.9     Identify a modern technological problem. Write a brief paper on the role of engineers and technology in solving this problem.

2.10    Select an engineering discipline that interests you and a particular job function within that discipline. Write a brief paper contrasting the different experiences an engineer in this discipline would encounter in each of three different industries.

2.11    Make a list of your own strengths and talents. Write a brief report on how these strengths are well matched with a specific engineering discipline.

2.12    Pick an engineering discipline that interests you. List and briefly describe the technical and design electives available in that discipline for undergraduates.

2.13 Select a consumer product you are familiar with (a stereo, clock radio, automobile, food product, etc.). List and briefly describe the role of all the engineering disciplines involved in producing the product.

2.14 Select an engineering discipline that interests you. Write a brief paper on how the global marketplace has altered this discipline.

2.15 Write a brief paper listing two similarities and two differences for each of the following engineering functions:
a) research and development
b) development and design
c) design and manufacturing
d) manufacturing and operations
e) sales and customer support
f) management and consulting

2.16 Find out how many engineering programs your school offers. How many students graduate each year in each discipline?

2.17 Which of the job functions described in this chapter is most appealing to you? Write a brief paper discussing it why it is appealing.

2.18 Write a paper about an engineer who made a significant technical contribution to society.

2.19 Report on the requirements for becoming a registered professional engineer in your state. Also report on how registration would be beneficial to your engineering career.

2.20 Make a list of general responsibilities and obligations you would have as an engineer.

2.21 Write a brief paper on the importance of ethical conduct as an engineer.

2.22 Select one of the industrial sectors within engineering and list five companies that do business in that sector. Briefly describe job opportunities in each.

2.23 Prepare a report on how the following items work and what engineering disciplines are involved in their design and manufacture:
a) CD player
b) CAT scan machine
c) Computer disk drive
d) Dialysis machine
e) Flat TV screen

2.24 Answer the following questions and look for themes or commonality in your answers. Comment on how engineering might fit into these themes.
a) If I could snap my fingers and do whatever I wanted, knowing that I wouldn't fail, what would I do?
b) At the end of my life, I'd love to be able to look back and know that I had done something about _____.
c) What would my friends say I'm really interested in or passionate about?
d) What conversation could keep me talking late into the night?
e) What were the top five positive experiences I've had in my life, and why were they meaningful to me?

2.25    Write a short paper describing your dream job, regardless of pay or geographical location.

2.26    Write a short paper describing how engineering might fit into your answer to 2.25.

2.27    List the characteristics a job must have for you to be excited to go to work every morning. How does engineering fit with those characteristics?

2.28    List the top ten reasons why a student should study engineering.

2.29    List the ten inappropriate reasons for a student to choose to study engineering.

2.30    Select an industrial sector and describe what you suppose a typical day is like for:
    a) Sales engineer                 d) Manufacturing engineer
    b) Research engineer            e) Design engineer
    c) Test engineer

2.31    Make a list of five non-engineering careers an engineering graduate could have and describe each one briefly.

2.32    Select one of the professional societies that was listed at the end of this chapter. Prepare a report on what the organization does and how it could benefit you as a practicing engineer.

2.33    Select one of the professional societies that was listed at the end of this chapter. Identify one of its technical divisions and prepare a report on that division's activities.

2.34    Identify one of the student branches of an engineering professional society on your campus and prepare a report on the benefits of involvement with that organization for an engineering student.

2.35    Contact one of the professional societies listed in this chapter and have them send you information. Prepare a brief presentation summarizing the material.

2.36    Write a letter to a $9^{th}$ grade class explaining the exciting opportunities in your chosen major within engineering. Include a short discussion of the classes they should be taking to be success in that same major.

2.37    Write a letter to a $9^{th}$ grade class describing the opportunities a bachelor's degree in engineering provides in today's society.

2.38    Write a letter to your parents detailing why you are going to major the field you have chosen.

2.39    Write a one-page paper describing how your chosen field of engineering will be different in 25 years.

2.40   Select one field of engineering and write a one page paper on how the advances in biology have influenced that field.

2.41   Select one field of engineering and write a one page paper on how the advances nanotechnology have influenced that field.

2.42   Select one field of engineering and write a one page paper on how the advances information technology influenced that field.

2.43   Select two fields of engineering and describe problems that span these two disciplines.

2.44   Write a brief paper on an emerging area within engineering.  Relate the area to your chosen engineering major.

2.45   Research the current spending priorities of the U.S. government in the area of technical research.  How will these priorities impact your chosen major?

2.46   For each grouping of engineering disciplines, describe applications or problems that span the disciplines.
  a.  Aerospace, Materials and Civil
  b.  Mechanical, Agricultural and Computer
  c.  Biological and Environmental
  d.  Industrial, Chemical and Electrical
  e.  Biomedical and nuclear
  f.  Agricultural and Aerospace
  g.  Materials and Biomedical
  h.  Civil and Computer

2.47   Select on professional organization and find out how they handle new and emerging technologies within their society (where do they put them and where can people working in emerging areas find colleagues?)

2.48   Interview a practicing engineering and a faculty member from the same discipline about their field.  Compare and contrast their views of the discipline.

2.49   Select two engineering majors and compare and contrast the opportunities available between the two.

2.50   Identify how your skills as an engineer can be used within your local community – either as full time work or as a volunteer.  Share your findings with the class by preparing a short oral presentation.

# 3. Profiles of Engineers

## 3.1 Introduction

This chapter contains a collection of profiles of engineering graduates to let you read first-hand accounts of what it is really like to be an engineer. Each engineer wrote his or her own profile.

Our intent is to provide you with a glimpse of the diversity of the engineering workforce. Engineers are people just like you, and had varying reasons for pursuing engineering as a career. Engineering graduates also take very diverse career paths.

In order to capture a flavor of this diversity, each engineer was asked to address three areas:

1. Why or how they became an engineer
2. Their current professional activities
3. Their life outside of work

This is not a comprehensive survey of the engineering workforce. To truly represent the breadth of engineering careers and the people in those careers would take several volumes, not just one chapter. This is meant to be only a beginning. It is also arranged simply to show you the wide range of engineering careers that are possible. To keep the wide-ranging feel, we've avoided ranking the profiles by subject, and instead present them alphabetically. This makes it easier to see the common bonds across all the disciplines.

We recommend that you follow up and seek out other practicing engineers or engineering students, to get more detailed information about the specific career path you might follow. As with the information we have collected in other chapters of this text, our goal is to assist you in finding that career path which is right for you. You are the one who must ultimately decide this for yourself.

The following Table 2.1 summarizes those who provided profiles.

TABLE 3.1  Summary of Profiled Engineers

| Name | BS Degree | Graduate Degree | Current Job Title |
|---|---|---|---|
| Sue Abreu | BS/E/BioMed | MD | Lt. Col and Medical Director |
| Moyosola Ajaja | BS/EE | | Software Engineer |
| Patrick Rivera Antony | BS/Aero | | Project Manager |
| Artagnan Ayala | BS/Aero | | Combustion Engineer |
| Sandra Begay-Campbell | BS/CE | MS/CE | Executive Director of AISES |
| Raymond C. Barrera | BS/EE | MS/Software | Computer Engineer |
| Linda Blevins | BS/ME | MS/ME & Ph.D. | Mechanical Engineer with NIST |
| Timothy Bruns | BS/EE | | Software Manager |
| Jerry Burris | BS/EE | MBA | General Manager |
| Bethany Fabin | BS/ABE | | Design Engineer |
| Bob Feldmann | BS/EE | MS/Comp, MBA | Director, Tactical Aircraft Systems |
| Steven Fredrickson | BS/EE | Ph.D. | Project Manager, NASA |
| Myron Gramelspacher | BS/ME | | Manufacturing Manager |
| Edgar Hammer | BS/ME/Mat | | President, SSI Consulting Group |
| Karen Jamison | BS/IE | MBA | Operations Manager |
| Beverly Johnson | BS/ME | MS/EM & MBA | Supervisor |
| James Lammie | BS/CE | MS/CE | Member of Board of Directors |
| Ryan Maibach | BS/CEM | | Project Engineer |
| Mary Maley | BS/AgE | | Product Manager |
| Jeanne Mordarski | BS/IE | MBA | Sales Manager |
| Mark Pashan | BS/EE | MS/EE & MBA | Director, Hardware Operations |
| Douglas Pyles | BS/E/Mgt | MS/CEM | Engineering Consultant |
| James Richard | BS/CEM | | Director of Mechanical Systems |
| Patrick Shook | BS/ME | MS/ME | Senior Engineer |
| David Swain | BS/AAE | | Executive Vice-President |
| Nana Tzeng | BS/ME | MS/ME | Design Engineer |
| Patrice Vanderbeck | BS/ME | | Electronics systems Engineer |
| Jack Welch | BS/ChE | MS/ChE & Ph.D. | Chief Executive Officer (retired) |
| Shawn Williams | BS/EE | | Product General Manager |
| Adel Zakaria | BS/ME | MS/IE & Ph.D. | Sr. VP Engr & Manufacturing |

## Profile of a Biomedical Engineer:
### Sue H. Abreu, Ft. Bragg, North Carolina

### Occupation
Lieutenant Colonel, Medical Corps, United States Army
Medical Director, Quality Assurance, Womack Army Medical Center

### Education
IDE (BSE, Biomedical Engineering), 1978
MD, Uniformed Services University of the Health Sciences, 1982

### Studying Engineering
As I started college, I was planning to be a teacher. Because of taking an elective class in athletic training, I developed an interest in sports medicine. I ended up taking most of my classes in aeronautical engineering so I could study the lightweight structures and materials that could be used to design artificial limbs or protective equipment for sports. Late in college, I decided to go to medical school and ended up graduating from college with an interdisciplinary engineering degree.

**Career Life**

After medical school, I specialized in nuclear medicine. In nuclear medicine we use small amounts of radioactive compounds to see how things work inside of people. By using special cameras that detect radiation and computers that help gather the data, we can watch how various organs function. We can do three-dimensional studies and quantify results. In nuclear medicine, I am a consultant to other physicians: I help them decide what tests might be helpful and discuss the meaning of the results of the nuclear medicine procedures we do for their patients.

I ended up in a field I never had heard of when I started college, but I found it as I kept exploring areas that intrigued me. I tried new classes and looked for opportunities that interested me, even if they didn't fit the paths most students followed. As a result, I found a specialty I enjoy, and I'm now doing a great deal of teaching within my specialty of nuclear medicine and in my current work in quality assurance.

So, be sure to follow your dreams—if you can take something you love doing and find a way to earn a living doing it, you will end up much happier than if you set money or prestige as your goals.

**Life Outside of Work**

Outside of work I enjoy skydiving. I volunteer as the team doctor for the U.S. Parachute Team and have traveled all over the world with them. I currently live in a large house on six acres in the country, not far from an airport with a parachuting center. I share the house with an airline pilot and an artist—both expert skydivers—who help make it a great place to live. Although I was married, I chose not to have children; but the dog and cats help keep us company here.

## Profile of a Computer Engineer:
## Moyosola O. Ajaja, Chandler, Arizona

### Occupation
Software Engineer at Intel Corporation

### Education
BS, Computer and Electrical Engineering, 1997

### Studying Engineering

I came into engineering the easy way—by excelling in math and physics in high school. Deciding to enroll at Purdue and pursue a dual degree in computer and electrical engineering was a little more complicated. I wanted to learn more about computers, and I wanted to seek my fortune in a distant land. I picked the U.S. and justified the 7000-mile journey from Lagos, Nigeria, where my family lived, to West Lafayette, Indiana, where Purdue is located, with the phrase "dual degree." (I understand that degree option is no longer available. Fellow adventurers will have to justify their journeys with a different explanation.)

During my first year at Purdue I set two goals for myself: first, find a scholarship to fund my education, and second, gain useful work experience. I applied for dozens of scholarships. I was partial to those offered by engineering firms that provided internships, since internships for first-year engineering students were very scarce. In addition, I attended every resume or interview preparation workshop offered during that year. My efforts paid off. I was invited to join the cooperative education program with a summer placement with Intel Corporation, and later I was awarded an Intel Foundation scholarship which paid my tuition.

### Career Life

Today, I work as a software engineer with Intel in Arizona. I develop hardware emulation units and validation test suites for new processors. What that means in plain English: I take descriptions of hardware features and functions and write software programs that behave like the hardware should. This is cheaper than fabricating silicon devices each time a change is made during design. As an Intel engineer with access to the latest and greatest technologies, I am constantly challenged to learn new things to remain at the leading edge of computer technology.

### Life Outside of Work

I have tried to maintain a balance between my work and my non-work activities. My weekend mornings are spent running with my dog or hiking up Camelback Mountain in Phoenix. The evenings are spent in classes like dog training, theology, or photography. My real passion is traveling, and through engineering school, internships, and my current assignment, I have met people who helped fuel my interest in increasingly diverse destinations.

I've discovered that engineering is a *discipline*, not just a major. The distinction here is that a discipline involves the development of the faculties through instruction and exercise, while a major is simply a field of study, an area of mental focus, or a concentration. For me this means the qualities of an engineer should be apparent in all I do. The guide I use is the Code of Ethics approved by the IEEE, which is presented in the Ethics chapter.

## Profile of an Aerospace Engineer:
### Patrick Rivera Antony

**Occupation:**
Project Manager, Boeing Space Beach

**Education:**
BS, Aerospace Engineering

## Studying Engineering
When Apollo 11 landed on the moon I was eight years old. I still remember to this day how excited I was and exactly where I was and how I felt when I saw Neil Armstrong step off the ladder of the lunar module and become the first man to walk on the moon. Math and science was always my favorite subject in school, so it gave me a good background to get started in my college studies.

Along with space travel, rockets and airplanes have always fascinated me. The whole concept of being able to fly and be free is amazing. I also knew that if I wanted to become an astronaut, I would have to learn to fly. So I got my pilot's license and found that the fundamental principles of flying are exactly what you learn in the courses taken for Aerospace Engineering.

Now, I would probably put more emphasis on my non-engineering classes. To be a good engineer now, you have to be able to communicate effectively, work in a team environment, and know different aspects of the business. Being good technically is not enough anymore. I would take more English, Business, and Social Science classes to make me a more well rounded individual and more valuable employee.

**Career Life**

Currently I am the Project Manager of the Performance Management and Continuous Improvement team for Boeing Space Beach, HB/SB Host Site Engineering. My responsibilities include developing and maintaining the management system used for organizational oversight, planning and operational assessment. I coordinate company policy deployment of business goals and strategies from the Vision Support Plan to the Engineering functional organization. Additionally, I interface with Program Management and Technical Managers to define performance requirements and develop innovative process and organizational solutions.

In my current management role, I spend a lot of time in planning meetings, so every day is different. I also do not often have an opportunity to use my engineering education, but as a leader of an organization, my engineering background allows me to communicate competently on technical matters related to the business.

**Life Outside of Work**

Most of my free time I try to spend with my wife and two children. But when I am not with them, I am studying. I am currently pursuing a Doctorate degree in Organizational Psychology with an emphasis in Change Management. With what little free time I have after that, I enjoy golf, hiking, fishing, and biking.

## Profile of an Aerospace Engineer:
### Artagnan Ayala, Gilbert, Arizona

### Occupation
Combustion Engineer , Diversity Organizations Manager, Honeywell

### Education
BS, Aeronautical and Astronautical Engineering, 1995
MS (in progress), Mechanical & Aerospace Engineering

### Studying Engineering
I have been interested in space since I was very young, I always wanted to be an Astronaut. This was my motivation to become an engineer. I figured that, if one day I was to climb on a rocket and go to outer space, I'd better know how it works. I am still working towards that goal.

My education has definitely met my expectations. I have applied what I learned in my job, and some of it to life in general. Engineering is not only a field you go into, but also a way of thinking. You are thought to solve problems, which can be applied to everything you do.

If I could start over, I would interact more with professors, take better notes, and learn more about statistics.

**Career Life**

Currently I am a Combustion Engineer, a Six Sigma Plus Black Belt, and the Diversity Organizations Manager for Society of Hispanic Professional Engineers at Honeywell Engines & Systems.

As a combustion engineer I design and develop combustion systems that are installed in Auxiliary Power Units and Industrial Power Generators. I have finished the development on one system, designed a technology demonstrator, and am currently designing a premixed fuel delivery system.

As a Six Sigma Plus Black Belt I apply statistical tools to improve all sorts of processes, from combustion system development and manufacturing.

As Diversity Organizations Manager, I am responsible for the company's contact and participation with the Society.

What I like the most about what I do is the diversity of my responsibilities. I get to apply my engineering skills every day, and I get to learn more skills. This has allowed me to receive my Black Belt certification, and participate in the Honeywell Quest for Excellence, a company event where teams with outstanding results present their work in a competition to win the Premier Achievement Award, the biggest team honor. I am particularly proud of my participation as presenter in 2 events, and making it to the finals in one of them.

**Life Outside of Work**

My wife Laura, a Graphic Designer, and recently we expanded our family with the arrival of our first daughter, Deanna Isabella. I devote most of my free time to my family and some to my studies.

Before we had a baby, I participated in a volleyball league, went dancing at clubs and concerts on weekends, and traveled outside of Arizona.

## Profile of a Civil Engineer:
## Sandra Begay-Campbell, Boulder, Colorado

### Occupation
AISES Executive Director

### Education
BSCE, 1987; MS, Structural Engineering, 1991

### Studying Engineering
I am a Navajo and the executive director of the American Indian Science and Engineering Society (AISES), which is a non-profit organization whose mission is to increase the number of American Indian scientists and engineers. I am the third executive director in the Society's twenty-year history and the first woman to serve in this position. I manage the Society's operations and educational programs. For more AISES information, check out www.aises.org.

In 1987, I received a BSCE degree from the University of New Mexico. I worked at Lawrence Livermore National Laboratories before I earned a MS, Structural Engineering degree from Stanford University. I also worked at Los Alamos National Laboratory and Sandia National Laboratories before accepting my current leadership position. Within AISES, I served as a college chapter officer, a national student representative, and board of directors member. I was the first woman AISES board of directors Chairperson.

In the sixth grade, I was very interested in architecture, but I knew I was not an artist. I also enjoyed math and solving problems so I looked into the engineering profession. I attended a "minority introduction to engineering" program as a high school junior and I discovered that civil engineers worked on a variety of interesting public projects, which included work with architects. This program solidified my decision to become an engineer.

### Career Life
One of the earliest challenges I faced was in continuing my structural engineering studies following the 1989 San Francisco Bay-Area earthquake. I was a first quarter graduate student at Stanford when the earthquake hit. Through prayer and reflection, I understood my unique role as an American Indian engineer. I must use my best knowledge to design structures for earthquake resistance, but my cultural heritage taught me the wisdom that engineers ultimately cannot control Nature and that we have to accept the consequences from natural phenomena.

### Life Outside of Work
Life outside of work is difficult to describe at this point in time. With the re-building of the AISES organization and relocation of the offices, I don't have much time for outside activities. I have also been commuting between Boulder, Colorado, and Albuquerque, New Mexico. In brief, my hobbies are watching college basketball, watching movies, and working on my home's backyard. My husband and I have two dogs and a cat.

## Profile of a Computer Engineer:
## Raymond C. Barrera, Gaithersburg, MA

### Occupation
Computer Engineer, Advanced Concepts and Engineering Division, Space and Warfare Systems Command Systems Center, San Diego

### Education
BS, Electrical and Computer Engineering 1989
MS, Software Engineering 1999

### Studying Engineering
I was very fortunate during high school to work for an archaeologist and her husband who were great mentors. To me archaeology is like detective work— finding bits of information here and there and putting them together to form the big picture. Dr. Bernice McAllister taught me the scientific methodology an archaeologist needs to base sound conclusions on evidence. I think I would be happy had I become an archaeologist, but I really enjoy building things. My dad's training as an electronic technician had gotten me interested in electronics when I was very young. That, with some encouragement from Dr. McAllister's husband, Capt. James McAllister, USN (ret) helped convince me to select Electrical Engineering as my specialty.

### Career Life
I work at a research, development, test and evaluation laboratory for the US Navy. I am involved in testing and system engineering of command and control systems. Command and control systems are used by tactical commanders for decision making and direction. I began working here in 1989 so I was here during Desert Storm. Perhaps even more important than the technical work is the ability to communicate. Not very many engineers work alone. A former Navy Admiral, Grace Hopper (who is said to have coined the computer term "bug") used a length of wire to describe a nanosecond to programmers. It was about a foot long, the distance that electricity could travel in one billionth of a second. But then she showed a microsecond—a coil of wire almost a thousand feet long. She was trying to convince programmers not to waste even a microsecond. Often the most difficult engineering challenge is to share an idea with others in oral and written presentations, but that is the only way these ideas can come to life.

### Life Outside of Work
My wife Martha and I spend most of our time outside of work with our new daughter Laura. I do have some flexibility on my work schedule so I can spend more time with her. I've been able to select job assignments that don't require too much travel. Since this is a research laboratory there are always new things to do. In the over ten years I've been here, no two have been the same. In this command alone there are engineers working with supercomputers, lasers, networking, marine mammals, 3-D displays, simulators, and sensors.

## Profile of a Mechanical Engineer:
### Linda G. Blevins, Gaithersburg, Maryland

### Occupation
Mechanical Engineer, National Institute of Standards and Technology

### Education
BSME, 1989; MSME, 1992; PhD, 1996

### Studying Engineering
During high school I discovered that I enjoyed mathematics. I learned about engineering when I participated in a six-week summer honors program at the University of Alabama before my senior year in high school. I took college calculus that summer, and I was hooked. I chose to study mechanical engineering because the course subjects are diverse and the industrial demand for mechanical engineers remains steady. As a co-op at Eastman Chemical Co., I worked on engineering problems in power and chemical plants. The concepts that I learned in classes came to life during the alternate semesters that I worked, and the money I earned helped pay for school. After earning a BS degree from the University of Alabama, I obtained an MS degree from Virginia Tech, and a PhD degree from Purdue University. I never would have set or achieved these goals without encouragement and advice from faculty members. Because these mentors played such valuable roles in my life, I would advise college students to get to know their professors well. These personal investments will be rewarding for years to come.

**Career Life**

I am a mechanical engineer in the Building and Fire Research Laboratory at the National Institute of Standards and Technology (NIST), a national research laboratory operated by the U.S. Department of Commerce, located in Gaithersburg, Maryland. Our goals are to study the ways that fires ignite, spread, and extinguish so that our nation can minimize the loss of lives and property to fires. My primary job function is to improve the accuracy of measurements made during fire research. A few things routinely measured are toxic gas concentration, temperature, and heat intensity. I spend my time developing laser-based instrumentation, devising computer (math) models of instrument behavior, designing laboratory equipment, tinkering with electronics, publishing papers, writing and reviewing research proposals, and presenting talks at conferences. In addition, I work on a project funded by the National Aeronautics and Space Administration (NASA) to study fires in space. Working in a research laboratory ensures that I am constantly learning and growing, and I realize every day how lucky I am to be here. My job is exciting, fun, and rewarding.

**Life Outside of Work**

During my free time, I enjoy hiking, rollerblading, and reading. I participate in a weekly bowling league and I manage a softball team each summer. I also volunteer as a member of the Mechanical Engineering Advisory Board at the University of Alabama. This allows me to travel home to Alabama (and visit my family) several times a year. Finally, I volunteer regularly to educate children and community members about the excitement of engineering.

## Profile of an Electrical Engineer:
### Timothy J. Bruns, St. Louis, Missouri

**Occupation:** Software Manager at Boeing Co.    **Education:** BSEE, 1983

### Studying Engineering

I became interested in electronics at a young age by building electronic kits from companies like Radio Shack and Heathkit. As a teenager, I became very active in local citizen's band (CB) radio groups. It was an easy decision for me to pursue a degree in engineering. The technology has changed so much since I graduated, and I have needed to stay current with the latest technology and to find ways to apply it to my line of work. If you are just starting out in engineering, I encourage you to apply yourself and do your very best in all your classes. When I arrived at Purdue I felt as if I was the least prepared of any of my classmates, but I worked hard and did very well. Some of the better prepared students did not apply themselves from the beginning and suffered as a result. One thing I would have done differently is to get to know my professors and teaching assistants better. In large universities and organizations it is easy to get lost in the crowd, and I wish that I had formed better friendships and relationships with my instructors.

### Career Life

I am the software manager for a team of 15 developers that is creating a Windows NT application. This application uses the latest technologies such as MFC, COM and ActiveX. A typical day is spent reviewing the technical work of the team, along with reviewing schedules and making estimates for future work. I often meet with customers of our product and suppliers of our software development tools. Since our program is just getting started, I have been spending a lot of time interviewing people who would like to join our team. It is difficult to say how I apply my engineering training directly to my current job. I know that my engineering degree has given me the ability to plan and organize the work of our team, and to solve the many problems that come up. The thing I like best about my job is the wide variety of assignments I have had in my 15 years with Boeing. Working in a large company gives me the ability to have several "mini-careers," all while working for the same company. A significant accomplishment that I have made while working at Boeing is the introduction of new tools and technology into the software development process. One tool that we have introduced automatically produces source code from a graphical representation. This tool enables us to bypass much of the labor-intensive and error-prone aspects of software design.

### Life Outside of Work

In the engineering field, particularly in electrical and computer engineering, you will find that the technology changes very rapidly. In my case, I stay abreast of the latest technologies by enrolling in evening computer classes through the local universities. I enjoy home "engineering" projects such as designing a new deck. My wife, Donna, and I keep very busy raising our two sons, Garrett and Gavin.

## Profile of an Electrical Engineer:
### Jerry W. Burris, Louisville, Kentucky

### Occupation
General Manager of Refrigeration Programs for General Electric Appliances

### Education
BSEE, 1985; MBA, 1994

### Studying Engineering
I have always had a curiosity about how things work (especially electronic devices). My parents recognized this at a very early age. They encouraged me to think about becoming an engineer. I was the child in the family who was always asked to fix the TV or electronic games. This continued through high school, where I excelled in math and science.

Purdue University was a natural choice for me, not only for its reputation for engineering excellence, but also due to the added bonus of having Marion Blalock and her Minority Engineering Program. This program has served as a recruiting magnet for Purdue and also has served as a mechanism for helping retain and matriculate students of color at Purdue.

While at Purdue, I was active in many extracurricular activities including leadership roles with NSBE and Kappa Alpha Psi fraternity. My early involvement in academics and extracurricular activities led to a full scholarship, which I received from PPG during my freshman year. Summer internships with PPG and IBM were invaluable in terms of giving me insight into what career path I wanted to pursue (design, manufacturing, or sales/marketing).

**Career Life**

I chose the technical sales and marketing route with General Electric's Technical Leadership Program. This premier program gave me advantages over direct hires in terms of exposure and training. After working six years I earned an MBA from Northwestern's Kellogg School of Management; I focused on global business, teamwork, and marketing.

My career has taken me from a role as a sales engineer, calling on industrial and OEM customers, to branch manager, with profit and loss responsibility, leading a team of nine people; to general manager of Refrigeration Programs at GE Appliances, where I now manage a $2 billion refrigeration product line.

I have been blessed with a lovely wife, who is also a Purdue and Northwestern graduate. We have two active children—Jarret, who is 7, and Ashlee, 4. We are managing dual careers at GE. This comes with significant challenges. However, GE has been very supportive of both of our careers.

**Life Outside of Work**

Life can not be all about work! You have to strive for balance. I have sought to keep God first in my life. I enjoy coaching my children in soccer, baseball and basketball, and I try to stay active with my own personal sporting activities. My favorite activities are listening to jazz music, traveling to exotic locations, managing our investment portfolio, and improving my golf game.

## Profile of an Agricultural Engineer:
## **Bethany A. Elkin Fabin,** Waterloo, Iowa

### Occupation
Design Engineer, 8000 Chassis Design Team – John Deere Waterloo Works

### Education
BS, Agricultural and Biological engineering

### Studying Engineering
When I began to explore career options, I was told that an engineering degree was the ticket to achieving success in a variety of fields. I investigated the Agricultural and Biological Engineering program at Penn State and discovered therein the opportunity to examine many aspects of engineering and agriculture under one discipline. I found my niche. This major provided the chance to "sample" many engineering topics and thus make knowledgeable decisions on what areas I wanted to pursue in future jobs. My Business Management minor also afforded many opportunities, and I would recommend that every engineer take at least a few business classes. I would also recommend getting involved in professional societies whenever possible. They provide many networking opportunities and a good preview of the job market. If I were to start my schooling over, I would take more of the hands-on classes. Also, I cannot begin to convey the importance of an internship or some kind of related work experience. Having the opportunity to work for a variety of companies in a variety of positions has helped me greatly in my career.

### Career Life
In my current position as a chassis design engineer for John Deere, I work with others to design parts for tractor frames, coordinate homologation and standard reviews for up-date programs, and coordinate projects with supporting teams. In the latter role, I de-velop general specifications to ensure that we meet customer requirements and imple-ment verification processes.

In a typical day of work, I spend a couple of hours working on Pro/E software designing and modeling parts. I also spend time working with suppliers and purchasing personnel to get parts quoted and ordered. In addition, I spend some time in our shop checking on prototype builds or test procedures, and some time in meetings working with different groups to keep people informed. The thing I like best about my job is the freedom I have to work on a variety of projects. It's nice to work for a company that has developed a strong name for itself and works diligently to stand behind their products.

### Life Outside of Work
Outside of work, I welcome every opportunity to travel with my husband and play host to out-of-state friends, relatives, and foreign exchange students. MBA classes are taking up much of my time off the job currently, but in my free time I find I enjoy music, sports, rowing, training my dog, remodeling my house, and gardening. My membership in the local chapter of American Society of Agricultural Engineers also keeps me busy with meetings and seminars.

## Profile of an Electrical Engineer:
### Bob Feldmann, St. Louis, Missouri

### Occupation
Director, Tactical Aircraft Mission Systems, The Boeing Company

### Education:
BSEE, 1976; MS, Computer Science, 1980; MBA, 1999

### Studying Engineering
My interest in engineering evolved naturally from my lifelong interest in science. Mathematics, while not my life's ambition, was interesting and satisfying. High school offered me the opportunity to enjoy learning about science, and I knew that I wanted to explore it even more in college.

In the mid-1970s, computers were not commonplace except in colleges, and I was hooked with my first FORTRAN programming class as a freshman. From that point on, I wanted to learn more about the hardware and software that made computers work. I oriented my electrical engineering curriculum toward digital electronics and used every elective I could to take software or software theory courses. While in college, I began a four-quarter stint as a cooperative engineering student at McDonnell Douglas in St. Louis. As a co-op, I was able to design software in the flight simulators (McDonnell was a world leader in simulation) and to work on a research design team for advanced flight control systems. Each semester when I went back to school, I would be at the library when Aviation Week magazine arrived, and I would read it cover to cover. When I graduated, I started my career as a software designer. My first day on the job, I was told that I would be responsible for the design of the software that controls the automatic carrier landing system on the F/A-18 aircraft. Ever since that first day, I have never been disappointed with the technical issues that have challenged me.

### Career Life
Today I am leading a team of over 800 engineers in the design and production of Mission Systems (also known as avionics) for the F-15, the F/A-18, the AV-8B, and the T-45 aircraft. My role as team leader for the organization is to ensure that the various product teams are providing outstanding value to our customers with the quality of our designs. I no longer write software, but I interact with the technical teams, coaching and guiding them through the difficult challenges of today's technically-exploding world. My proudest recent accomplishment was leading a team of engineers in the design and flight test of a reconnaissance system for the F/A-18. That system will provide the United States with its first manned tactical reconnaissance capability since the mid-1980s.

### Life Outside of Work
My life outside the office centers around outdoor activities and my family. My wife and I have three sons, all of whom play soccer and baseball (I have coached each one at various times). I really enjoy golf, bike riding, and other outdoor activities. My wife and I receive great pleasure from watching our sons grow up.

## Profile of a Computer Engineer:
## **Steven E. Fredrickson,** Houston, Texas

### Occupation
Project manager of the Autonomous Extravehicular Robotic Camera for NASA; Electronics Engineer, NASA Johnson Space Center, 1995 – present

### Education
BS, Computer and EE 1992; PhD, Engineering Science, 1995

### Studying Engineering
To prepare for a leadership role in the emerging information society, I studied electrical and computer engineering as an undergraduate. At Purdue I supplemented engineering studies with non-engineering courses and extracurricular activities, and sought experiences to develop practical business skills. One highlight was the Cooperative Education Program. Three "Co-op" tours at NASA introduced me to software design, robotic control systems, and neural networks. This early work experience intensified my interest in advanced study of electrical engineering and robotics. To simultaneously satisfy my desires to engage in advanced academic research and to gain personal international experience, I pursued an engineering doctorate program in the Robotics Research Group at Oxford University.

I am extremely pleased with the universities I attended and the fields of study I completed to prepare for my current career. I would offer three recommendations to anyone pursuing an engineering path: 1) participate in Co-op or similar programs, 2) develop effective oral and written communications skills, 3) explore opportunities to study abroad.

**Career Life**

When I returned to NASA as a robotics research engineer, I transitioned from specialized research in artificial neural networks to broadly focused applied engineering. As project manager of the Autonomous Extravehicular Robotic Camera (AERCam) project, I have led a multidisciplinary team of engineers in development of a free-flyer robotic camera to provide "bird's eye" views of the Space Shuttle or International Space Station. Despite this deliberate transition to a project leadership role, it has been imperative for me to maintain my core technical skills. To ensure continued technical proficiency, I participate in several training courses and technical conferences every year.

**Life Outside of Work**

As much as I enjoy working at NASA, I believe it is essential to maintain outside interests. For me, that starts by spending time with my wife, Becky. Since Becky is pursuing a joint engineering and medical career, it can be demanding at times. The key for us has been to develop outside activities that we can enjoy together. Currently these include teaching Sunday school, participating in Bible study, attending concerts and plays, jogging, lifting weights, climbing at an indoor rock gym, and traveling. In addition, we allow each other time to pursue individual interests, which for me include reading, aviation, and golf.

## Profile of a Mechanical Engineer:
## **Myron D. Gramelspacher,** Hartland, WI

### Occupation
Manufacturing Manager, General Electric

### Education
BSME, 1989

### Studying Engineering
I started at Purdue University in August 1985 in the engineering program. My interests in math and science were what really drove me to initially pursue opportunities in the field of engineering. My initial focus was in civil engineering, since I liked the concept of being able to work on roads, bridges, and outdoor structures. By learning more about the various engineering disciplines through seminars during my freshman year, I changed my mind and decided to pursue mechanical engineering. I felt that a degree in mechanical engineering would allow me more versatility and options in the workplace. Looking back on my college days, I wish I had taken courses in both business and foreign language to supplement my technical background.

**Career Life**

I graduated from Purdue University in 1989 with a degree in mechanical engineering, and started with General Electric (GE) as part of the Manufacturing Management Program. This program provided me with an opportunity to have six-month rotational assignments in two GE businesses. My first year was with GE Transportation Systems in Grove City, Pennsylvania, followed by a year with GE Aircraft Engines in Cincinnati, Ohio. In 1991, I transferred to the GE Medical Systems division in Milwaukee, Wisconsin. Since that time, I have held various positions in the Sourcing group, including supplier quality engineer, buyer, and team leader of the mechanical sourcing department. I also had the opportunity to live in Paris, France, for a year, heading up an Eastern European initiative. During that time, my efforts focused on the identification and qualification of suppliers in Eastern Europe. This position required that I travel throughout Europe, making it possible for me to experience different cultures and surroundings. This was a truly challenging and rewarding experience, both for my wife and me.

I currently hold the position of a Black Belt in GE's Six Sigma quality program. I utilize the Six Sigma tools and methodology to drive both process and product improvements that reduce costs, and ultimately impact our customers. The analytical skills and systematic problem solving techniques that I gained through my undergraduate engineering courses have greatly contributed to the many opportunities and successes I have had in my professional career.

**Life Outside of Work**

I now am attending Marquette University, working toward my MBA. An MBA will complement my technical background and enable me to strengthen my overall business knowledge. Outside work, I enjoy making landscaping improvements around the house and tackling various wood-working activities. My wife, Kim, and I enjoy traveling in our spare time. I also enjoy playing golf, tennis, and softball.

## Profile of an Engineer:
### Edgar T. Hammer, Greensburg, Pennsylvania

### Occupation
President, SSI Consulting Group

### Education
BS Mechanical/Metallurgical Engineering

### Studying Engineering
I enjoy solving problems, and once classes started in my engineering program, I never considered pursuing any other major. It wasn't always easy, but solving problems is what engineers do and I love it. My hardest choice was selecting between mechanical and metallurgical engineering majors. After switching back and forth a few times, I discovered that many colleges would allow you to design your own "interdisciplinary major." I did just that and now probably have the distinction of being the only graduate from Grove City College with a BS in Mechanical/Metallurgical Engineering.

### Career Life
I started working as a practicing metallurgical engineer doing testing in a quality assurance department. I picked up industrial statistics along the way, which came easily with my engineering math background. All engineers learn a little computer programming (when I started, personal PCs had 4 Kb—not Mb—of RAM). I was able to write some useful applications, combining my engineering, statistics and programming knowledge, that caught my employer's attention, and eventually I was promoted to head of the department.

I found that I really enjoy programming and there is a nice market for technical software. In engineering you'll write equations, create flow charts, and draw charts and graphs to illustrate your data. That all translates directly to software development. My engineering training has literally equipped me for technical programming better than computer science would have.

Today, I have found my niche in the quality assurance industry as a consultant, trainer, and software developer. I am president of my own company and enjoy the perks of being my own boss. I rarely practice my pure engineering disciplines, but I rely on the foundations of them daily.

### Life Outside of Work
More important than my career is my family and deep faith in God. Through engineering, I have been blessed with a career that lets me keep my family a priority, allowing me to spend considerable time with them and to provide well for them. At times, engineering has also been a stumbling block to my faith. Mankind sometimes becomes prideful in his ability to explain his world. But focus on what we don't know and you will see that we understand only a tiny portion of the universe. I am convinced that the complexity and order we see reveals the hand of the master engineer Himself. I guess that puts us in good company!

## Profile of an Industrial Engineer:
### Karen Jamison, Dayton, Ohio

### Occupation
Operations Manager, Jamison Metal Supply, Inc.

### Education
BSIE, 1988; MBA, 2000

### Studying Engineering
I didn't grow up knowing I wanted to be an engineer, but luckily my high school guidance counselor recognized my science and math abilities and encouraged me to try engineering. I firmly believe that engineering is a wonderful career in and of itself, and that it can be an excellent stepping stone for any other career you may wish to pursue in the future.

I chose industrial engineering because I am highly interested in improving the processes people use to do their work. Industrial engineering provides both technical challenges and the opportunity to work with all kinds of people.

If you are just starting to think about engineering or are trying to choose a specific discipline, talk to as many practicing engineers and professors as you can. Become involved in organizations on campus that will let you interact with other engineering students and practicing engineers.

I also highly recommend the co-op program. I had over two years of work experience when I graduated, and I knew what types of work I would enjoy. It is definitely to your advantage during interviews to know what type of job will best suit you, and to be able to speak intelligently on that subject.

Finally, remember that grades aren't everything but that your education is invaluable. If I were to do one thing differently, I would study to truly learn and understand the content instead of with the goal of getting a good grade in the class.

### Career Life

Until last year, I was a consultant focusing on process improvement and business process re-engineering. Now I am learning to run Jamison Metal Supply, which is a business my parents founded 25 years ago. My job includes anything and everything that needs to be done. My primary responsibilities are overseeing operations to ensure quality products and timely deliveries, ordering steel for inventory and special orders, and pricing the material we sell.

I use my engineering training in all kinds of ways. I am working on updating our physical inventory system to better utilize warehouse floor space; I schedule customer orders to meet promised delivery times; and I am updating our computer system. Most importantly, engineering has taught me how to approach solving a problem and how to manage my time.

### Life Outside of Work

My time outside of work is concentrated on completing my MBA degree, but I do find time for having fun as well. One of my favorite hobbies is crewing for a hot air balloon. I also teach a sign language class at the University of Dayton, and am vice president of the Purdue Club of Greater Dayton, Ohio. I think engineering is a very flexible field that allows individuals to prioritize their lives any way they wish.

## Profile of a Mechanical Engineer:
## Beverly D. Johnson, Waterloo, Iowa

### Occupation
Supervisor in Wheel Operations at John Deere Waterloo Works

### Education
BSME; MS, Engineering Management

### Studying Engineering
My education includes a BS in Mechanical Engineering from the United States Military Academy, an MS in Engineering Management from the University of Missouri, Rolla, and my current study in the Executive Master's Degree Program at Northwestern University, Evanston, IL.

I think engineering is a very rewarding career because you can see the results of your effort every day. Engineering offers opportunities to create, build, design, and sometimes even destroy. Also, the analytical tools you develop in your engineering coursework make studying other subjects easier, and they are applicable to everyday life.

I truly enjoy my career in engineering. It is a dynamic career field that has taken me to many different jobs and many different places. I have done everything from constructing buildings and roads in Germany and the Hawaiian Islands to my current work as a supervisor in the wheel operations for the John Deere Waterloo Works.

### Career Life

I have been with the John Deere Waterloo Works for two years, working in various engineering assignments such as quality engineering, project management, and process redesign. My current assignment as a supervisor in Wheel Operations is focused in production. I am responsible for the assembly processes pertaining to the tires and wheels for the 7000 and 8000 series tractors. I am also responsible for the daily supervision of the wage department personnel. Although my job is sometimes hectic, it is also very rewarding as I watch what our department is able to accomplish every day.

Prior to joining John Deere I spent nine years as a military officer in the U.S. Army Corps of Engineers. My primary responsibilities included the construction of buildings and roads, and the development and training of other engineers. My work with the military allowed me to live in, and travel throughout, Europe and the Pacific Islands.

### Life Outside of Work

Although I chose engineering over journalism, my favorite pastimes are reading and writing. I also exercise regularly and compete in sports. I volunteer my time to the Boys and Girls Club of Waterloo, the American Red Cross, and a local university. However, my most important responsibility, and the most enjoyable, is the time I commit to the care and development of my two children, Colbert, 6, and Randy, 3.

## Profile of a Civil Engineer:
## James L. Lammie, New York

### Occupation
Board of Directors, Parsons Brinckerhoff Inc.

### Education
BS, Civil Engineering, 1953; MS, Civil Engineering, 1957

### Career Life
When I grew up, my father worked in a steel mill in Pittsburgh, the City of Bridges. I was fascinated with the many different bridges and what could be done with steel. I knew that I wanted to build things. I was fortunate to win an appointment to West Point, which was founded as the first engineering school in the U.S.

After graduation, I spent 21 years in the Army Corps of Engineers working on a wide variety of military and civil engineering projects all over the world. After retiring from the Army I knew I wanted to be a Project Manager on big projects, so I joined Parsons Brinckerhoff, Inc. and spent seven years as a consultant Project Manager for design and construction on the Metropolitan Atlanta Rapid Transit project (MARTA), the most rewarding period of my professional career. Today, my grandchildren ride what I helped build—a most rewarding feeling.

After MARTA, I had the pleasure of serving as the CEO of Parsons Brinckerhoff, Inc., the largest transportation design firm in the U.S., for the next fourteen years. Today, as a member of the Board of Directors of our employee-owned firm, I am still involved in some of our mega projects: the Central Artery Highway project in Boston, the new Taiwan High-Speed Rail system, the Bay Area Rapid Transit extension to the San Francisco Airport, and many others. The high point of my job is getting involved in critical project decisions and being able to "kick the tires" of work under construction.

**Life Outside of Work**

Thanks to my varied career in engineering, construction and management, I am also able to participate in a variety of outside activities: the Transportation Research Board (TRB), the Institute for Civil Infrastructure Systems (ICIS), the Engineering Advisory Board at Purdue University, and the National Academy of Engineering (NAE). I also teach and lecture on Project Management, Leadership, and Engineering Ethics. During my career, the high points have been presenting proposals and winning major jobs, election to the National Academy of Engineering, and receiving an Honorary Doctorate at Purdue University.

I always enjoyed participating in a variety of sports, until my knees gave out. The most personally rewarding aspect of my life over the years has been the companionship of my wife, three children (all in the medical profession, thanks to my wife's nursing career), and my eight grandchildren (with three going to colleges close by, permitting frequent visits).

## Profile of an Electrical Engineer:
## Ryan Maibach, Farmington, Michigan

### Occupation
Project Engineer at Barton Malow Company

### Education
BS–CEM (Construction Engr. and Management), 1996

### Studying Engineering
I think I have construction in my blood. My family has been involved in construction for four generations, and I was always surrounded by it while growing up. I can remember walking along beams with dad at a very young age, much to the irritation of my mother. So when it came time to choose a career, construction seemed to be the obvious answer.

Knowing what I wanted to study before I entered college was very nice, but it did not make my freshman year any easier. The first year of engineering school requires that you gut-out the tough prerequisites. During my college days, two extra-curricular activities were particularly rewarding to me. The first was my involvement in student organizations, which helped me to develop my leadership skills. The second was taking advantage of summer job internships, which provided practical experience by bringing to life the theoretical classroom education. Looking back, I should have taken more technology-related classes. I have found that technological skills are highly sought after in the job market.

**Career Life**

After graduation, I went to work for Barton Malow Company, a national design and construction services firm, and was made a field engineer on a basketball arena construction project. Presently, I am a project engineer working on a hospital expansion project. In my position I have a great deal of flexibility in terms of how I use my time throughout the day. I have worked on design development, purchasing contracts, scheduling, and subcontractor supervision.

Construction has been just as exciting as I had hoped. Every day when I leave my project, I can see what new accomplishments have been made—always, more steel has been hung or walls erected since I arrived. An interesting aspect of construction work is learning about our clients' businesses. It helps to have an understanding of our their industries in order for projects to be successful. For example, in order to make the present hospital expansion successful, our project team must understand how the hospital functions, overall. Having the opportunity to work on a variety of projects allows people in the construction industry to learn a great deal about various industries throughout their career.

**Life Outside of Work**

The spirit of camaraderie common to a construction site fosters friendships with coworkers that often continues outside of the work place. I have participated in a variety of community activities with my co-workers, ranging from speaking to school groups to refurbishing old homes for the elderly. Individuals looking for a challenging career which offers a lifetime of learning experience will find the construction industry very rewarding.

## Profile of an Agricultural Engineer:
## **Mary E. Maley,** Battle Creek, Michigan

### Occupation
Product Manager, Kellogg Company

### Education
BS, Agricultural Engineering (food engineering)

### Studying Engineering

Math and science were always my favorite subjects, with the best part being the story problems where the concepts were applied. The idea of using scientific principles to solve a problem is what led me to choose engineering as a major. I would get to learn some more about math and chemistry as well as do something useful with that knowledge. That happened at college and continues to happen in my job.

### Career Life

Today at Kellogg Company, the work I do is varied from day to day. My role is to make sure our manufacturing facilities have all the information they need, at the right time, to bring new products to market. That means coordinating the work from many different departments and gaining a consensus on the critical tasks to meet the timeline. You might ask, "What does that have to do with engineering?" Primarily, I bring together a myriad of details into one final outcome, just as all engineers do in combining the known facts to reach a solution. I just get to add some more unknowns and assumptions, such as dealing with people and changing requirements. The biggest challenge is getting the project accomplished to meet the needs of the consumer (that's you) before any of our competitors do.

Since Kellogg Company is a global company, my work affects the entire world. These days I work on projects for North America, Mexico, and Southeast Asia. I have had the opportunity to learn about other cultures and adapt our food products to fit their lifestyles. With manufacturing being located outside the U.S. as well, I encounter the varying work procedures and government regulations of each country. It makes my job challenging and enjoyable.

### Life Outside of Work

Certainly, working at Kellogg's is not all that I do. My job is just one part of life. I find I need outlets for creative activities and for making contributions for the betterment of our world. Through sailboat racing I find a time of total concentration and a chance to apply aerodynamic principles. This also provides a fun way to have some competition. On the creative side, I participate in the handbell choir at my church. For me, music is a way to use my whole brain in the interpretation of notes into an emotional song of praise.

Perhaps more important is how I can give back to the community and the world in which I live. To promote cultural activities in my town, I serve on the board of directors of the Art Center. Through the Presbyterian Church, I lead the Benevolence Committee in disbursing funds and educating the congregation about the outreach in our community, the nation, and the world.

## Profile of an Industrial Engineer:
## Jeanne Mordarski, Albuquerque, NM

### Occupation
Sales Manager, LightPath Technologies, Inc.

### Education
BSIE; MBA

### Studying Engineering
The engineering workload at Purdue was quite a shock to me. My first two years were a struggle and I was afraid to get involved in extracurricular activities. By my junior year, I became more concerned that I was missing out on the "college experience" than I was about my grades. I became an active member in several campus organizations—the best decision I ever made. I was forced to balance my studies and personal life. I broadened my network of friends, developed leadership skills, and learned to manage my time more effectively. As a bonus, my grades improved tremendously.

My emphasis within the IE curriculum was on Production and Manufacturing Systems. I accepted a production supervisor position with Corning Inc. after graduation. This role put me in the middle of the action, and taught me to think on my feet and make sound decisions. For eight years I worked at Corning in various engineering and manufacturing capacities. During this time, I was able to land a one-year tenure in Japan implementing Process Management Systems at our facility in Shizuoka.

For three years I took evening classes working toward an MBA from Syracuse University. In my course work, I realized how much I enjoyed the business side of things. Upon completion of my degree, I accepted a sales manager position at Corning in the telecommunications market.

### Career Life
I recently left Corning to work as a sales manager for a start-up company, LightPath Technologies, Inc. Working for a large company directly from college gave me invaluable experience. The structure enabled me to work more effectively. However, as I progressed through the ranks at Corning, I realized that this same structure was limiting my ability to contribute because of the many management layers. In my current role at LightPath, we are introducing new products to the telecommunications market.

### Life Outside of Work
I have an eclectic mix of interests outside of work. I truly enjoy exercise and the outdoors. On weekends, you'll find me skiing, camping, hiking, rock climbing, or biking. I love international travel and scuba diving, and take every opportunity I can to participate in both. I have recently taken up Latin social dance and kickboxing. I am also involved with the Purdue Alumni Association in Albuquerque. I like to keep busy and have worked very hard to strike a balance between my career and personal life. I seldom work more than 40 hours a week. I made it a goal to be more productive during work hours to minimize overtime and unnecessary stress. It usually works.

## Profile of an Electrical Engineer:
### Mark Allen Pashan, Red Bank, New Jersey

### Occupation
Director of Hardware Development, Lucent Technologies

### Education
BS; MS; MBA

### Studying Engineering
When I was in high school trying to decide which career to pursue, I had a number of criteria: I wanted a job that I'd look forward to each day, that offered continuous learning, and that offered a reasonable level of financial stability. Engineering satisfied those criteria for me. I enjoyed math and science (the foundations of engineering) in high school, but engineering is more than number crunching. The field of engineering rewards creativity, the ability to find a better way to solve a problem. If I had to do it over again, I'd still choose engineering, but I'd also have bought more shares of Wal-Mart, Lucent, and Yahoo when they were first offered.

### Career Life
In my career, I have advanced through a number of levels of technical management, and currently have about 130 engineers reporting to me. My job is no longer at the level of designing integrated circuits. I guide my team's progress on a number of new product development activities. I work to make sure we have the right people working on the right things at the right time. I set priorities among the competing needs of the business, and evaluate new business opportunities. To do my job, I use a combination of business and technical judgment: what are the future customer needs, what are the available and soon-to-be-available technologies, what are my competitors doing and what may they do next, who can do the work and work well together, and can we get the work done in time and at a reasonable cost. The end results are new products introduced into the marketplace that turn a profit for the business. That goal can only be achieved through others. A good part of my job is getting my teams to achieve more than they thought possible.

This is the best time in history to be an engineer. There are more available alternatives than ever—from startup companies to large established firms, from full-time to part-time work hours. There are more opportunities for continuing education and there is the potential for significant financial reward for those willing to take a risk.

My organization is spread across three states and I have customers and suppliers all over the world. My job requires travel and long hours, and I couldn't do my job and have a family without the support of my wife, Reem (see following profile). But we do it together and the kids are a joy (even when they don't always obey). I enjoy a number of activities outside of work such as basketball, traveling, and dining out.

## Profile of an Engineer:
## Douglas E. Pyles

### Occupation
Engineering Consultant, Ingersoll-Rand

### Education
BS, Interdisciplinary Engineering

### Studying Engineering

I enjoyed science in high school, especially physics. I would describe myself during that time as a loner, a trait that became a weakness for me when I went to college. If I were to give advice to a student just beginning to study engineering I would say the following: 'Don't live in a shell, become active in one or two student organizations, make friends in your classes, get help right away and when you need it, don't be afraid!' If I had it to do over again I would try to follow this advice.

### Career Life

I work for Ingersoll-Rand, at the Tool & Hoist Division Business Development Center. My title is Engineering Consultant, which is the first level on Ingersoll-Rand's technical career ladder. The career ladder is for people who, after progressing through the ranks, want to continue pursuing areas of technical expertise rather than managing people.

My primary responsibility at this time is supporting New Product Development Teams in the areas of mechanisms, design, testing, and manufacturing. I am personally responsible for insuring that portions of a system meet all performance, cost, manufacturability, and life targets. During the course of a project we travel extensively in a team-wide effort to learn the needs of the customer. I also participate with and help guide the team as we develop customer specifications for the final product.

### Life Outside of Work

As I write this, I am also running an analysis on a Unix workstation that sits next to my PC. In the past, we have had to rely on experience and a process of building and testing to perfect our products. Now we have the advantage of computer analysis. I hold several patents for designs that were developed in this way. I enjoy being able to answer complex questions and make decisions about designs before making prototypes. I also really enjoy working on projects that have the potential for increasing market share and profits for our division. This type of analysis is only one of the many tools we use to develop high-quality products for an ever more demanding and changing marketplace.

When I am not slaying dragons at work, I am likely to be with my wife and three kids. I still maintain that anyone with children has no spare time, and has to schedule all their fun. On Wednesday nights, I play music with a different group of people. We have several guitars, a bass and a keyboard, as well as dedicated vocalists. We are responsible for leading music at church most every Sunday morning. I play the twelve-string guitar. When I was in school, I never thought I would ever find myself up in front of 300 people singing and playing music! It shows me that whatever stage of life I am in I must always look for opportunities to reach and to grow.

## Profile of a Construction Engineer:
## **James T. Richard, PE,** Buffalo Grove, Illinois

### Occupation
Director of Mechanical Systems Service, Siemens Building Technologies, Inc.

### Education
BS, 1984 and MS, 1995 in Construction Engr and Management

### Studying Engineering
When I entered college, I had no idea in which engineering field I wanted to major. After learning more about the different types of engineering majors and opportunities during my freshman year, I selected construction engineering. This program gave me the opportunity to combine civil and mechanical engineering into a single concentrated major.

Equally as important as choosing your field is taking advantage of campus activities. Social skills are critical for your career and make the rest of your life more fulfilled. I was a member of a number of student organizations, including the Purdue Varsity Glee Club, which performed throughout the United States and Europe and at President Reagan's Inaugural Ball.

**Career Life**

After receiving my master's degree, I became director of Mechanical Systems Services for Siemens Building Technologies, Landis Division. I am responsible for developing the mechanical service business in North America and assisting approximately 100 branch offices. I am very proud of my department and the way we have grown.

When working with 100 branch offices, there is no typical day. What's exciting about my work is starting from ground zero on a project and giving it life. Our effort to build a world-class service organization pumps me up daily. It's exciting to answer questions such as "How can we do that better?" and "How can we make it better?" I am constantly applying the problem-solving techniques engineering teaches to take us from point A to point B.

**Life Outside of Work**

My biggest accomplishment is coordinating work, home, marriage and fatherhood (two children)—and I love them all. It's quite a juggling act, but provides tremendous satisfaction. My wife is an executive with Motorola and has an equally demanding schedule. Consequently, we both put a premium on family activities.

I also try to stay involved with my profession outside of work. I'm a registered professional engineer and a member of a number of professional organizations including the American Council for Construction Education, Association of Mechanical Engineers, Society of Professional Engineers and the American Society of Heating, Refrigeration and Air Conditioning Engineers.

## Profile of a Mechanical Engineer:
### **Patrick J. Shook,** Columbus, Indiana

### Occupation
Senior Engineer, New Product Development, Cummins Engine Company

### Education
BSME, 1992; MSME, 1994

### Studying Engineering

Mr. Myers, my high school chemistry teacher, had a discussion with me one day about Purdue's co-operative education program. He could see my interest in math and science and pointed me toward a field which I knew very little about—engineering. I investigated, and with high expectations, made the decision to attend Purdue to study and to prepare for what seemed to be a very interesting career.

By the end of the first semester during my freshman year, I had decided to pursue a mechanical engineering degree. This was after many discussions with junior and senior engineering students as well as with my father and a few professors. I had grown up in a family which owned a general contracting business (house construction, remodeling, etc.) and the broad variety of topics of study within mechanical engineering seemed to fit my desires. I also signed up to become a co-op student in order to obtain valuable work experience as well as to help pay for my education.

After graduating with my BSME, I entered into a research assistantship at Purdue for an intense (but extremely rewarding) two years on the way to obtaining my MSME. The most exciting task given to me by Prof. Fleeter was to build and operate a helicopter engine compressor test stand.

After graduating with my MSME, I hired on at Cummins Engine Company and worked for four years as a mechanical development engineer. This time was filled with designing abuse tests for semi-truck engines and determining how to improve the components that wore out during those tests. Without a doubt, learning how to work with people is easily 50-percent of my job. Since I have been at Cummins, I have learned that being clear with people concerning the goals of a plan is extremely important. As in a football huddle, everyone on the "field" needs to know what the "play" is and how to execute it.

Since July 1998, I have been working in a new position which focuses on engine cycle simulation. This has been primarily computer work and has re-sharpened my skills in fluid mechanics, thermodynamics, and heat transfer. The variety in my job has been enjoyable: from defining customer requirements for specific components to maximizing work processes within the structure of a large company.

### Life Outside of Work

Outside of work, my wife and I do our best to serve the Lord and our church. In the past, we have both taken and taught classes on what the Bible says about marriage. We wanted to build on a good foundation and have enjoyed our marriage more and more with each year.

## Profile of an Aeronautical Engineer:
## David O. Swain, St. Louis, Missouri

### Occupation
Executive vice president, The Boeing Company

### Education
BS, Aeronautical Engineering, 1964

### Studying Engineering
I had demonstrated strong mathematics and physics skills in high school and was drawn to engineering because I understood it was a well-paid profession. I was very interested in the space program following the Soviet Union's launch of Sputnik in 1957 and was drawn to aerospace. The field has met all my expectations.

**Career Life**

If I could do everything over again, I would do everything the same. My advice for beginning students is to focus on learning and not worrying so much about their specific career path.

I joined McDonnell Douglas in 1964 as an engineer on the manned Gemini project. In 1972 I moved to Tactical Missile programs and worked on the Tomahawk and Harpoon/Standoff Land Attack Missile. In 1987 I became vice president and general manager for strategic business development for McDonnell Douglas Astronautics Company. I was named senior vice president and C-17 program manager in 1991. This was one of my most significant accomplishments in that I led the C-17 program through difficult and troubled times. It is now on the right track.

I currently am executive vice president for Phantom Works, the research and development organization of The Boeing Company. I have overall responsibility for Phantom Works and its operations in Southern California, Seattle, Mesa, St. Louis, and Philadelphia. Some of the exciting projects we are involved with include: Joint Strike Fighter, Future X Space Vehicle, Canard Rotor Wing, Space-Based Laser, Space-Based Radar, Common Support Aircraft, Uninhabited Combat Aircraft, Blended-Wing Body, Solar Orbit Transfer Vehicle, and Deepwater Ship. In a typical day, I meet with customers, suppliers, peers, direct reports, and employees. I lead, coach and teach. My normal work hours are 11-12 hours a day. I travel frequently to California, Washington state, and Washington, D.C.

**Life Outside of Work**

I use my engineering training in my current job by applying logic and asking penetrating questions. I also enjoy learning from other people. What I like best about my job is the fact that I am working on projects that have significant value.

I enjoy doing things with my family. I have four children and seven grandchildren. I also like racquetball and golf. I am a trustee on the St. Louis Science Center Board and serve as an advisor for Purdue University. I also find a great deal of reward from providing good advice to interested individuals.

## Profile of a Mechanical Engineer:
### Nana Tzeng, Seattle, Washington

### Occupation
Design Engineer, The Boeing Company

### Education
BSME, 1997; MSME, 1998

### Studying Engineering
What I enjoyed most about being an engineering student was making stuff—in other words designing and fabricating mechanical parts. While I was at Purdue, I participated in the Solar Racing Club. As one of the few mechanical engineers on a team dominated by electrical engineers, I helped improve the braking system, performed computer aided stress analysis on the chassis, learned to weld, machined rotors and other parts, and got to be driver of the solar car. The experience was not only rewarding, it also helped me relate what I read about in textbooks with applying that knowledge. College offers many extracurricular opportunities and I would encourage any engineering student to become involved in hands-on activities and research projects.

### Career Life
My career really is rocket science! I currently work in the Instrumentation Development and Design group at Rocketdyne, the division of The Boeing Company that designs and develops rocket engines. The team I work with is responsible for all the sensors and electrical components on the Space Shuttle Main Engine. My latest project is the re-design of the spark ignition system. This involves the design of components and tooling, creating and updating of drawings, and working with the manufacturing team to improve the fabrication process. I also help the members of my team analyze sensor data from hot fire tests and space shuttle flights. Because the nature of my work is highly technical, I regularly use the knowledge and skill I gained as an engineering student. Now that I am familiar with the complexity of rocket engines and the detailed work that goes into building one, it's even more amazing when I see everything come together during launch.

### Life Outside of Work
Ever since I finished school, I have been able to develop other interests and hobbies, some of which are golf, photography and snowboarding. I also often enjoy hiking, camping, rollerblading, shopping, concerts, clubs, etc. The best advice I have to offer to students of any discipline is to keep an open mind and take advantage of your opportunities.

## Profile of a Mechanical Engineer:
## **Patrice Vanderbeck,** Cedar Falls, Iowa

### Occupation
Electronics systems engineer for the 6000 and 7000 series John Deere tractors

### Education
BSME, 1982

### Studying Engineering
I like to understand how things work. Math was my favorite subject in high school, followed closely by the sciences. Engineering seemed like the appropriate career for me, based on my interests. My college advisor recommended I study mechanical engineering after assessing my capabilities and interests. It sure was the right direction. The engineering curriculum can be difficult, but an engineering degree gives you many options. You can go into design, management, marketing, sales, law, manufacturing, or research, to name just a few areas. If you decide you want a change, it makes it easier for you to move on in a new direction. The engineering degree will open doors for you.

Given the opportunity, I would have changed two things about the path I took. First, I would have developed better study skills in high school, or earlier in college. I had to do some backtracking because of this. Also, I would have worked at a company that manufactures a product (like tractors) directly out of college instead of starting at a consulting firm. I learned it was important for me to work where a product is manufactured.

**Career Life**

I have worked at three companies since graduating in 1982. I am currently the electronics systems engineer for the 6000 and 7000 series John Deere tractors. I make sure that the different engineering teams within the electronics and vehicle groups are communicating with each other. My job combines design, program management, negotiating, and communicating. I deal with current tractor-related issues and the designs for new tractor programs at the systems level. I love my job because it is never boring. I am always learning. I work with many talented, dedicated people. When I was in a previous position at John Deere, I had design responsibility for a device for left-hand control of the forward and reverse movement of a tractor. It had 38 subassemblies. Between the supplier and me, we designed these subassemblies into a very small package. It took a lot of development effort to get the assembly to work perfectly within the entire system of the tractor. In the end, the device was well received by our customers for its function and reliability.

**Life Outside of Work**

I enjoy biking, skiing, weaving, entertaining, and reading. I belong to an investment club and a reading club. I am married to another engineer. We have a vacation home on the Mississippi River where we do a lot of boating and entertaining. We love to travel. My husband and I do volunteer work with Habitat for Humanity and with our church. Our engineering jobs allow us to live a comfortable life and to enjoy fun things like travel and boating.

## Profile of a Chemical Engineer:
## **Jack Welch,** Fairfield, Connecticut

### Occupation
CEO of General Electric (Retired, 2002)

### Education
BSChE; MSChE; PhD

### Career Life

The man called "CEO of the century" by the editor-in-chief of *Time* magazine is an engineer. Jack Welch, who led General Electric's transformation over the past two decades into a global technology and services giant, started with the company as an engineer in Pittsfield, Mass. He had earned his BS ChE from the U of Mass in 1957, and followed that with an MS and PhD from the U of Illinois in 1958 and 1960.

In high school he had captained the hockey and golf teams and earned the distinction of being voted "Most Talkative and Noisiest Boy" by his classmates. "No one in my family had ever gone to college, but I had that ambition," Welch recalls. "Of course, believe me, my mother had that ambition for both of us."

"Life is a series of experiences, a series of steps if you will," he continues. "Every time you're reaffirmed, every time someone tells you you're Okay, you can go on to the next step, the next challenge. Well, my teachers in the Engineering Department told me I was Okay. In fact, they told me I was really good. A couple of them practically adopted me, and told me I had what it took to go on to graduate school. I had never even thought of graduate school. But they really believed in me."

After earning his PhD, Welch returned to Massachusetts and GE's Chemicals Division for his first job as a development specialist. It was on that first job that he demonstrated many of the leadership traits that characterize him to this day.

"I was an entrepreneur in a small business outside the mainstream of GE—the plastics business. My technician and I were partners working on the same thing. We had two people, then four people, then eight people, then 12. Today, GE Plastics is a $6 billion business. But it started that way. Everyone's involved. Everyone knows. Everyone's got a piece of the action. The organization's flat. All these things are from when I was 26 years old."

Welch's rapid rise in GE continued, and in 1981 he became the eighth chairman and CEO of the company that was founded in 1892. Although he recently retired, the organization he led was named "Most Admired" by Fortune magazine and "Most Respected" by the Financial Times.

Yet he described his job running a company with 1998 revenues of approximately $100 billion as "not rocket science." Instead he saw his key role as allocating both human and financial resources in a way that will continue GE's growth. "My job is allocating capital, human and financial, and transferring the best practices. That's all. It's transferring ideas, putting the right people in the right jobs and giving them the resources to win," he says.

Welch, now the father of four and recently a grandfather for the fourth time, continues on the golf course his winning ways that began in high school. He's twice won his club championship and has even bested well-known pros in friendly play.

# Profile of an Electrical Engineer:
## Shawn D. Williams, Twinsburg, Ohio

### Occupation
Product General Manager at GE Appliances

### Education
BSEE, 1985

### Studying Engineering

The opportunity to further my math and science abilities attracted me to engineering. I really enjoyed calculus, chemistry, and physics in high school, so I attended Marion Blalock's Target Cities Luncheon in Chicago. At the luncheon, Purdue engineering students spoke about engineering, and their comments opened my eyes to engineering as a possible career.

As a freshman in engineering at Purdue, I queried several juniors and seniors in various engineering disciplines on their course work. I concluded that electrical engineering would allow me to leverage my foundation in mathematics and physics. In addition, computers were becoming more popular, and I concluded that EE would provide insights into computers. My advice to students is that you remain persistent and disciplined in pursuing an engineering degree.

### Career Life

Currently, I am a regional sales manager for General Electric Industrial Systems. My primary job responsibilities include delivering top-line sales growth on an annual sales volume of $130 million. I lead a team of 70 employees (mostly engineers) throughout six states in the Midwest as they execute business strategies in their local trading areas.

In the course of a day, I may do such things as interview for an open sales engineer position, review a trading area strategy with a general manager, expedite a delivery with a factory for a distributor principal, and provide coaching to a new sales engineer on a project. The best aspect of my job is the variety of strategic and tactical tasks in which I engage. My engineering degree provided the solid foundation for me to understand the technical nature of the products, but the ability to handle multiple tasks is key in executive management.

### Life Outside of Work

Obtaining an engineering degree has assisted me in providing a lifestyle for my family that I never thought possible while growing up. It is imperative for me to balance work and family life. My wife and my son and I spend weekends visiting the zoo, seeing friends, playing board games, and attending church.

On a personal note, I am the primary provider for my family, so keeping in shape physically, mentally, spiritually, and intellectually are keys to a successful life. I accomplish this by running 3 to 4 miles three times a week; consistently working on lowering my golf handicap of 13 by playing golf with customers and friends weekly; and incessantly learning about life through Bible reading, management books, and by mentoring and coaching employees, students and family members. I can honestly say that without my engineering experiences, the life I have now would still be just a dream!

## Profile of a Mechanical Engineer:
## Dr. Adel A. Zakaria, Waterloo, Iowa

### Occupation

Senior VP, Engineering and Manufacturing, Worldwide Agricultural Equipment Division

### Education:

BSME, Egypt, 1967; MSIE, 1971; PhD, Industrial Engineering, 1973

### Studying Engineering

Growing up in a developing country, I viewed engineering as an instrument for making genuine progress. It has offered me an opportunity to be part of moving things forward in our generation. In particular, I was interested in how things are made. Manufacturing offered a unique way to work with three domains: things, people, and systems. In today's highly technical society, engineering will offer a student an ideal foundation for several careers. A co-op or summer job experience can be very helpful in getting early exposure and making one's study a lot more interesting and meaningful.

### Career Life

Within a $7 billion Worldwide Agricultural Equipment Division, I guide the work of 20,000 employees who engineer, manufacture, and provide product support to our customers.

A typical day for me in the last year might include:

- visiting with key customers and their servicing dealer in Nebraska
- chairing a Worldwide Combine Product Council in Germany
- visiting a new joint venture site in India to review progress of the factory construction
- reviewing the results of a business improvement team

I enjoy the breadth of my job, working with a variety of people, the disciplines they represent, and the constant challenge of leading change. Not only have I been able to apply engineering training, but I have had to constantly augment it by learning and expanding my knowledge in new, evolving areas.

Significant accomplishments in my career include the early development of cellular manufacturing, the development of computer aided design and manufacturing tools (CAD/CAM), guiding the development and introduction of our company's first worldwide tractor product platform, and the breakthrough in forging a new win/win labor strategy with our unions.

### Life Outside of Work

My hobbies outside of work include racquetball and photography. Our family (including our two daughters) has traveled together in over 30 countries. We enjoy learning about the world's people and cultures. We also have camped in most of the U.S. and Canadian national parks. At various times I have been active in a number of national engineering societies. I also have participated in a number of community volunteer groups, including a hospital board, United Way, and the Boys & Girls Club.

# Exercises and Activities

3.1    Select five of the engineers profiled in this chapter. Prepare a one-page summary of their current responsibilities.

3.2    Select five other engineers profiled in this chapter. What were some of the common factors that got them interested in engineering or helped them succeed in school?

3.3    Of the engineers profiled in this chapter, describe three who have unique, non-engineering jobs.

3.4    Make a list summarizing the current jobs and responsibilities of five of the engineers in this chapter.

3.5    What types of career paths are available to engineers? Make a list of seven different career paths taken by engineers in this chapter.

3.6    Select a historical engineering figure. Prepare a one-page profile on this person using the format used in this chapter.

3.7    Interview a practicing engineer and prepare a profile on him or her, similar to the format used in this chapter.

3.8    Select five of the engineers profiled in this chapter and list some of the challenges they had to overcome during their education and/or career.

3.9    Prepare a one-page paper summarizing some of the challenges engineers face today.

3.10    Select one of the engineers profiled in this chapter and prepare an oral presentation on that engineer's company or organization.

3.11    Which career path presented in this chapter sounds most appealing to you?

3.12    Which of the engineers' stories can you relate to the most? Why?

3.13    What are your goals for your life outside of work? In this chapter, did you read about anyone with similar goals and/or interests? Explain.

3.14    Imagine yourself five years from now. Write your profile using the same format as presented in this chapter.

3.15    Imagine yourself ten years from now. Write your profile from that perspective.

3.16    Imagine yourself 25 years from now. Write your profile from that perspective.

3.17    What are your thoughts about the success and fulfillment found by engineers who obtained an advanced degree outside the field of engineering?

3.18    Based on the profiles and your own experience, what are the advantages and disadvantages of getting a graduate degree in engineering?

3.19    Prepare a matrix of the bachelor's degrees the engineers in this chapter received and their current positions. How many work in jobs traditionally associated with bachelor's degrees?

3.20    Select your preferred major and prepare a list of potential career paths it could lead to. Provide a brief explanation for each option.

3.21    Today's engineering workforce is truly diverse, with both men and women from all ethnic backgrounds working together. This was not always the case. Select a historical figure who was a woman or minority engineer and prepare a one-page profile which discusses the difficulties she or he had to overcome.

3.22    Research a company you would be interested in working for and list the engineers in their upper-level management structure. Identify the highest level of management currently held by an engineer in that company.

3.23    Prepare a one-page paper on how some of the engineers profiled have addressed family responsibilities amidst an engineering career.

3.24    A stereotype of engineers is that they are boring loners who only care about numbers and technology. Based on the personal-interest sections of the profiles, prepare a report refuting this stereotype.

# 4. A Statistical Profile of the Engineering Profession

## 4.1 Statistical Overview

Many students who choose a major in engineering know little about the profession or the individuals who work in the field. This chapter will answer such questions as: How many people study engineering? What are their majors? What is the job market for engineers like? How much do engineers earn? How many women and minorities are studying engineering? How many practicing engineers are there in the U.S.?

The picture of the engineering profession presented in this chapter will assist you in better understanding the field. This information has been gathered from a variety of surveys and reports that are published on a regular basis by various organizations. As with many types of data, some are very current and others not as current. We have endeavored to provide commentary that explains the information presented.

## 4.2 College Enrollment Trends of Engineering Students

As observed in Figure 4.1, the number of first-year students enrolled in engineering programs has fluctuated greatly during the period from 1955 to 2000, according to the 2000 Engineering Workforce Commission's annual survey of enrollment patterns. During the 1950s and 1960s the number of first-year students pursuing bachelor's degrees in engineering programs ranged between 60,000 and 80,000. During the 1970s there was a significant drop in student interest, and freshman enrollment fell to a low of 43,000. This was followed by the largest long-term climb in student enrollment, which eventually peaked in the early 1980s at about 118,000 students. From that point until 1996 there was a steady decline in student enrollment, with a low of about 85,000 first-year students. This was followed in 1997 to 2000 with a modest increase in student enrollment, which was projected to last for several years.

Why has enrollment in engineering (and higher education in general) fluctuated so greatly over the last few decades? The primary factor influencing enrollment is the number of high school graduates in a given year. Of course, those numbers are directly correlated to the number of births 18 years previously. Given these facts and the information in Figure 4.2, it can be determined how the birth rates from 1960 to 2000 have influenced, and will influence, the enrollment of college-bound students. Likewise, Figure 4.3 provides data concerning high school graduations from 1988 to 2008. This shows that high school graduations started to climb in 1992 and will increase in 1999, but could slow after 2005. Projections indicate that there could be modest increases in 2006 through 2008.

Obviously, there are other factors which influence a student's choice of a major, including the general economy and the relative popularity of various fields in any given year.

It also should be noted that the total number of students pursuing graduate engineering degrees increased significantly from the late 1970s to 1992. Since then, there was a relatively steady decline until 1997, when enrollment started increasing again, with 2000 numbers approaching those of the early 1990's.

Figure 4.1 Engineering enrollments: selected indicators, 1955-2000
(Sources: Through 1978, Table 23, "Engineering Enrollments and Degrees in the Placement of Engineering and Technology Graduates," 1979 (New York: Engineering Manpower Commission); for 1979-1998, Engineering and Technology Enrollments, Fall 1979-1998, Engineering Workforce Commission of the American Association of Engineering Societies, 1979-2000, New York and Washington.)

Figure 4.2 Births in the U.S., 1960-2000

(Source: U.S. Department of Health and Human Services, Centers for Disease Control and Prevention National Center for Health Statistics, 2001.)

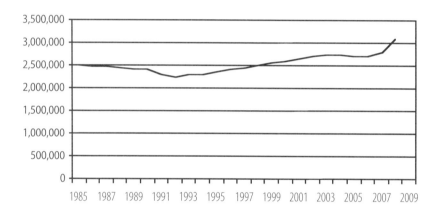

Figure 4.3 High school graduations, 1985-2008

(Source: National Center for Education Statistics, 1998. Projected data are for the years 1996-2008.)

# 4.3 College Majors of Recent Engineering Students

Table 4.1 provides specific 2000 enrollment data for undergraduate and graduate students in several broad engineering fields. This information shows that over 390,000 students were majoring in one of the undergraduate engineering disciplines, with another 118,000 pursuing graduate engineering degrees. The largest field of study at all levels is the electrical and computer engineering discipline, followed by mechanical and aerospace. It is interesting to note that over 57,000 students are in "other engineering disciplines" (which includes the many smaller, specialized fields of engineering) and almost 34,000 are in pre-engineering programs or have not yet declared a major.

Table 4.1  Fall 2000 Engineering Enrollments, By Broad Disciplinary Groups

| | All Engineering Disciplines | Electrical and Computer | Mechanical and Aerospace | Civil and Environ-mental | Chemical and Petroleum | Industrial, Management, Manufacturing | All Other Disciplines | Pre-Engineering or Undecided |
|---|---|---|---|---|---|---|---|---|
| **Full-Time Undergrads** | | | | | | | | |
| First Year | 101,773 | 32,194 | 15,918 | 7,193 | 5,292 | 1,573 | 18,476 | 21,127 |
| Second Year | 76,706 | 26,217 | 14,938 | 7,369 | 5,258 | 2,286 | 11,921 | 8,717 |
| Third Year | 74,055 | 27,953 | 16,198 | 8,875 | 6,131 | 2,912 | 10,134 | 1,852 |
| Forth Year | 92,502 | 34,551 | 20,470 | 12,448 | 8,545 | 4,341 | 11,618 | 529 |
| Fifth Year | 8,082 | 2,692 | 1,824 | 1,419 | 761 | 578 | 782 | 26 |
| | | | | | | | | |
| **Full-Time Under-grad** | 353,118 | 123,607 | 69,348 | 37,304 | 25,987 | 11,690 | 52,931 | 32,251 |
| Part-Time Undergrad | 37,685 | 17,496 | 7,685 | 3,533 | 1,921 | 1,397 | 4,167 | 1,486 |
| Total: all Undergrad | 390,803 | 141,103 | 77,033 | 40,837 | 27,908 | 13,087 | 57,098 | 33,737 |
| | | | | | | | | |
| **Candidates for M.S.** | | | | | | | | |
| Full Time | 48,904 | 20,121 | 7,652 | 5,842 | 2,427 | 2,924 | 9,921 | 17 |
| Part Time | 29,668 | 11,705 | 3,876 | 3,431 | 933 | 1,765 | 7,670 | 288 |
| Total all M.S. | 78,572 | 31,826 | 11,528 | 9,273 | 3,360 | 4,689 | 17,591 | 305 |
| | | | | | | | | |
| **Candidates for Ph.D.** | | | | | | | | |
| Full Time | 32,917 | 11,694 | 5,623 | 3,424 | 3,602 | 1,185 | 7,389 | ---- |
| Part Time | 6,768 | 2,601 | 1,092 | 795 | 383 | 455 | 1.442 | ---- |
| Total: all Ph.D. | 39,685 | 14,295 | 6,715 | 4,219 | 3,985 | 1,640 | 8,831 | ---- |
| | | | | | | | | |
| Full-Time Grads | 81,821 | 31,815 | 13,275 | 9,266 | 6,029 | 4,109 | 17,310 | 17 |
| Part-Time Grads | 36,436 | 14,306 | 4,968 | 4,226 | 1,316 | 2,220 | 9,112 | 288 |
| Total: all Grads | 118,257 | 46,121 | 18,243 | 13,492 | 7,345 | 6,329 | 26,422 | 305 |

(Source: Engineering Workforce Commission of the American Association of Engineering Societies.)

# 4.4 Degrees in Engineering

When examining the number of degrees awarded to engineering graduates during the years from 1991-2000, there is a close correlation between enrollment and degrees. Since 1986, when over 78,000 engineering graduates were produced, to 1997 when approximately 65,000 degrees were awarded, there was a steady decline (with the exception of 1993) until 1995, and then a slight increase until 1997. However, there have been declines in 1998 and 1999, and then a modest increase in 2000, dropping to a total of 63,600. Table 4.2 documents the number of bachelor's degrees awarded by field of study for the decade, 1991 to 2000. It is interesting to note that for many of those years, the greatest number were earned by students in electrical engineering, until 1995, when more were awarded to graduates in mechanical engineering. The large drop in electrical engineering degrees probably reflects the decision by some students to study computer engineering or computer science (which at many schools are in the same department, or are closely aligned with electrical engineering programs).

Table 4.2  Bachelor's Degrees in Engineering, by Discipline, 1991-2000

| Curriculum | 1991 | 1992 | 1993 | 1994 | 1995 | 1996 | 1997 | 1998 | 1999 | 2000 |
|---|---|---|---|---|---|---|---|---|---|---|
| Aerospace | 2898 | 2915 | 2707 | 2311 | 1789 | 1722 | 1372 | 1291 | 1221 | 1274 |
| Agricultural | 267 | 325 | 323 | 392 | 503 | 588 | 770 | 691 | 621 | 624 |
| Architectural | 355 | 445 | 486 | 548 | 563 | 490 | 508 | 514 | 536 | 557 |
| Biomedical | 665 | 686 | 756 | 745 | 831 | 971 | 922 | 1000 | 939 | 1172 |
| Ceramic | 310 | 289 | 286 | 238 | 186 | 178 | 179 | 163 | 157 | 135 |
| Clinical | 3612 | 3849 | 4674 | 5299 | 6153 | 6539 | 6830 | 6492 | 6195 | 6044 |
| Civil | 7748 | 8413 | 9196 | 9842 | 10267 | 10920 | 11119 | 10475 | 9748 | 8750 |
| Computer | 4447 | 4574 | 4675 | 4916 | 4897 | 5577 | 6194 | 7361 | 8192 | 9816 |
| Electrical & Electronic | 19858 | 18337 | 17588 | 15935 | 14966 | 13859 | 13466 | 12495 | 12423 | 12643 |
| Engineering Science | 1110 | 998 | 1089 | 1046 | 1011 | 1 104 | 1041 | 1027 | 966 | 949 |
| Environmental | 168 | 261 | 388 | 537 | 692 | 827 | 882 | 791 | 750 | 697 |
| General Engineering | 1115 | 1120 | 1087 | 1144 | 1224 | 1099 | 1057 | 1046 | 1089 | 115 |
| Industrial & Manufacturing* | - | - | - | - | - | - | - | - | | |
| - Engineering Management* | 373 | 338 | 279 | 244 | 255 | 290 | 304 | 339 | 404 | 454 |
| - Manufacturing Engineering* | 232 | 204 | 168 | 213 | 221 | 206 | 211 | 211 | 218 | 196 |
| - Industrial: All Others* | 3690 | 3541 | 3242 | 3065 | 3044 | 3157 | 3113 | 3037 | 3224 | 3133 |
| Marine, Naval Architectural & Ocean | 410 | 442 | 453 | 474 | 439 | 507 | 448 | 411 | 414 | 379 |
| Materials & Metallurgical | 856 | 875 | 944 | 845 | 872 | 863 | 870 | 889 | 834 | 901 |
| Mechanical | 14626 | 14737 | 15109 | 15569 | 15293 | 14687 | 14136 | 13418 | 12913 | 12989 |
| Mining & Mineral | 182 | 207 | 293 | 292 | 305 | 377 | 428 | 419 | 421 | 373 |
| Nuclear | 261 | 262 | 257 | 258 | 258 | 233 | 182 | 147 | 113 | 130 |
| Petroleum | 235 | 237 | 316 | 297 | 281 | 241 | 236 | 220 | 228 | 250 |
| Systems | 368 | 382 | 396 | 449 | 400 | 471 | 503 | 393 | 442 | 413 |
| Other | 200 | 216 | 289 | 287 | 299 | 361 | 320 | 432 | 452 | 605 |
| Total | 63986 | 63653 | 65001 | 64946 | 64749 | 65267 | 65091 | 63262 | 62500 | 63635 |

*Detailed subcategories for industrial engineering programs adopted by EWC in 1991.

(Source: Engineering Workforce Commission of the American Association of Engineering Societies.)

# 4.5 Job Placement Trends

The 1999-2000 job market for graduating engineers was excellent. In fact, experts agree that it was probably the hottest job market in over 30 years. According to a survey of 320 companies done by the Collegiate Employment Research Institute at Michigan State University, employers expected a 10% to 15% increase in job openings for 2000-2001 engineering graduates. This increase is considerably higher than that in the overall job market for new college grads.

One of the biggest factors driving the job market is the shortage of engineering graduates to fill this expanding pool of new job openings. This trend is expected to continue for the next year or two because of the drop in the birth rate years ago. As a result of these factors, competition is intense among employers seeking to fill vacant positions. This is reflected in a noticeable increase in starting salaries and extra inducements such as stock options, tuition reimbursement and signing bonuses.

The Bureau of Labor Statistics, in its 2000 report, shows that over 2 million people work as engineers, which is approaching a record high for the profession. Figure 4.4 illustrates the recent trends in engineering employment from 1994 to 2000. These trends involve some modest fluctuation, perhaps due to corporate downsizing, and some engineers changing employers. However, the overall picture is one of stable employment growth.

Figure 4.5 illustrates the low unemployment in the engineering profession. In 1994 the unemployment rate of engineers was about 3.5%. Since that time, there has been a steady decrease, with the 2001 unemployment rate around 1.7%.

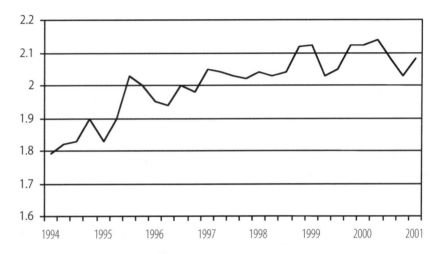

Figure 4.4  Numbers of employed engineers (in millions). Source: Bureau of Labor Statistics.

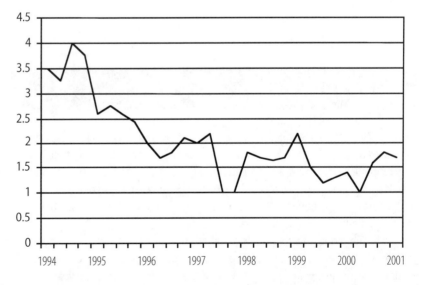

Figure 4.5  Engineering unemployment: recent trends. (Source: for Figs. 4.4 and 4.5 Bureau of Labor Statistics Current Population Survey.)

# 4.6 Salaries of Engineers

Salaries of engineering graduates consistently have been among the highest for all college graduates over the past several decades. To some extent, this has been due to the shortage of engineering candidates to fill available jobs. However, even in "down" years and leaner economic periods, engineers tend to do better in the job market than do their counterparts from other majors.

As reflected in Table 4.3, recent data indicate that average starting salaries for 2000 graduates in selected engineering fields have increased significantly. In a departure from the mode of the past number of years, computer engineering and computer science graduates have now joined the traditional leader, petroleum engineers, at the top of the salary list. This trend may reflect the consistent shortage of trained graduates in those fields.

Table 4.4 provides starting salary data for 2001 graduates compared to 2000 graduates by engineering curriculum and some of the types of employers that hire students in those fields. Starting salaries are higher in virtually every category for 2001 graduates.

One question often raised by engineering students concerns the long-term earning potential for engineers. In other words, does the high starting salary hold up over the length of one's engineering career? Figure 4.6 presents 2000 median salaries for all engineers based on the number of years since their bachelor's degree. It is apparent that an engineer's relative earnings hold up well for many years following graduation.

One significant difference in salaries is shown in Figure 4.7 and 4.8. These data relate the 2000 salary difference between engineers who are placed into supervisory positions and those who remain in non-supervisory roles. Over the course of their career, supervisors tend to earn significantly higher salaries than those in non-supervisory positions.

Figure 4.9 shows how an engineer's income is influenced by the level of education achieved. As might be expected, those with master's and doctoral degrees have higher compensation levels. Length of experience in the engineering field is also an income determinant. Figure 4.9 demonstrates a consistent, regular growth in an engineer's income level as experience is gained. Generally, the average increase is at about 4.6% per year of experience.

Table 4.5 demonstrates 2000 median salaries by region of employment. This compares graduates of 1980-1983 with those of 1990-1991. Salaries are highest in the Pacific Region, followed closely by those in the Western South Central. The lowest reported salaries are for engineers in the New England and Mid-Atlantic Regions. (Geographic strongpoints have shifted dramatically in recent years.)

Figure 4.10 provides the 1999 median annual income by major branch of engineering for all working engineers. Petroleum engineers, nuclear engineers and computer engineers show the highest income levels, with aerospace being bumped from its former top-four position.

When viewing the average starting salaries of engineers by discipline and job function in Table 4.6, it is surprising to observe that there is great variance within discipline and function. The highest reported starting salaries are for computer science engineers in the testing function, with aerospace again having to step down. The lowest is for bio-engineers in the management trainee function.

It is also interesting to examine median annual income by industry or service. Table 4.7 compares the 2000 median salaries of graduates from 1980-1983 with those of 1990-1991. The data show those involved in manufacturing-related industries are at the top of the salary list. The lowest from these classes are involved in public administration.

Table 4.3  Starting Salary Offers in Engineering (Reported, September, 2001)

### Bachelors Degrees

| Curriculum | Average | 75th Percentile | 25th Percentile | % average increase since 2000 | % average increase since 1999 |
|---|---|---|---|---|---|
| Aerospace | 47,085 | 51,000 | 44,000 | 8.0 | 15.8 |
| Agricultural | 44,392 | 50,280 | 38,000 | 18.5 | 14.8 |
| Architectural | 41,403 | 43,800 | 39,000 | 7.1 | 13.9 |
| Bioengineering | 46,678 | 53,000 | 43,888 | 4.0 | 12.0 |
| Chemical | 51,255 | 55,000 | 49,000 | 4.8 | 9.2 |
| Civil | 40,979 | 43,200 | 38,000 | 8.0 | 13.6 |
| Computer Engr. | 53,653 | 57,500 | 50,000 | 6.9 | 17.5 |
| Computer Sci. | 52,473 | 56,800 | 49,000 | 7.0 | 17.5 |
| Electrical | 52,092 | 56,000 | 48,500 | 7.2 | 15.3 |
| Environmental | 42,795 | 49,000 | 37,500 | 8.1 | 16.2 |
| Industrial / Manufacturing | 48,234 | 52,500 | 44,720 | 4.9 | 11.9 |
| Materials | 49,384 | 53,000 | 45,000 | 0.7 | 13.7 |
| Mechanical | 48,588 | 52,000 | 45,000 | 5.7 | 12.3 |
| Metallurgical | 47,772 | 51,300 | 45,000 | 5.5 | 11.0 |
| Mining | 42,208 | 45,000 | 38,400 | 13.0 | 6.6 |
| Nuclear | 47,914 | 51,000 | 45,000 | 1.8 | 11.5 |
| Petroleum | 54,761 | 59,000 | 50,000 | 8.7 | 8.6 |

### Masters Degrees

| Aerospace | 57,224 | 65,376 | 51,000 | 4.8 | 5.5 |
|---|---|---|---|---|---|
| Chemical | 57,317 | 62,000 | 53,500 | 5.2 | 10.1 |
| Civil | 44,234 | 47,000 | 40,000 | -0.8 | 4.7 |
| Computer Engr. | 60,974 | 69,996 | 52,000 | 6.8 | 3.9 |
| Computer Sci. | 62,312 | 68,000 | 55,000 | 1.5 | 21.0 |
| Electrical | 64,188 | 69,996 | 60,000 | 5.5 | 12.3 |
| Industrial / Manufacturing | 56,875 | 63,000 | 47,000 | 1.5 | 14.0 |
| Mechanical | 56,565 | 61,200 | 52,000 | 5.6 | 9.0 |

### Doctoral Degrees

| Aerospace | 66,769 | 78,000 | 55,000 | * | * |
|---|---|---|---|---|---|
| Chemical | 74,153 | 80,000 | 70,000 | 8.8 | 10.1 |
| Civil | 61,606 | 72,000 | 51,000 | * | 5.2 |
| Computer Engr. | 71,720 | * | * | * | 24.8 |
| Computer Sci. | 84,033 | 96,000 | 72,000 | 17.0 | * |
| Electrical | 79,383 | 87,000 | 73,200 | 6.7 | 12.0 |
| Industrial / Manufacturing | 62,121 | * | * | * | * |
| Mechanical | 70,124 | 78,500 | 62,000 | 0.4 | 9.1 |

Source: NACE (National Association of Colleges and Employers) Salary Surveys, September, 2000 & September, 2001

Table 4.4  Average Starting Salary by Curriculum in 2000 and 2001 and Some Common Types of Employers for Each Field

| Curriculum    Type of Employer | 2001 | 2000 |
|---|---|---|
| **Aerospace / Aeronautical Engineering** | | |
| Aerospace Industry | 48,796 | 41,845 |
| **Agricultural Engineering** | | |
| Agribusiness Manufacturing | 49,720 | 31,992 |
| Engineering Services | 38,600 | 34,500 |
| **Architectural Engineering** | | |
| Architectural Services | 41,619 | 34,663 |
| Building Materials & Construction | 41,695 | 38,338 |
| Consulting Services | 46,722 | 40,712 |
| **Bioengineering / Biomedical Engr.** | | |
| Consulting Services | 55,700 | 49,013 |
| Medical Equipment & Supplies | 50,063 | 46,367 |
| **Chemical Engineering** | | |
| Chemicals and Allied Products | 52,859 | 49,736 |
| Computer & Electronic Products | 51,678 | 50,616 |
| Consulting Services | 52,330 | 48,218 |
| Food and Beverage | 50,281 | 49,146 |
| Household / Personal Care Products | 55,182 | 52,043 |
| Petroleum | 54,902 | 50,964 |
| Pharmaceutical | 52,844 | 49,308 |
| **Civil Engineering** | | |
| Building Materials & Construction | 42,487 | 38,899 |
| Consulting Services | 40,690 | 38,336 |
| Federal Government | 37,662 | 33,835 |
| Local / State  Government | 36,495 | 33,132 |
| Transportation | 38,279 | 36,843 |
| **Computer Engineering** | | |
| Communications | 54,677 | 51,702 |
| Computer & Electronic Products | 56,370 | 51,392 |
| Computer Systems/Design/Programming | 54,227 | 48,684 |
| Consulting Services | 55,167 | 51,484 |
| Electrical Equipment/Components | 55,659 | 51,288 |
| **Electrical / Electronics Engineering** | | |
| Aerospace | 51,986 | 46,664 |
| Communications | 52,761 | 47,710 |
| Computer & Electronic Products | 54,937 | 51,274 |
| Computer Systems/Design/Programming | 54,244 | 50,923 |
| Consulting Services | 52,729 | 48,587 |
| Electrical Equipment/Components | 52,911 | 50,128 |
| Utilities | 49,325 | 46,704 |
| **Environmental Engineering** | | |
| Consulting Services | 35,620 | 40,278 |
| Environmental / Waste Management | 43,500 | 37,250 |
| **Industrial / Manufacturing Engineering** | | |
| Aerospace | 45,653 | 44,924 |
| Consulting  Services | 53,333 | 49,472 |
| Electrical Equipment/Components | 41,943 | 45,313 |
| Manufacturing / miscellaneous | 45,773 | 43,131 |
| **Materials Engineering** | | |
| Aerospace | 48,667 | NA |
| Computer & Electronic Products | 54,455 | 54,575 |
| Manufacturing / miscellaneous | 51,520 | 45,680 |
| Metals / Metal Products | 46,780 | 45,800 |
| **Mechanical Engineering** | | |
| Aerospace | 48,429 | 44,663 |
| Computer & Electronic Products | 52,601 | 48,728 |
| Consulting Services | 50,422 | 47,971 |
| Electrical Equipment/Components | 49,175 | 49,943 |
| Manufacturing / miscellaneous | 48,206 | 44,869 |
| Metals & Metal Products | 47,068 | 43,081 |
| Petroleum | 50,441 | 47,458 |
| Transportation Equipment Manufacturing | 50,020 | 47,297 |

Source: NACE (National Association of Colleges and Employers) Salary Surveys, July, 2000 & September, 2001

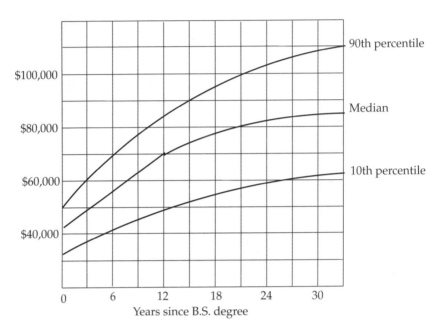

Figure 4.6  Current Salaries of Engineers in All Industries (All Engineers, All Degree Levels) Based on Years Since Baccalaureate Degree

(Source: "Engineers' Salaries: Special Industry Report, 2000." Engineering Workforce Commission of the American Association of Engineering Societies, Inc.)

Figure 4.7  Salary curves for engineering supervisors, all degree levels

(Source: "Engineers' Salaries: Special Industry Report, 2000" Engineering Workforce Commission of the American Association of Engineering Societies, Inc.)

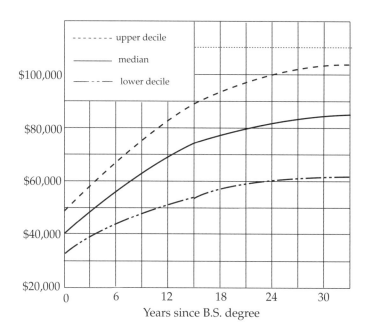

Figure 4.8 Salary curves for engineering non-supervisors, all degree levels.

(Source: "Engineers' Salaries: Special Industry Report, 2000" Engineering
Workforce Commission of the American Association of Engineering Societies, Inc.)

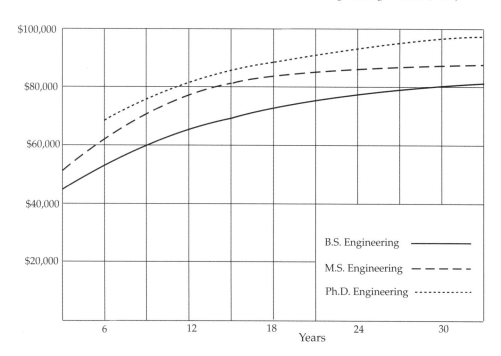

Figure 4.9  Current Income of All Engineers in All Industries by Highest Degree
Earned and Length of Experience.

(Source: "Engineers' Salaries: Special Industry Report, 2000"   Engineering
Workforce Commission of the American Association of Engineering Societies, Inc.)

Table 4.5 Income by Region, All engineers – All degree levels, Graduates of 1990-91 versus Graduates of 1980-1983, 2000 Median Salaries

| Region | Graduates of 1990-91 | Graduates of 1980-1983 |
| --- | --- | --- |
| New England (CT, ME, MA, NH, RI, VT) | 57,379 | 66,389 |
| Middle Atlantic (NJ, NY, PA) | 57,737 | 69,471 |
| East North Central \(IL, IN, MI, OH, WI) | 68,798 | 78,560 |
| West North Central (IA, KS, MN, MO, NE, ND, SD) | 55,854 | 67,439 |
| South Atlantic (DE, DC, FL, GA, MD, NC, SC, VA, WV) | 56,334 | 69,297 |
| East South Central (AL, KY, MS, TN) | 60,557 | 69,109 |
| West South Central (AR, LA, OK, TX) | 70,588 | 81,478 |
| Mountain (AZ, CO, ID, MT, NM, NV, UT, WY) | 60,869 | 72,315 |
| Pacific (AK, CA, HI, OR, WA) | 64,578 | 81,561 |

Source: 2000 Engineer's Salaries: Special Industry Report, Engineering Workforce Commission of the American Association of Engineering Societies, Inc.

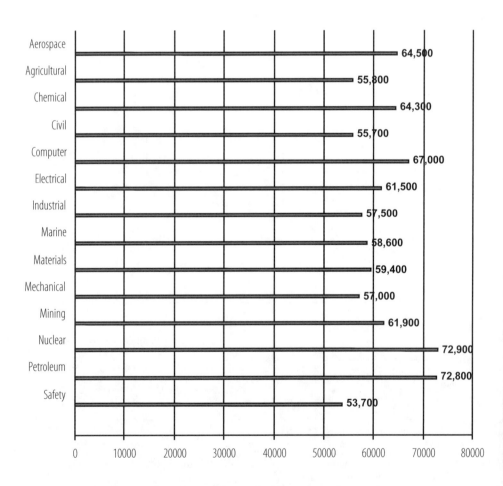

Figure 4.10  Income by engineering discipline, 1999. Source: Bureau of Labor Statistics, National Occupational Employment Estimates, 1999

Table 4.6  Average Starting Salaries by Curriculum and Job Function

| Curriculum | Research & Development | Testing | Design or Construction | Manufacturing | Sales | Consulting | Management Trainee |
|---|---|---|---|---|---|---|---|
| Aerospace | 49,375 | 48,500 | 49,380 | 48,500 | * | 45,000 | * |
| Agricultural | 50,000 | 51,936 | 48,355 | 50,280 | * | * | * |
| Bioengineering | 37,200 | * | 51,000 | 53,000 | * | 54,333 | 35,500 |
| Chemical | 52,557 | 45,963 | 45,389 | 51,749 | 47,263 | 52,582 | 50,517 |
| Civil | 39,869 | 34,020 | 40,308 | 45,000 | 57,000 | 41,232 | * |
| Computer Engr. | 51,833 | 55,144 | 52,138 | 52,380 | 53,263 | 52,941 | * |
| Computer Sci. | 52,882 | 57,330 | * | 50,000 | 50,000 | 52,514 | 40,000 |
| Electrical | 52,330 | 51,358 | 53,272 | 51,480 | 53,298 | 53,136 | 47,925 |
| Industrial/ Manufacturing | 45,667 | * | 39,900 | 47,248 | 38,534 | 54,454 | 46,557 |
| Materials | 51,029 | 43,500 | 40,000 | 50,815 | 45,500 | * | * |
| Mechanical | 48,737 | 48,494 | 47,398 | 47,995 | 49,949 | 50,012 | 47,200 |

* Data not available

(Source: NACE (National Association of Colleges and Employers) Salary Survey, September, 2001.)

Table 4.7  Income by Industry Sector, All engineers – all degree levels, Graduates of 1990-91 versus Graduates of 1980-1983, 2000 Median Salaries

| Industry Sector | Graduates of 1990-91 | Graduates of 1980-1983 |
|---|---|---|
| All Manufacturing Industries | 63,782 | 74,254 |
| Computer/Electrical / Electronics Manufacturing Industries | 60,893 | 70,167 |
| Electrical Equipment Manufacturing Industries | 62,403 | 72,369 |
| Fabricated Metals Manufacturing Industries | 61,221 | 71,795 |
| Machinery Manufacturing Industries | 62,126 | 72,644 |
| Metals Manufacturing Industries | 61,002 | 70,813 |
| Primary Metals Manufacturing | 59,684 | 70,105 |
| Other Manufacturing | 65,180 | 75,920 |
| All Non-Manufacturing | 57,566 | 68,919 |
| Architectural, Engineering, and Related Services | 58,317 | 73,094 |
| Research and Development | 64,167 | 79,389 |
| Professional, Scientific, and Technical Services | 60,870 | 76,323 |
| Public Administration | 53,911 | 56,991 |
| Utilities | 65,375 | 72,080 |
| Other Non-Manufacturing Industries | 57,566 | 68,919 |

(Source: 2000 Engineer's Salaries: Special Industry Report, Engineering Workforce Commission of the American Association of Engineering Societies, Inc.)

# 4.7 The Diversity of the Profession

For many years, engineering was a profession dominated by white males. In recent years, this has been changing as more women and minority students have found engineering to be an excellent career choice. Unfortunately, the rate of change in both enrollment and degrees awarded to these students has been slower than expected. This is especially discouraging at a time when population statistics show a significant increase in the number of college-bound women and minority students.

Table 4.8 provides information concerning degrees awarded to women and minority groups over a recent 10-year period. While the numbers have been increasing, the growth rate has been slower than expected. Since recent demographic data show that the number of minority students attending college is growing, we would expect to see a larger percentage of these students earning engineering degrees. So far, this is not the case—especially at the graduate level, where increases have been modest.

Table 4.9 shows enrollment data for women and minorities in all undergraduate engineering programs. The percentage of women in engineering is less than 20%, while the percentage of under-represented minorities is about 15%. These figures have stayed fairly steady for the past few years, despite increased demand for engineers and increased recruitment efforts.

Table 4.8  Degrees in Engineering, by Level and Type of Student, 1988–1997

| Level and Type | 1991 | 1992 | 1993 | 1994 | 1995 | 1996 | 1997 | 1998 | 1999 | 2000 |
|---|---|---|---|---|---|---|---|---|---|---|
| **Bachelor's Degrees** | 63986 | 63653 | 65001 | 64946 | 64749 | 65267 | 65091 | 63262 | 62500 | 63635 |
| Women | 10016 | 9972 | 10453 | 10800 | 11303 | 11737 | 12160 | 11796 | 12360 | 13140 |
| African Americans | 2304 | 2374 | 2637 | 2769 | 2897 | 3120 | 3203 | 3144 | 3171 | 3150 |
| Hispanic Americans | 2663 | 2708 | 2845 | 3045 | 3409 | 3557 | 4005 | 3938 | 4073 | 4124 |
| Native Americans | 146 | 163 | 175 | 207 | 230 | 263 | 265 | 247 | 328 | 347 |
| Asian Americans | 6305 | 6479 | 6764 | 6881 | 7056 | 7333 | 7625 | 7131 | 7226 | 7529 |
| Foreign Nationals | 4540 | 4389 | 4604 | 4908 | 4893 | 5042 | 5017 | 5083 | 5052 | 5048 |
| **MS/PE Degrees** | 27754 | 28540 | 31104 | 31943 | 32235 | 31012 | 30574 | 30212 | 30229 | 30453 |
| Women | 4117 | 4414 | 4876 | 5131 | 5381 | 5495 | 5800 | 6125 | 6205 | 6431 |
| African Americans | 482 | 516 | 638 | 675 | 712 | 749 | 765 | 778 | 836 | 778 |
| Hispanic Americans | 498 | 527 | 629 | 679 | 708 | 769 | 835 | 856 | 805 | 851 |
| Native Americans | 37 | 40 | 57 | 54 | 48 | 72 | 65 | 81 | 46 | 96 |
| Asian Americans | 2344 | 2423 | 2512 | 2624 | 2829 | 3132 | 2926 | 2929 | 2826 | 2613 |
| Foreign Nationals | 8262 | 9054 | 10122 | 10385 | 10697 | 9955 | 10208 | 10524 | 11010 | 11680 |
| **Doctoral Degrees** | 5680 | 5958 | 6198 | 6458 | 6611 | 6811 | 6986 | 6567 | 5833 | 5929 |
| Women | 551 | 592 | 600 | 711 | 802 | 854 | 861 | 810 | 858 | 937 |
| African Americans | 43 | 39 | 46 | 63 | 68 | 78 | 99 | 95 | 114 | 96 |
| Hispanic Americans | 51 | 62 | 55 | 48 | 68 | 114 | 90 | 102 | 95 | 85 |
| Native Americans | 6 | 10 | 2 | 4 | 6 | 13 | 8 | 8 | 12 | 11 |
| Asian Americans | 345 | 394 | 383 | 432 | 599 | 618 | 511 | 467 | 408 | 397 |
| Foreign Nationals | 2898 | 3088 | 3306 | 3376 | 3240 | 3253 | 3381 | 3117 | 2627 | 2961 |

Table 4.9  Women and Minorities in the Engineering Pipeline, 2000

|  | Percentage of all Full-Time Undergraduates | |
|  | First-Year Full-Time B.S. Students | All Full-Time B.S. Students |
|---|---|---|
| Women | 18.92% | 19.68% |
| African Americans | 8.04 | 6.57 |
| Hispanic Americans | 7.44 | 7.90 |
| Native Americans | 0.62 | 0.59 |
| Asian Americans | 10.90 | 11.45 |

Source: Engineering Workforce Commission of the American Association of Engineering Societies

# 4.8 Distribution of Engineers by Field of Study

According to the Bureau of Labor Statistics (see Figure 4.11), approximately 25% (almost 360,000) of all practicing engineers are employed in the electrical engineering discipline. This is followed by about 230,000 mechanical engineers, and 200,000 civil engineers. It is interesting to note that "other engineers" (which includes the smaller, specialized fields of engineering) include about 415,000 engineers. Comparing this information to previous data on enrollment and degrees, one would expect these numbers to remain somewhat steady, although there probably will be a slight increase in overall mechanical engineering employment and a slight decline for electrical engineering.

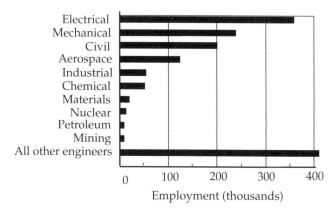

Figure 4.11  Practicing engineers in the U.S. in 1999.

## 4.9 Engineering Employment by Type of Employer

According to the Bureau of Labor Statistics, in 1998 the largest sector of engineering employment, almost 50%, was in manufacturing industries. These include electrical and electronic equipment, industrial machinery, transportation equipment, and instruments.

In 1998, over 720,000 engineering jobs were in service industries which includes engineering and architectural services, research and testing services, and business services in which firms designed construction projects or did other engineering work on a contractual basis. Engineers were also working in communications, utilities and computer industries.

Government agencies at the federal, state, and local levels employed approximately 166,000 engineers in 1998. The largest number were employed by the federal government in the Departments of Defense, Transportation, Agriculture, Interior, Energy, and the National Aeronautics and Space Administration (NASA). The majority of state and local workers are usually involved in local transportation, highway and public works projects.

In 1996, almost 50,000 engineers were self-employed, typically as consultants.

## 4.10 Percent of Students Unemployed or in Graduate School

Though 2000-2001 showed one of the strongest job markets in years for engineers, what about the number of students who are still unemployed and those who go on to graduate school? As indicated in Table 4.10 and referred to earlier in Figure 4.1, the total number of students pursuing graduate engineering degrees increased significantly between the late 1970's and 1992. Since then, there has been a steady decline. Although there was that slight increase in 1997, a steady and significant decline continued for 1998 and 2000. The early period covered by Table 4.10 includes a period of time when jobs for bachelor-level graduates were not as plentiful, and many were opting for graduate school, which may explain the rapid rise in graduate school enrollments. However, in the current job market, recruiters are competing directly with colleges and universities for qualified candidates. This has contributed to the continuing decline in graduate school enrollment.

Table 4.10 also shows the low unemployment rate for graduating engineers between 1990 and 2000. The figures for 1992 and 1994 reflect a tighter job market than that of 1990. It should be noted that, in general, the unemployment rate for new engineering graduates is far lower than for any other major.

If this strong engineering job market continues, one might expect that the unemployment rate for engineers will decrease, and the number going to graduate school will also continue to fall.

Table 4.10  Engineering Graduate School Full and Part-time Enrollment Trends and percent of Unemployed Engineers

|  | 1990 | 1992 | 1994 | 1998 | 2000 |
|---|---|---|---|---|---|
| **Aerospace** | 3,563 | 3,935 | 3,550 | 2,780 | 3,042 |
| **Chemical** | 6,657 | 6,926 | 7,292 | 6,683 | 6,714 |
| **Civil** | 11,625 | 13,900 | 14,164 | 11,009 | 11,151 |
| **Electrical** | 32,095 | 32,635 | 29,524 | 25,302 | 27,764 |
| **Industrial** | 9,713 | 11,552 | 11,401 | 9,428 | 5,234 |
| **Mechanical** | 17,115 | 18,566 | 17,310 | 14,032 | 15,201 |
| **TOTAL** | 80,750 | 87,510 | 83,230 | 69,220 | 69,106 |
|  |  |  |  |  |  |
| **% Unemployed Engineers** | 2.9% | 5.2% | 4.3% | 1.7% | 1.5% |

(Sources: "Engineering Enrollments and Degrees", Engineering Manpower Commission of the American Association of Engineering Societies, Inc. 1990, 1992, 1994, and Engineering Workforce Commission of the American Association of Engineering Societies, Inc., 1998, 2000.)

# 4.11  A Word from Employers

As discussed earlier, the current job market for graduating engineers is excellent. Despite this favorable situation for engineering students, there are important issues that employers want students to consider. In a 1999 survey of 320 employers done by the Collegiate Employment Research Institute at Michigan State University, recruiters were asked to provide commentary on any special concerns they had about new engineering graduates.

One interesting conclusion to the study found that "Employers want the *total package* when they hire their next engineering graduates. Not satisfied with academically well-prepared graduates, employers want individuals who possess and can demonstrate excellent communication and interpersonal skills, teamwork, leadership, and computer/technical proficiency. A willingness to learn quickly and continuously, to problem solve effectively, and to use their common sense is also desired. New employees must be hard-working, take initiative, and be able to handle multiple tasks." In addition, employers are looking for students with new emerging skills and aptitudes such as the ability to understand e-commerce, computer capabilities that include programming skills, and the ability to adapt to constant change. In the new global economy, increasing competition requires a strategy to respond quickly.

Therefore, it is important to realize that just earning an engineering degree is no longer sufficient to prepare students for the fast-paced, technologically changing global economy. Students must devote time in college developing those critical competencies that employers are seeking in order to be well prepared for the challenges that await them.

# Exercises and Activities

4.1    If an administrator used only the data from 1975 to 1980 to predict future engineering enrollment, how many first-year, full-time students would be expected in 1990? What would the percentage error have been?

4.2    What is the percentage increase in projected high school graduations in 2006 compared to those in 1998?

4.3    Using the data of Fig. 4.4, estimate the number of employed engineers in 2004.

4.4    Using the starting salary data from Table 4.3, and assuming an annual increase at the current cost of living (currently about 3%), calculate what you could expect to receive as a starting salary in your chosen field of study if you graduate three years from now.

4.5    Using the information from Figure 4.9 estimate your salary after 30 years, assuming you have: a) a bachelor's degree, b) a master's degree, and c) a doctoral degree.

4.6    Using the information from Figures 4.7 and 4.8, estimate your salary after 30 years, assuming you are: a) a supervisor, and b) a non-supervisor.

4.7    If you were a supervisor in the upper decile (10%) rather than a non-supervisor in the upper decile, what would be your percentage increase in salary 20 years after graduation?

4.8    Calculate the percentage increase of the highest paid Bachelor's engineer in 2000 compared to the lowest paid Bachelor's engineer in 2000.

4.9    Suggest at least one reason for the relatively low income in the New England Region of Table 4.5.

4.10    Suggest at least one reason for the relatively low income of government workers compared to those in the private sector.

4.11    Project the number of women and minorities in engineering in the year 2005 using the data of Table 4.8. Do not include foreign nationals.

4.12    Estimate the percentage of engineers who are chemical engineers using the data of Fig. 4.11.

4.13    Based on the current economic situation, do you expect the employment demand for graduating engineers to increase or decrease? Explain the basis for your answer.

4.14   If we experience a significant economic downturn, what do you think will happen to enrollment in graduate engineering programs? Support your answer by referencing the appropriate figures or tables.

4.15   Review the information presented in Figures 4.2 and 4.3 and Table 4.8. Make a prediction about the future enrollment of women and minorities in engineering. Explain your answer.

4.16   What factors do you think influence how employers determine hiring needs in any given year?

4.17   Attend a campus career fair. Talk with at least 3 recruiters from different firms. Make a list and discuss those factors which employers are stressing in their hiring decisions.

4.18   Based on the current economic situation, do you expect the salaries of engineers to continue to increase at the recent pace? Why or why not? Using other tables in this chapter, explain the basis for your answer.

4.19   Review the enrollment data in this chapter. What do you think high schools, colleges, and the engineering profession can do to increase the number of students choosing to pursue an engineering career?

4.20   What factors do you think contribute to the fact that entry level engineering salaries are usually higher than any other major on campus?

# 5. Succeeding in the Classroom

## 5.1 Introduction

As an engineering student, an important goal is to succeed in the classes needed to attain an engineering degree. How this is done depends to a great deal on the individual student and the style, temperament and strengths of that student. This chapter presents strategies proven to be effective in helping students succeed. As with any tips on success, you need to look at the techniques being outlined and decide if they can be applied to you and your own individual style. If you are unsure about some of the suggestions, give them a try. You may find yourself in the same situation as one freshman engineering student after being given a book on study skills.

*Success:*
*"The achievement of something desired, planned or attempted."*

The student enters a professor's office very excited and asks for a short conference with the professor:

| | |
|---|---|
| *Student:* | *Professor! It worked!* |
| *Professor:* | *That's great. What worked?* |
| *Student:* | *The book!* |
| *Professor:* | *That's great. What book?* |
| *Student:* | *That study skills book.* |
| *Professor:* | *So you read it?* |
| *Student:* | *Yes, I read it like you suggested. It sure sounded like a lot of unworkable stuff at first. But then I decided, what the heck. I'll try it for a week. Couldn't hurt, I figured. Well, it really did work! It even worked for physics lab. I got an A on the last lab. I still can't believe it.* |
| *Professor:* | *That's great. I am glad it helped.* |
| *Student:* | *Could I get another copy of that book? I have some friends who want to read it now…* |

Just like that student, we are often reluctant to try new things. We are comfortable in what we do. You may have been very successful in high school with your study methods and wonder why you need to develop different study methods and techniques. The answer is that your environment has changed. College is very different from high school. The courses are structured differently and require significantly more preparation outside of class. The expectations of the instructors are different and the grading is different. Modifying your habits to become more effective is a key to succeeding in engineering studies. Learning to adapt to meet new challenges is a skill that will serve you well not only in college, but for the rest of your life.

The methods for excelling in the classroom outlined in this chapter have proven effective for numerous students. We suggest that you consider all of these ideas and adapt the ones that fit best with your own personality. If you are unsure whether they will work, try some out for a trial period and see. If those don't work, move on to others until you find something that fits with your own style.

*The three components to succeeding in your academics are your ability, your attitude and your effort.*

The three components to succeeding in your academics are your ability, your attitude and your effort. There is no book that can address the first item. While there is unquestionably a level of ability that is needed to succeed in engineering, there are numerous students who had the ability but failed at engineering because of poor attitudes or poor or inefficient effort. There are also many students who lacked natural ability but because of their positive attitude and extraordinary effort are now excellent engineers. This chapter focuses on techniques to improve these two components for success that are totally under your control.

## 5.2 Attitude

Your attitude is the first thing that you have control over when beginning any challenge, including studying engineering. You can expect success or expect failure. You can look for the positives or dwell on the negatives.

*"Ability is what you're capable of doing. Motivation determines what you do. Attitude determines how well you do it."*

*—Lou Holtz, South Carolina football coach*

Approaching your classes, professors and teaching assistants with a positive attitude is the first key to succeeding in your engineering studies. Each class should be looked at as an opportunity to succeed. If there are difficulties along the way, learn to deal with those and move ahead. Many students will decide that a certain class is too difficult or that the professor is a poor teacher and will expect to do poorly. Many fulfill their own prophecy and actually fail. Other students look at the same situation and overcome the hurdles and excel.

It is a distinct possibility that you may have professors you don't like or who you think teach poorly. You may even have an experience like the following example.

## Example 5.1

A physics professor of mine discovered the overhead projector during a semester and moved his lecture from the blackboard to overheads. The problem was that he was very tall and stood so that he blocked the overhead. It shined on his coat and tie but we couldn't see what he was writing. When we told him in class that we couldn't see, he moved out of the way and looked at the overhead. He then adjusted the focus and asked if we could see now. After we responded yes, he moved back to write more and blocked the overhead again. When another student tried to explain that he was blocking his view of the overhead, the professor explained that no matter where he stood, he would block someone's view. He suggested that the student move to the other side of the room. A third student from the other side of the room chimed in that he couldn't see from that side either. Unfortunately he was in the back of the lecture hall, so the professor thought that was the reason. He moved out of the way of the projector and asked one of the students in the front row if she could see. She said, "Yes, but..." The professor cut her off before she could finish and asked the guy in the back of the room if he had glasses. When he said no, the professor recommended he get his eyes checked because the woman in front said she could see. At this point we all gave up taking notes for the rest of the lecture.

Later, I complained to an advisor who was the director of the university honors program. Rather than a strategy for communicating with the professor, I received a story about a professor he had had. It seems that he had an English professor who was about 108 years old (or *seemed* that old). He couldn't stand up without leaning on something, so he used the blackboard. His shoulder made contact with the blackboard, which wasn't a problem in itself except that he wrote at shoulder level. The result is that he erased what he wrote as he moved along the board. So if you weren't in a seat where you could see the three or four words he wrote before he erased them with his shoulder, you couldn't take notes.

*Attitude is totally under your control.*

I am still not sure what lesson I was supposed to learn from that story. It struck me, however, that even with professors like that, the advisor earned his Ph.D., became a professor and eventually ran the university's honors program. He had developed strategies to cope and kept a good attitude. Attitude is totally under your control.

# 5.3 Goals

As freshmen, there were a number of us who were in an honors program. One in the group set a goal to be a Rhodes Scholar from the first day of his freshman year. The rest of us were too busy adjusting to college and taking it all in. In our senior year, when we were interviewing for jobs and deciding whether to go to graduate school, the friend who had set the goal was being toured around the state as the newest Rhodes Scholar from our university. The difference over our college careers was astounding, simply because he had established clear, elevated goals. The rest of us did okay, but he excelled, in large part because he had defined a standard and kept striving for it.

The lesson that I took away from that friendship in college was that setting clearly defined goals early is a key to success in school, in an engineering career or in life as a whole.

In setting goals, there are two essential elements—"height" and time. You might ask, "How high do I set my goals?" This will vary from individual to individual. Goals should be set high, but attainable. A common management practice is for a team to decide on a reasonable goal to accomplish. Then the bar is raised and a stretch goal is defined that encourages the team to produce more. This stretch goal is designed to push the group farther than they would have gone with only the original goal. Sometimes this stretch goal looks reasonable and sometimes it looks totally unreachable. What often happens, though, is that the stretch goal is met, realistic or not. As an individual, setting stretch goals helps us grow. When setting goals for yourself, look at what you think you can do and set base goals. Then establish stretch goals a little higher than the base goals. You may just find yourself meeting those higher goals.

The second key element in goal setting is time. When will the goal be accomplished? It is important to have long-term goals and short-term goals. Long-term goals help guide where you are headed. For example, what do you want to do with your engineering degree? An appropriate long-term goal as a freshman could involve determining the kind of entry-level job you desire. This will help answer the question "Why am I doing this?" when your classes get tough.

Short-term or intermediate goals should be stepping-stones to the achievement of your long-term goals. An advantage of an academic setting is that there are natural breaks, semesters or quarters, for the establishment of goals within specific time frames. Goals can be set for the academic year, and for each semester or quarter. These provide intermediate goals.

Short-term goals need to be set and rewarded. Rewarding short-term goals is essential to keeping motivation high in the quest for long-term goals. Many students lose sight of their long-term goals because they didn't set short-term, "rewardable" goals. Short-term goals may be an "A" on an upcoming quiz or test, or it may be completing a homework set before the weekend.

*"Do you feel the world is treating you well? If your attitude toward the world is excellent, you will receive excellent results. If you feel so-so about the world, your response from that world will be average. Feel badly about your world and you will seem to have only negative results from life."*

—*Dr. John Maxwell, founder of INJOY, Inc.*

*"If you aim at nothing, you will hit it every time."*

—*Joseph McRae Mellichamp, professor emeritus of management science, University of Alabama*

## Example 5.2

| Long-term goal | Job as a design engineer for a major manufacturer of microprocessors |
|---|---|
| Intermediate goals | 3.7 GPA after freshman year<br>3.5 GPA at graduation<br>Hold an office in the local chapter of IEEE<br>Internships after sophomore and junior years |
| Short-term goals | 'B' on first calculus test<br>'A' in calculus<br>3.7 GPA after first semester<br>Join IEEE and attend meetings<br>Complete chemistry lab by Friday |

In the list of short-term goals above, there are opportunities for rewards. Perhaps there's a compact disk you want. Set a goal of a certain grade on that calculus test. If you meet the goal, you can go buy the CD.

*Clearly define your goals by writing them down.*

Clearly define your goals by writing them down. It is much more powerful to write down your goals than to simply think about them. Some people post them where they can see them daily. Others put them in a place where they can retrieve and review them regularly. Either method is effective. For short-term goals, establish weekly goals. Select a day, probably Sunday or Monday, and establish the goals for the week. This is a great time to establish the rewards for your weekly goals. Plan something fun to do on the weekend if you meet those goals.

At the end of each semester, examine your progress toward your long-term goals. This is an advantage to being in school. Every semester, there is an opportunity to assess your progress and start fresh in new classes. Take advantage of this. Examine your long-term goals and make any changes if necessary.

## Determining Goals

It can be intimidating or difficult for students early in their careers to decide on appropriate goals. Fortunately, there are numerous resources on campus to help. Academic advisors can provide criteria for grade-point goals. Things to consider include:

1. What is the minimum GPA to stay enrolled in school?
2. What grades do you need to continue in engineering or to enter engineering?
3. What grades are needed to be eligible for co-op or intern positions?
4. What GPA is required to make the dean's list or honor list?
5. What GPA is needed to be eligible for the engineering honorary organizations?
6. What grades are needed for admittance to graduate school?
7. What grades are needed for scholarships?

There are also people on campus who work with the placement of graduates. They are excellent resources to get information on what employers are looking for. What credentials do you need to be able to get the kind of job you desire? If you know this as a freshman, you can work toward it. You can also go directly to the employers and ask these same questions of them. If companies come to your campus, talk to them about what credentials are needed to get a job. Job fairs are great times to do this. Approaching company representatives about establishing goals is also a great way to show initiative and distinguish yourself from your classmates.

Goals are important for outside the classroom too. It is important to have goals in all areas of your life. Later in this chapter we will discuss such goals.

# 5.4 Keys to Effectiveness

Once goals have been set regarding grades, the next step is to achieve those grades. Effort and effectiveness may be the most important components of your success as a student. There are numerous cases of well-equipped students who end up failing. There are also students who come to the university poorly prepared or start off poorly but eventually succeed. The difference is in their effort and their effectiveness. The following suggestions contain strategies to improve personal effectiveness in your studies.

### Take Time to Study

One of the strongest correlations with a student's performance is the time he or she spends studying.

 **A general rule is that it takes a minimum of two hours of study outside of class for each hour of class lecture.**

According to the reminder above, if you are taking 15 hours, that means that you need to spend 30 effective hours a week studying outside of class. In a typical high school, the time requirement is much less than this. Developing the study habits and discipline to spend the needed time studying is a prime factor in separating successful and unsuccessful students.

### Go to Class

*Missing class only delays the expenditure of time you will need to master the material.*

There is also a high correlation with class attendance and performance. This may seem obvious, but you will run into numerous students who will swear that it doesn't matter if you go to class. Many instructors in college do not take attendance, so no one but you knows if you were there, which is different from high school. The classroom is where important information will be presented or discussed. If you skip class, you may miss this information. Also, by skipping class, you miss an opportunity to be exposed to the material you will need to know, and you will have to make up that time on your own. Missing class only delays the expenditure of time you will need to master the material. Most often, it may take much more than an hour of independent study to make up an hour's worth of missed class time. It is possible for you to never miss a class in your undergraduate years, except for illness or injury, serious family problems or trips required by groups to which you belong. Never missing a class could be one of your goals.

Also, you may want to schedule classes at 8 a.m. and 4 p.m. This will get you up and going and keep you there! You will seldom, if ever, study at 8 a.m. and at 4 p.m. Do your studying in the middle of the day. This will help you to use your time effectively.

## Make Class Effective

The first component of making class effective is sitting where you can get involved. If the classroom is large, can you see and hear in the back? If not, move up front. Do you need to feel involved in the class? Sit in the front row. Do you fall asleep in class? Identify why. Ask yourself, "Am I sleeping enough at night?" If the answer is no, get more sleep.

The second step for making class time effective is to prepare for class. Learning is a process of reinforcing ideas and concepts. As such, use the class time to reinforce the course material by reviewing material for the lecture beforehand. Most classes will have a textbook and assigned reading for each lecture. Take time before the lecture to skim over the relevant material. Skimming means reading it through but not taking time to understand it in depth. This will make the class time more interesting and more understandable, and it will make note taking easier.

> **"After class, reread the assigned sections in your textbook. Concentrate on the sections highlighted by the lecture.  By doing so, you will not waste time trying to understand parts of the book that are not critical to the course."**

By following the above advice, you will have had three exposures to the material, which helps you remember and understand the material better.

## Keep Up with the Class

Class is most effective if you keep up. An excellent short-term goal is to master the lecture material of each course before the next lecture. A course is structured in such a way that you master one concept and then move on to the next during the next lecture. Often in science and engineering classes the concepts build on each other. The problem with falling behind is that you will be trying to master concepts that depend on previous lectures, and you are going to hear lectures that you will not understand.

There is a very practical reason that a three-credit class meets three times per week. During a 15-week semester, there are only 45 hours of class meeting time. That amount of class time could be held in a week of consecutive nine-hour days. Arranged that way, a whole semester could be taken in a month. Then it would only take a year to get a bachelor's degree. So why don't we do it that way?

*The current educational model is designed to introduce a concept and give you, the learner, time to digest it.*

The reason is that the current educational model is designed to introduce a concept and give you, the learner, time to digest it. It is an educationally sound model. In engineering study, digesting means reviewing notes, reading the text and working several problems related to the topic. If you are still confused about an aspect of the previous lecture, the next class provides a great opportunity to bring it up with the professor. Get your questions answered before going on to the next topic.

A note of caution should be given here. Friends of yours may be studying subjects where they can study from test to test rather than from class to class. They may insist that you don't need to keep up so diligently because they don't. Some majors cover material that is more conceptually based. That is, once you understand the concept, you have it. Engineering, math and science courses are not this way. To be mastered they require extensive study and preparation over an extended period of time. In some courses there are only a few basic concepts, yet the entire course is dedicated to the application of these few concepts. An example is Statics, a sophomore-level course, in which there are only two fundamental equations germane to the course: the sum of forces equals zero, and the sum of moments equals zero. The entire course involves applying these two equations to various situations, and using the information in engineering applications.

## Take Effective Notes

A main component in keeping up with your classes is taking effective notes for each lecture. Effective notes capture the key points of the lecture in a way that allows you to understand them when you are reviewing for the final exam three months later. Suggested note-taking strategies follow:

1. Skim the assigned reading prior to class to help identify key points.
2. Take enough notes to capture key points but don't write so much that you fail to adequately listen to the presentation in class.
3. Review your notes after class to annotate them, filling in gaps so they will still make sense later in the semester.
4. Review your notes with other students to ensure that you captured the key points.
5. Review your notes early in the semester with your professor to be certain you are capturing the key points.

In class, your job is to record enough information to allow you to annotate your notes later. You don't have to write everything down. Getting together with classmates after class is a terrific way to annotate your notes. That way you have different perspectives on what the main points were. If you really want to make sure you captured the key points, go and ask your professor. Doing so early in the semester will not only set you up for successful note taking throughout the semester, it will also allow you to get to know your professor.

## Work Lots of Problems

Because the applications of the concepts are the core of most of your classes, the more problems you can work, the better prepared you will be. In math, science or engineering courses, doing the assigned homework problems is a minimum. Search for additional problems to work. These may be from the text or from old exams. Your professor should be able to steer you to appropriate problems to supplement the homework.

## Caution Using Solution Manuals and Files

For some classes, there are homework files or solution manuals available. While these tools may help get the homework done quicker, they very often adversely affect exam performance. The problem is that most professors don't allow the solutions to be used during the test. Using them becomes a crutch and can impair your ability to really learn the material and excel on exams. If they are used at all, use them only after exhausting all other possibilities of working out the answer yourself. If it takes a lot of work to figure out a homework problem, that concept will be learned well and you will have a higher probability of demonstrating your knowledge on the next test. Taking the easy way out on homework has a very consistent way of showing up on test results. Minimal learning results from copying down a solution.

## Group Studying

A better model for studying than using solutions is to study in a group. Numerous studies have shown that more learning takes place in groups than when students study by themselves. Retention is higher if a subject is discussed, rather than just listened to or read. If you find yourself doing most of the explaining of the ideas to your study partners, take heart! The most effective way to learn a subject is to teach it. Anyone who has taught can confirm this anecdotally; they really learned the subject the first time they taught it.

If you aren't totally convinced of the academic benefits of group studying, consider it a part of your engineering education. Engineers today work in groups more than ever before. Being able to work with others effectively is essential to being an effective engineer. If you spend your college years studying in groups, working in groups will be second nature when you enter the workforce.

*"The most important single ingredient to the formula of success is knowing how to get along with people."*
—*Theodore Roosevelt*

Studying with others will also make it more bearable, and even fun, to study. There are numerous stories about students preparing for exams and spending the whole day on a subject. It is difficult to stick to one subject for an entire day by yourself. With study partners you will find yourself sticking to it and maybe even enjoying it.

Studying with others makes it easier to maintain your study commitments. Something we all struggle with is discipline. It is much easier to keep a commitment to study a certain subject if there are others who are depending on you. They will notice if you are not there and studying.

Group studying is more efficient if properly used. Chances are that not everyone in the group will be stuck on the same problems. So right away, you will have someone who can explain the solution. The problem areas common to everyone before the group convenes can be tackled with the collective knowledge and perspectives of the group. Solutions are found more quickly if they can be discussed. More efficient study time will make more time for other areas of your life!

*"I will pay more for the ability to deal with people than any other ability under the sun."*
—*John D. Rockefeller*

Choosing a group or partner may be hard at first. Try different study partners and different size groups. Ray Landis, Dean of engineering at California State University, Los Angeles and a leading expert on student success, suggests studying in pairs. "That way each gets to be the teacher about half the time." Larger groups also can be effective. The trick is to have a group that can work efficiently. Too large a group will degenerate into a social gathering and not be productive in studying. Whichever group size you choose, here are some basic tips for group studying.

1. Prepare individually before getting together.
2. Set expectations for how much preparation should be done before getting together.
3. Set expectations on what will be done during each group meeting.
4. Find a good place to convene the group that will allow for good discussion without too many distractions.
5. Hold each other accountable. If a group member is not carrying his or her weight, discuss it with that person and try to get them to comply with the group's rules.
6. If a member continues to fail to carry their own weight or comply with the group's rules, remove that person from the group.

In a subsequent section, differences in learning styles will be discussed. Depending on your own learning style and personality, you may need a certain kind of environment to

## Define a Study Spot

study efficiently. Some resources attempt to describe the "perfect study environment." In reality, there is no one perfect study environment or method of studying. What is crucial is that you find what you need and find a place that meets your needs. For some, it involves total quiet, sitting at a desk. Others may prefer some noise, such as background music, and prefer to spread out on the floor. Whatever you decide is the best environment, pick one where you can be effective.

Cynthia Tobias describes the conflict she has with her husband's view of a proper study environment in her book *The Way They Learn*:

*What is crucial is that you find what you need and find a place that meets your needs.*

> I have always favored working on the floor, both as a student and as an adult. Even if I'm dressed in a business suit, I close my office door and spread out on the floor before commencing my work. At home, my husband will often find me hunched over books and papers on the floor, lost in thought. He is concerned.
>
> "The light is terrible in here!" he exclaims. "And you're going to ruin your back sitting on the floor like that. Here, here! We have a perfectly clean and wonderful rolltop desk." He sweeps up my papers and neatly places them on the desk, helps me into the chair, turns on the high-intensity lamp, and pats my shoulder.
>
> "Now, isn't that better?" he asks.
>
> I nod and wait until he is down the hall and out of sight. Then, I gather all my papers and go back down on the floor… It does not occur to him that anyone in their right mind could actually work better on the floor than at a desk, or concentrate better in 10-minute spurts with music or noise in the background than in a silent 60-minute block of time.

Different people have different ways in which they can be most efficient. You need to discover yours so that you can be efficient. If you are unsure of what works best for you, test some options. Try different study environments and keep track of how much you get done. Stick with the one that works best for you.

If you and your roommate have different styles and needs, then you will need to negotiate to decide how the room will be used. Both of you will have to be considerate and compromise, because it is a shared space.

One item that is a distraction for almost everyone is television. Because it is visual, it is almost impossible to concentrate on homework with a TV on, although many students swear they are still "productive." A few may be, but most are not. Watching TV is one of the biggest time-wasters for students. It is hard enough for first-year students to adjust to college life and the rigors of pre-engineering classes. We strongly recommend **not** having a television set in your residence hall room or apartment for at least your first year. It is not even good for study breaks, which should be shorter than 30- or 60-minute programs. This may be one temporary sacrifice you have to make to be successful in your studies.

## 5.5 Test-taking

Most courses use written, timed exams as the main method of evaluating your performance. To excel in a course, you need to know the material well and be able to apply it in a test situation. The early part of this chapter provided tips on how to improve your understanding of the course material. The second part is test-taking skills.

Ask any student who has been away from school for a few years. Taking tests is a skill and needs to be practiced. Just like a basketball player who will spend hours practicing lay-ups, it is important to practice taking tests. A great way to do this is to obtain past exams. A couple of days before the test, block out an hour and sit down and "take" one of the old tests. Use only the materials allowed during the actual test. For instance, if it is a closed-book test, don't use your book. In this way, you are doing two things. The first is assessing your preparation. A good performance on the practice test indicates that you are on the right track. The second thing you are doing is practicing the mechanics of taking the test. Again, with the basketball analogy, players will scrimmage in practice to simulate game conditions. You should simulate test conditions. What do you do when you reach a problem you can't do? Most people will start to panic, at least a little bit. It is critical that you stay calm and reason your way beyond whatever is blocking you from doing the problem. This may mean skipping the problem and coming back to it, or looking at it in another way. In any case, these are the types of things you have to do on tests, and by practicing them under test conditions, you will do better on the actual tests.

Before the first test in the course, visit your professor and ask him or her how to assemble a simulated test. It may be that old exams are the way to go, or possibly certain types of problems will be suggested. Professors don't like the question "What is going to be on the test?" but are much more receptive to requests for guidance in your preparation. The professor's suggestions will also help you focus your studying, to make it more efficient.

## Taking the Test

There are some general guidelines for taking any test. The first one is to come prepared to take the test. Do you have extra pencils just in case? Can you use your calculator efficiently? Breaking your only pencil or having your calculator fail during the test can be very stressful and prevent you from doing your best, even if you are given a replacement. Proper planning for the test is the first step toward succeeding.

The second step is to skim over the entire test before beginning to work the problems. This is partially to make sure that you have the complete test. It also gives you an overview of what the test is going to be like so you can plan to attack it efficiently. Take note of the weighting of the points for each problem and the relative difficulty.

The next step is to look for an easy problem or one that you know you can do. Do this problem first. It will get you into the flow of doing problems and give you confidence for the rest of the test. Also, doing the problems you know you can do will ensure that you have made time for the problems for which you should get full credit.

Always keep track of the time during a test. Tests are timed, so you need to use the time efficiently. Many students will say they "just ran out of time." While this may prevent you from finishing the test, it should not prevent you from showing what you know. Look at the number of points on the test and the time you have to take the test. This will tell you how long to spend on each problem. If there are 100 points possible on the test and 50 minutes to complete it, each point is allotted 30 seconds. So a 10-point problem should only take 5 minutes. A common mistake is wasting too much time struggling through a difficult problem and not even getting to others on the test. Pace yourself so you can get to every problem even if you don't complete them. Most instructors will give partial credit. If they do, write down how you would have finished the problem and go on to the next when the allotted time for that one is up. If there is

*A common mistake is wasting too much time struggling through a difficult problem and not even getting to others on the test.*

time at the end, you can go back to the unfinished problem. But if not, the instructor can see that you understood the material and can award partial credit.

Remember that the goal of a test is to get as many points as possible. A professor can't give you any credit if you don't write anything down. Make sure you write down what you know and don't leave questions blank. Another common mistake occurs when students realize they have made a mistake and erase much of their work. Often, the mistake they made was minor and partial credit could have been given, but none is awarded because everything was erased.

Concentrate on the problems that will produce the most points. These are either problems you can do relatively quickly and get full credit for, or ones worth a large portion of the total points. For instance, if there are four problems, three worth 20 points and one worth 40 points, concentrate on the 40-point problem.

Leave time at the end, if possible, to review your answers. If you find a mistake and there is not enough time to fix it, write down that you found it and would have fixed it if you had time. If you are taking a multiple-choice test, be careful about changing your answers. Many studies find that students will change right answers as often as they change wrong answers. If you make a correction on a multiple-choice test, make certain you know that it was wrong. Otherwise, your initial answer may have been more likely to have been correct.

Think. This may sound basic, but it is important. What is the question asking? What does the professor want to see you do? When you get an answer, ask yourself if it makes sense. If you calculate an unrealistic answer, comment on it. Often a simple math error produces a ridiculous answer. Show that you know it is ridiculous but that you just ran out of time to find the error.

**After the Test**

After you get your test back, look it over. It is a great idea to correct the problems you missed. It is much easier to fill in the holes in what you missed as you go along in the semester than to wait until you are studying for the final exam. Most final exams cover the material from all the semester tests. If you can't correct the problems yourself, go and see the professor and ask for help. Remember, you *will* see the material again!

## 5.6 Making the Most of Your Professors

One of the most under-utilized resources at a university is the faculty. Many students do not take full advantage of the faculty that they are paying to educate them. The professor is the one who decides on course material, what is important and what is not, and how you will be evaluated. Yet few students take the time to get to know their professors. Here are some reasons to get to know your professors:

1. They are professionals and have experience in the fields you are studying and, therefore, have perspectives that are valuable to you.
2. Every student will need references (for scholarships, job applications or graduate school applications) and professors can provide them.
3. They are the ones in charge of the course and can help you focus your studying to be more effective.
4. They are the experts in the field and can answer the hard questions that you don't understand.

5. They assign your grades, and at some point in your college career, you may be on the borderline between grades. If they know you and know that you are working hard, they may be more likely to give you the higher grade.
6. They are likely to know employers and can provide job leads for full-time positions or internships.
7. They may be aware of scholarship opportunities or other sources of money for which you could apply.
8. They can provide opportunities for undergraduates to work in labs, which is great experience. You may even get paid!

It is interesting to examine the history of the university. In centuries past, a student did not go to a university to get "trained" to do a job. They went to be mentored by great scholars (the faculty). In modern universities, many students come looking primarily for the training needed to get a good job, and getting mentored by faculty is low among their priorities. Consider this: a school could send you the course materials at home and have you show up for one day to take your exams. Yet students still come to a university to study. Why? A main reason is for the interaction with the people at the university—faculty, staff, and other students.

*Students come to a university for the interaction with the people at the university—faculty, staff, and other students.*

If there are so many reasons to get to know professors, why don't most students do so? Many professors don't encourage students to come see them, or don't present a welcoming appearance. Professors may remind students of their parents, people who seem to be out of touch with college students. It may also be that many students don't really understand the advantages of getting to know their professors.

Here are some things to keep in mind when getting to know professors. The first is that they have spent their careers studying the material you are covering in your courses and, therefore, they enjoy it. So it would be a very bad idea to start by telling your calculus professor that you find math disgusting. This is insulting and suggests he has wasted his career.

The second is that professors teach by choice, especially in engineering. Industry salaries are higher than academic salaries for engineering Ph.D.'s, so professors teach because they want to. They also probably think that they are good at it. If you are having a problem with an instructor's teaching style, approaching him or her with a problem solving strategy such as the following can be very constructive: "I am having trouble understanding the concepts in class. I find that if I can associate the concepts with applications or examples I understand them better. Can you help me with identifying applications?"

*Faculty members have many demands on their time (research, securing grants, writing, committee assignments, etc.) in addition to your class.*

As a group, professors also love to talk, especially about themselves and their area of expertise. They have valuable experience that you can benefit from. Ask them why they chose the field they did. Ask them what lessons they learned as students. You will get some valuable insights and also get to know someone who can help you succeed in your class, and possibly in your career. Steven Douglass, in his book *How to Get Better Grades and Have More Fun* suggests asking each professor, "What is the main objective in the course?"

Faculty members are very busy. They have many demands on their time (research, securing grants, writing, committee assignments, etc.) in addition to your class. These other demands are what keep the university running. So respect their time. It is okay to get to know them but don't keep popping in just to shoot the breeze. Also, when you come with questions, show them the work you have been doing on your own; you only need to get past a hurdle. A busy faculty member won't mind helping a hard-working student clear up an idea or concept. However, showing up and asking how to do the homework can get interpreted as "Do my homework for me." This would not make the positive impression you want.

# 5.7 Learning Styles

In the last several decades scientists have made many discoveries regarding the complexities of the human brain. While the brain is an important organ in each person's body, we now know that each person's brain is unique to him or her. Even before birth our genetic code, as well as our environment after birth, shapes our developing brain. Proper nutrition, stress, drugs and alcohol are some of the factors that can affect a developing brain.

Each person is born with all the brain cells, or neurons, that they will ever have. (Estimates reach as high as 180 billion neurons.) Neurons differ from other cells in that they do not reproduce. However, they do form connections with other neurons. As they are used, neurons can grow tentacle-like protrusions called dendrites which can receive signals from other neurons. In the first five years of life, with normal stimulation, a multitude of connections are created as the brain learns to move the body, speak, analyze, create, and control. However, people can continue to create more neurological connections as they continue to use their brain and learn new material. With the exception of diseases, like Alzheimer's, or physical injury, the brain is capable of creating new connections throughout our lifetime.

This is great news! None of us is ever too old or too dumb to learn something new. Another benefit of brain research is that it has identified several different ways that people think and memorize. Memorizing refers to how people assimilate, or add, new material to existing knowledge and experience. It also includes how we accommodate, or change our previous way of organizing material. Thinking refers to how we see the world, approach problems and use the different parts of our brain. One method does not fit all when it comes to how we learn best, so time spent identifying your own strengths and weaknesses will prove very beneficial. Once you identify the aspects of your own preferred learning style, information can be manipulated so that you learn more information in less time with higher rates of recall.

**Memory**
**Languages**

Place a check mark by all the statements that strongly describe your preferences.

### Auditory
_____I enjoy listening to tapes of lectures.
_____I often need to talk through a problem aloud in order to solve it.
_____I memorize best by repeating information aloud or explaining it to others.
_____I use music and/or jingles to memorize.
_____I remember best when information fits into a rhythmic pattern.
_____I would rather listen to the recording of a book than read it.
_____I remember names of people easily.

## Visual

_____I follow pictures or diagrams when assembling something.

_____I am drawn to flashy, colorful, visually stimulating objects.

_____I prefer books that include pictures or illustrations with the text.

_____I create mental pictures to help me remember.

_____I usually remember better when I can see the person talking.

_____I remember faces of people easily.

_____I prefer to follow a map rather than ask for directions.

## Kinesthetic

_____I can memorize well while exercising.

_____I usually learn best by physically participating in a task.

_____I almost always have some part of my body in motion.

_____I prefer to read books or hear stories that are full of action.

_____I remember best when I can do something with the information.

_____I solve problems best when I can act them out.

You probably checked some in each of the three areas; however, most people have a preferred modality for learning new material. See which area you checked most often and develop ways to study within that area. For example, if you are an auditory learner, buy a small tape recorder and record your lectures to listen to over and over again—perhaps while going to and from class. In lecture, sit where you can hear the professor well. Try focusing on what is being said during the class and taking notes on the material from your tape recording later. Ask the professor questions during class and office hours. Read your assignments out loud to yourself, or into your tape recorder. Keep visual distractions in your study place to a minimum so you can focus on active listening, to your own voice or to a tape. Create jingles, songs, poems, or stories to help you remember. When studying with others, have them drill you on what you're learning, or spend time explaining it to them.

If you are a visual learner, you will benefit from a different set of strategies. During class make sure you sit where you can see the professor and the board or screen clearly. Write notes during lecture with plenty of pictures and meaningful doodles. Rewrite notes later in a more organized fashion, highlighting the main ideas emphasized in class. Create diagrams, lists, charts, and pictures to help clarify material. Write out questions that you have to ask the professor, either in class or during office hours. Highlight main ideas in your books as you read, and don't be afraid to write notes in the margins. Use more than one color to emphasize different things. Use mental images to help you remember how to solve problems. When studying with others, have them give you written practice problems and tests. Use mental maps to illustrate how ideas fit together or to plan upcoming papers or projects. Use video material to reinforce concepts and events.

## Example 5.3

I sat down next to a Ph.D. student for a lecture. As we got ready for class, she prepared a little differently. Instead of taking out one writing instrument, she had four—she held four different-colored pens in her left hand. She would take one at a time and use them with her right hand to write. Different ideas had different colors. As I watched, she drew diagrams between points with the different colors. Her notes looked like a work of art. She was one of the most visual learners I have ever seen. Early in her academic career she struggled until she developed methods for succeeding. And she has succeeded, all the way through her Ph.D. program in engineering.

Kinesthetic learners may have the biggest disadvantage, because most formal college learning environments lack kinesthetic activities. The exception to this would be classes with labs. Take advantage of these as often as possible, since doing something with the information you are learning is easiest for you. During lecture, make connections between what is being said and similar things you have done in the past. Talk with your professor during office hours about ways to be more hands-on with the material. Perhaps you could volunteer in your professor's lab, or use models and experiments on your own at home. Try memorizing while exercising, especially repetitive activities like jogging, running stairs, jumping rope, or swimming laps. When studying with others, try acting out the material. Shorter, more frequent study times may be more productive than a single marathon session.

## Thinking Skills

The way the brain works is much more complex than simply by adding new material to old. Thinking refers to how we see the world, approach problems and use the different parts of our brain. As science has rapidly improved its ability to measure the brain's activity, our understanding of brain functions has dramatically changed. As research continues, our understanding of this wondrously complex organ is continuing to evolve.

Models have been developed to try to categorize how different individuals think. Early models focused on the specialization of the right and left hemispheres of the brain. Further research has unlocked other complexities and required the development of more complex models. A common current model was developed by Ned Herrmann and shows the brain as four quadrants. Originally, these were four physiological components of the brain. As research continued, it was determined that brain function was too complex to be precisely represented in this format. However, the model has proven effective as a representation of our preferred thinking style and the four quadrants are currently used as a metaphorical model of how we prefer to process information. Currently, the four quadrants are used as a metaphorical model.

Figure 5.1 shows the evolution of the quadrant model from the hemispherical models. Quadrants A and B represent the "left brain" and quadrants C and D represent the "right brain." Herrmann's complete model is detailed in his book *The Creative Brain*. These quadrants illustrate personal preferences and a dominant thinking style. However, most people exhibit characteristics of each quadrant to some degree. Successful people are aware of their strong areas and can capitalize on them while compensating for their weaker areas, and possibly building them up.

Tables 5.1 – 5.4 highlight examples of the various thinking skills, preferred learning activities and practice activities for each quadrant. Examine the lists and try to place yourself in one or more quadrants.

The most successful students pick and choose a variety of strategies that work best for their learning style and specific classes. Observe what other successful students are doing and learn from their example. Remember, your one-of-a-kind brain is very capable of learning new material. Spend some time learning how you learn best and you've paved the way for successful learning.

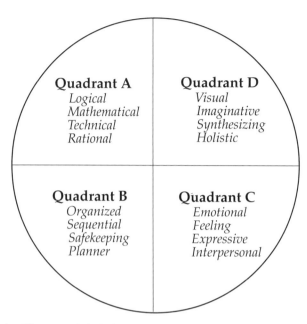

Figure 5.1  Herrmann's "whole brain model" describing four quadrants of the brain.
(Copyright Herrmann International.)

Table 5.1  Quadrant A

| Thinking | Preferred Learning Activities | Practice Activities |
|---|---|---|
| Factual<br>Analytical<br>Quantitative<br>Technical<br>Logical<br>Rational<br>Critical | Collecting data and information<br>Listening to lectures<br>Reading textbooks<br>Studying example problems and solutions<br>Using the scientific method to do research<br>Doing case studies<br>Dealing with things rather than people | Develop graphs, flow charts and outlines<br>    from info and data<br>Do a library search on a topic of interest<br>Join an investment club<br>Do logic puzzles or games<br>Play "devil's advocate" in a group decision<br>    process<br>Break down a machine and identify the<br>    parts and their functions |

Table 5.2  Quadrant B

| Thinking | Preferred Learning Activities | Practice Activities |
|---|---|---|
| Organized<br>Sequential<br>Controlled<br>Planned<br>Conservative<br>Structured<br>Detailed<br>Disciplined<br>Persistent | Following directions<br>Doing repetitive, detailed homework<br>    problems<br>Doing step-by-step lab work<br>Using programmed learning and tutoring<br>Listening to detailed lectures<br>Making up a detailed budget<br>Practicing new skills through repetition<br>Making time management schedules<br>Writing a "how to" manual | Cook a new dish following a complicated<br>    recipe<br>Organize a desk drawer or closet<br>Prepare a family tree<br>Be exactly on time all day<br>Assemble a model kit by instructions<br>Develop a personal budget<br>Learn a new habit through self-discipline<br>Set up a filing system for your paper work |

Table 5.3  Quadrant C

| Thinking | Preferred Learning Activities | Practice Activities |
|---|---|---|
| Sensory<br>Kinesthetic<br>Emotional<br>Interpersonal<br>Symbolic | Sharing ideas with others<br>Learning by teaching others<br>Using group study opportunities<br>Enjoying sensory experiences<br>Taking people-oriented field trips<br>Respecting other's rights and views<br>Study with background music | Study in a group<br>Share your feelings with a friend<br>Get involved in a play or musical<br>Volunteer in your community<br>Explore your spirituality<br>Become a penpal to someone from<br>　another culture |

Table 5.4  Quadrant D

| Thinking | Preferred Learning Activities | Practice Activities |
|---|---|---|
| Visual<br>Holistic<br>Innovative<br>Metaphorical<br>Creative<br>Imaginative<br>Conceptual<br>Spatial<br>Flexible<br>Intuitive | Getting actively involved<br>Using visual aids in lectures<br>Leading a brain storming session<br>Experimenting<br>Doing problems with many answers<br>Relying on intuition, not facts<br>Thinking about the future | Daydream<br>Play with clay, Legos, Skill Sticks<br>Focus on big picture, not the<br>　details of the problem<br>Create a logo<br>Invent a new recipe and prepare it<br>Imagine yourself 10, 20, 50 years in<br>　the future<br>Use analogies and metaphors in<br>　writing |

*In your professional career, you will have to deal with all different styles effectively to be successful.*

The model can also be used to help understand professors who have a different thinking style. Most people teach like they would like to be taught. If you are in a class where your professor has a very different style than you do, use the model to problem-solve ways to get past this hurdle. It is great experience for later in life. In your professional career, you will have to deal with all different styles effectively to be successful. Being aware of the differences and being able to work with them will be a tremendous asset.

If you would like to know which quadrants apply to you, you can take an assessment profile called the Herrmann Brain Dominance Instrument. An academic advisor or testing center on campus should be able to provide you with information about the HBDI. You may also contact Herrmann International at 794 Buffalo Creek Road, Lake Lure, North Carolina 28746, or at their Web site at *www.hbdi.com.*

# 5.8 Well-Rounded Equals Effective

*Developing the habits that make you truly effective during your college years will set you up for success in life.*

Being an effective person goes beyond being a good student. Developing the habits that make you truly effective during your college years will set you up for success in life. Part of being effective is functioning at full capacity. To do this, you must have the various dimensions of your life in order. One analogy that is frequently used is to look at a person as a wheel. A wheel will not roll if it is flat on one side. Similarly, people cannot function optimally if there is a problem area in their life. Five key areas are graphically represented in Figure 5.2. It is important to maintain each area to become as effective as you can be.

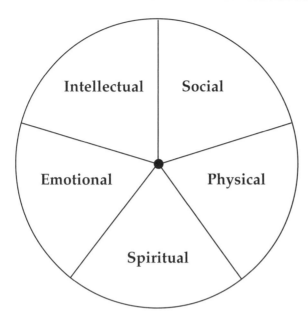

Figure 5.2  Areas of wellness for students.

## Intellectual

One of the main purposes of going to college is to expand the intellectual dimension of your life. Take full advantage of the opportunities at your institution. Besides engineering courses, schools will require general education courses (languages, humanities or social science classes). These non-technical classes are a great way to broaden yourself. Many students find these classes a refreshing break from their engineering classes.

In addition to course work, colleges have a wide range of activities to enhance your intellectual development. These range from student organizations to seminars to special events. Planning some non-engineering intellectual activities will help to keep you motivated and feeling fresh for your engineering classes. Some students find that reading a novel or two during the semester also provides a refreshing, intellectual break from engineering studies.

## Social

The social aspect of college is one that most students seem to master. Planning appropriate social activities will make your college experience more fun. There is a strong correlation between enjoying your college experience and being successful. Also, establishing a group of friends to socialize with will provide a support structure to help you when times get tough. College friends also provide a network that can be very beneficial later in your professional life.

If you have trouble identifying appropriate social activities, there are campus resources to help you. The first place to go is your residence hall staff. Also, most campuses have a central office for student activities. This office can provide you with a list of student organizations and activities.

## Physical

All the studying and preparation is a waste of time if you are not physically able to perform when the tests or quizzes come. Also, you can waste time if you are not able to study efficiently. There are three physical areas that are essential to becoming an effective student. They are fitness, sleep and nutrition.

People are more productive when they are physically fit, and students are no exception. Exercise is an activity which will not only help you study better but will allow you to live a longer, more productive life. College is a great time to develop a fitness habit. Fitness activities can include aerobics, jogging, biking, walking or participating in intramural sports. On every campus there are people who can help you develop a fitness plan that is right for you.

Sleep is an area that many students abuse. Finally they are away from their parents and can go to bed any time they want. The problem is that classes still come at the same time regardless of when you fall asleep. Sleep deprivation reduces a person's productivity, which includes studying. Each person needs a different amount of sleep. Typically, most adults need six to eight hours of sleep. It is important to schedule enough time for sleep or you won't be able to be effective in the classroom. Studies have shown that student performance is actually reduced by staying up very late the night before a test to study. You cover more material the night before the test but are so tired at test time that you can't retrieve the material. Honestly evaluate how much sleep you need. Do you fall asleep in class? If so, you probably aren't getting enough rest. Another test for sleep deprivation is to not set an alarm one morning and see when you wake up. If you are on a schedule, most people will wake up at the same time without the alarm. If you sleep several hours past when the alarm would have gone off, you very possibly are sleep-deprived and need more rest.

The third physical area is diet. Again, Mom and Dad are not there to make you eat that broccoli, so you don't. Our bodies are designed to work properly with the right input. While you don't have to become a diet fanatic, eating a sensible diet will enhance your ability to succeed in your academics.

A final note on the physical dimension pertains to non-dietary substances, such as alcohol. The number one problem on most college campuses is alcohol abuse. Every year, thousands of very capable students fail due to alcohol. If you choose to indulge, make honest assessments of the impact on your studying. Don't let alcohol become the barrier to your success.

## Spiritual

Few would disagree that humans are spiritual beings and function much better and more efficiently with a balanced spiritual component in their lives. Stephen Covey, in his book *The Seven Habits of Highly Effective People*, puts it this way:

> *The spiritual dimension is your core, your center, your commitment to your value system. It's a very private area of life and a supremely important one. It draws upon the sources that inspire and uplift you and tie you to the timeless truths of all humanity. And people do it very, very differently.*
>
> *I find renewal in daily prayerful meditation on the Scriptures because they represent my value system. As I read and meditate, I feel renewed, strengthened, centered and recommitted to serve.*

Search for the spiritual activities that keep you renewed. This may be through a small group study, attending campus worship services, meditating or immersing yourself in good music for a time of personal reflection. If you aren't sure where to start, try something which is close to what you were used to before college. This will probably be the most comfortable for you. If it doesn't seem to be the right thing for you, branch out from there.

Bill Hybels, pastor of a suburban Chicago church, explained the need to attend to this dimension when he addressed a gathering in 1996 in Detroit. He is an avid runner, eats well, and is physically sound. He has many casual friends and is socially active. He has written a number of books and articles and is sound intellectually. Yet he reached an emotional crisis that prevented him from being effective in his job and with his family.

Hybels describes emotional reserves like a gas tank. We need to put reserves into our tank so we have the ability to take it out when we need it. There will be times in your college career, and beyond, when you need those reserves. The lesson Hybels shared is the need to watch the gas gauge and not to let it run out. Students are no different.

A positive step to take to keep the emotional dimension in check is to set up a support network. This network is what you use to make deposits into your emotional tank. Friends are an integral part of the network. Do you have a person or people whom you can talk honestly with about important issues in your life? This might be a close friend, a parent, a relative, an advisor, a professor or a counselor. Cultivate at least one friendship where you feel the freedom to share honestly.

The second aspect to maintaining emotional reserves is to schedule time that is emotionally neutral or energizing. Many students get themselves into a lot of activities that are emotionally draining and don't schedule time for themselves for recovery. Eventually this catches up with them and adversely impacts their effectiveness.

A final note on this area is that on each campus there are advisors and counselors who are available to talk to you. Don't feel that you have to have a severe psychological problem to simply talk with them. These people are trained to help you and will just listen if that is what you need. They are a great resource for maintaining the emotional dimension and thereby helping to keep your academic effectiveness high.

**Emotional**

# 5.9 Your Effective Use of Time

You may be wondering how to find time to maintain all these areas every day: studying two hours for each hour spent in class, keeping a balanced lifestyle, etc. And what happened to all that fun time you've heard about? Aren't the college years supposed to be the best years of your life?

The answer is effective time management. Time management is a skill that will help you succeed as a student and will serve you well throughout your entire career. Mastering time management will set you ahead of your peers. It is interesting that when engineers and managers are surveyed and asked why they don't take the time to manage their time, the most common response is inevitably that they don't have time. Students have the same feelings.

*Time management is a skill that will help you succeed as a student and will serve you well throughout your entire career.*

If we consider the academic requirements with a 15-credit load, we will have 45 hours per week of class and study time. That could be done Monday through Friday from 8 a.m. until 6 p.m., assuming you take an hour for lunch every day. Think about it. That would give you every evening and every weekend free to have fun or to do other activities.

Many students don't believe they can do this, but the hours add up. The key to making such a schedule work is to make every minute count. Treat school like a full-time job, working similar hours. This will not only help you to effectively manage your life during your academic years, but it also will provide excellent preparation for your future work life.

Table 5.5  Sample Class Schedule

|  | Mon | Tues | Wed | Thurs | Fri |
|---|---|---|---|---|---|
| **8:00** | English |  | English |  | English |
| **9:00** | Chem 101 |  | Chem 101 |  | Chem 101 |
| **10:00** |  | Chem Lab |  | CS Lab |  |
| **11:00** | CS 101 | Chem Lab | CS 101 | CS Lab |  |
| **12:00** |  | Chem Lab |  |  |  |
| **1:00** |  |  |  | Chem Rec |  |
| **2:00** | Math | Math | Math |  | Math |
| **3:00** |  | Engr 101 |  | Engr 101 |  |
| **4:00** |  |  |  |  |  |
| **5:00** |  |  |  |  |  |
| **6:00** |  |  |  |  |  |
| **7:00** |  |  |  |  |  |
| **8:00** |  |  |  |  |  |
| **9:00** |  |  |  |  |  |

*It is much more efficient to decide in advance what you will study and when.*

Let's assume that you have a schedule like the one in Table 5.5. This student has Chemistry, Math, English, Computer Science and an Introduction to Engineering course. On Monday, the day starts at 8 a.m. with two classes back to back. Then there is an hour break before Computer Science. If you are on the job, you need to be working, and we'll define working as attending class or studying. You need to find a place to study during this hour break. This is where many students go astray. They will waste an hour like this between classes. Similarly, there is a two-hour time block before Math class. This is time to grab lunch and get some more studying done.

It will take some planning to make your day effective because you will need to find suitable places to study. You can use these times to do your individual studying or meet with a study partner or group. It is much more efficient to decide in advance what you will study and when. It can be a waste of precious time to try decide which subject you will study during breaks between classes. Plan enough time for each subject so you can study from class to class and not fall behind in any way. For example, you can't do all your studying for Chemistry on Monday, Math on Tuesday and Computers on Wednesday or you will fall behind the lectures.

*Extra time blocks will allow you to keep up and give you the flexibility to take time off with a friend for something fun but still stay on your weekly schedule.*

Another thing that scares many students away from effective time management is that it sounds too rigid. One of the reasons for going to college is to have fun and meet people. How can you "go with the flow" if everything is scheduled? The answer is responsible flexibility. If you have your schedule planned out, you can decide when it's a good time to interrupt your "work time" to have some fun. A criterion for making this decision is recovery time. In your schedule, is there enough non-work time to convert to work time before deadlines arrive in your classes? For instance, taking a road trip the day before an exam would be irresponsible. However, right after you have midterms, there might be a window of time to take such a break and still recover.

Scheduling two hours of study per class period is usually enough to keep up during your first year. However, there are classes or times in a semester when this won't be enough. Once you fall behind, it is hard to catch up. To avoid this, a weekly make-up time is an excellent idea. This is like an overtime period at work. At an aerospace company where I worked, Saturday morning from 6 to 10 was my favorite time to schedule this overtime work. This didn't interfere with anything else in my normal schedule and I could get the work done and still have a full Saturday to enjoy. For some reason, 6 a.m. on Saturday isn't a very popular time for students. However, an evening or two or Saturday afternoon will work, too. These extra time blocks will allow you to keep up and give you the flexibility to take time off with a friend for something fun but still stay on your weekly schedule.

## Scheduling Work and Other Activities

No matter how good you get at squeezing in study time between classes, there are still only 24 hours in each day. As a result, there is a limit to what anyone can do in one week. Planning a schedule complete with classes, study time (including make-up time), eating, sleeping and other activities is essential to deciding how much is too much. If you need to work while you're in school, schedule it into your week. If you join a student organization, plan those hours into your schedule. You need to see where it all fits. A common pitfall with students is that they over-commit themselves with classes, work and activities. Something has to give. It is easier to drop an activity than to have to repeat a class.

*A schedule can help guide you in determining what is manageable.*

Ray Landis, from California State University at Los Angeles, describes a 60-hour rule to help students decide how much to work. Fifteen credit-hours should equate to 45 hours of effort. That leaves 15 hours for work, totaling 60 hours. Students may be able to work more than this, but each individual has a limit. You need to learn for yourself what your limit is.

If you need to work to make it through college financially, you may need to look at the number of credits you can handle effectively each semester. A schedule can help guide you in determining what is manageable. Before deciding to take fewer credits so you can work more hours, examine the value of the money you can make at your job compared with the starting salary you would expect by graduating a semester earlier. In some cases, it will make sense to work more while taking fewer classes. In others, it will make financial sense to take out a loan, graduate sooner, and pay the loan back with your higher salary.

## Calendars

It is important to have a daily and weekly calendar. Another critical item in effective time management is a long-range schedule. For students, this is a semester calendar. This will help you organize for the crunch times, like when all five of your classes have midterms within two days, or you have three semester-projects due on the same day at the end of the semester. It is a fact of student life that tests and projects tend to be bunched together. That is because there are educationally sound times to schedule these things, and most professors follow the same model. Given this fact, you can plan your work schedule so that you spread out your work in a manageable way. This is the part you have control over.

*There are educationally sound times to schedule these things, and most professors follow the same model.*

Once, while I was in school, we complained to a professor about this on the first day of class. He offered to give the final exam in the next class period and have the semester project due the second week of the semester to free us up for our other classes. We declined his offer and decided it was best to learn the material first.

## Organizers: Paper or Electronic?

Calendars and task lists can be kept in numerous forms, and you will need to determine which is most effective for you. There has been a proliferation of electronic tools and gadgets that are designed to supplement or replace paper calendars and organizers. Among the most common electronic options are PDA's (Personal Digital Assistants) which can function like a personal organizer as well as a cellular phone, fax sender, and Internet portal. PDA's began with pen-based input but now incorporate handwriting recognition features as well as keyboard inputs. PDA's are also called palmtops, hand-held computers, and pocket computers. Most of the traditional time management companies, such as Franklin Covey, have created electronic versions of their products to be used with PDA's.

Among my colleagues, there are mixed feelings about the use of electronic tools to replace paper tools. Some, including me, have found them useful. Others, however, prefer the traditional calendars or a combination of the two. In a "race," my colleagues who continue to use paper organizers and calendars typically can locate an open date for a meeting more quickly than those who use an electronic organizer.

The media you use to organize your schedule is a personal choice that you have to make. For any system to be effective, it must fit your own needs and style. Our recommendation is that you explore the options before settling on what is best for you, and stay flexible. As new tools are developed, there may be options that enhance your effectiveness and are worth exploring.

## Organizing Tasks

A common method for organizing one's day is to make a "To Do" list each morning (or the evening before). This is a good way to ensure that everything needing attention gets it. Things not done on one day's list get transferred over to the next.

The one drawback to making a list is determining what to do first. You can approach the tasks in the order you write them down, but this ignores each task's priority. For instance, if you have a project due the following day for a class and it gets listed last, will it get done? A popular planning tool is the Franklin Planner produced by the Franklin Covey Co. (Material here used courtesy of Franklin Covey.) They stress the importance of prioritizing your life' activities according to your values, personal mission statement, and goals. They suggest using a two-tier approach to prioritizing activities. The first is to use A, B and C to categorize items as follows:

A = Vital (needs attention today)
B = Important (should be taken care of soon)
C = Optional (no one will notice if it is not done today)

Keep your long-term and intermediate goals in mind when assigning priority. And be sure to write simple statements for each task on your list, beginning each item with an action verb.

Once you have categorized the items, look at all the A priorities. Select the item to do first and give it a 1. This item now becomes A1. The next becomes A2, etc. Now you can attack the items on your list starting with A1 and working through the A's. Once you have finished with the A's, you move on to B1 and work through the B's. Often, you will not get to the C priorities, but you have already decided that it is okay if they don't get done that day.

Table 5.6 shows a sample list for a typical student's day. If you are starting a time management routine, include your list. This keeps you reminded of it, and it gives you something to check off your list right off the bat!

Some lessons can be learned by looking at how this student has scheduled activities. One is that paying the phone bill is not necessarily a top priority, but it requires little time to address. A helpful approach is to look at the quickly done A items and get them out of the way first. Then move on to the more time-consuming items. Also note that a workout was planned into the schedule, right after the chemistry lab. The workout is a priority that comes after some extensive studying. This can help motivate the student to finish a difficult task and then move on to something enjoyable.

Table 5.6 Sample Prioritized List of Tasks

| Priority | Task |
|---|---|
| A3 | Finish chemistry lab (due tomorrow) |
| A5 | Read Math assignment |
| B3 | Do math homework problems |
| B4 | Outline English paper |
| A4 | Workout at Intramural Building |
| B2 | Debug computer program |
| A6 | Annotate Math notes |
| B1 | Annotate chemistry notes |
| A7 | Spiritual meditation |
| C1 | Call Stacy |
| C2 | Write parents |
| A2 | Mail phone bill |
| C3 | Shop for CD's |
| A1 | Plan Day |

Looking at this schedule, you may think that poor Stacy is never going to get called or the letter to the parents written. It often happens that the more fun things are, the more likely they end up a C priority. But it is okay to put relationship-building or fun activities high in priority. People need to stay balanced. Maybe today Stacy is a C, but tomorrow might have that call as an A priority.

The Franklin Covey people provide additional help for identifying priorities. They rate tasks by two criteria, urgency and importance, as shown in Table 5.7. Franklin Covey stresses that most people spend their time in Quadrant I and III. To achieve your long-term goals, Quadrant II is the most important. By spending time on activities that are in II, you increase your capacity to accomplish what matters most to you. The urgency is something that is usually imposed by external forces, whereas importance is something you assign.

You may well want to use a schedule that shows a whole week at a time and has your goals and priorities listed. An general example is shown in Table 5.8.

A weekly schedule allows you to see where each day fits. It also keeps your priorities in full view as you plan. Franklin Covey offers many versions of their Panner, including the Collegiate Edition. Contact them at 1-800-654-1776 or visit their website at www.franklincovey.com.

**Stay Effective**

There are two final notes on time management that can help you stay effective. The first is to plan to tackle important activities when you are at your best. We are all born with different biological clocks. Some people wake up first thing in the morning, bright-eyed and ready to work. These people are often not very effective in the late evening. Others would be dangerous if they ever had to operate heavy equipment first thing in the morning, especially before that first cup of coffee. However, once the sun goes down, they can be on a tear and get a lot done. Know yourself well enough to plan most of your studying when you are at your best. It is a common tendency among students to leave studying until late in the evening. This is often because they haven't done it earlier in the day. There are always stories about how the best papers are written at 3:00 a.m. Some are, but most are not. Know yourself well enough to plan effectively.

*Know yourself well enough to plan most of your studying when you are at your best.*

And finally, pace yourself. You are entering a marathon, not a sprint. Plan your study time in manageable time blocks and plan built-in breaks. A general suggestion is to take a 5-10 minute break for every hour spent studying. This will help keep you fresh and effective.

Table 5.7   Franklin Covey's 'Time Matrix' for Daily Tasks (copyright 1999 Franklin Covey Co.)

| | Urgent | Not Urgent |
|---|---|---|
| **Important** | **I** • Crises<br>• Pressing deadlines<br>• Deadline-driven projects, meetings, preparations | • Preparation<br>• Prevention<br>• Values clarification<br>• Planning<br>• Relationship building<br>• Recreation<br>• Empowerment  **II** |
| **Not Important** | • Interruptions, some phone calls<br>• Some mail, some reports<br>• Some meetings<br>• Many pressing matters<br>**III** • Many popular activities | • Trivia, busywork<br>• Some phone calls<br>• Time-wasters<br>• 'Escape' activities<br>• Irrelevant mail<br>• Excessive TV  **IV** |

# 5.10 Accountability

The single most important concept to being successful is to establish accountability for your goals and intentions. The most effective way to do this is with an accountability partner. This is someone whom you trust and from whom you feel comfortable receiving honest feedback. Ideally, you will meet regularly with this person and share how things are going. Regularly means about once per week. Throughout a semester you can fall behind quickly, so weekly checks are a good idea. The person you choose can be a peer, a mentor, a parent, a relative, or any one you feel comfortable with, respect and trust.

*The single most important concept to being successful is to establish accountability for your goals and intentions.*

Your accountability could also be to a group of people. Bill Hybels describes the change in his life when he entered into an accountability group with three other men. These men held each other accountable in all areas of their lives—professional, familial and spiritual. You can do the same as a student.

The first step is to give your accountability partners the standard by which you want to be held. An easy way to do this is to share your goals with them. Let them see and hear from you where you want to be in five years, in ten years and beyond that. Let them also see the short-term goals you have set to achieve those longer-term objectives. Then, weekly, share with them how you are doing in meeting your goals. You will find that it is much easier to stay on course with the help of at least one close friend. Success will come much easier.

Table 5.8 Sample Generic Weekly Schedule and Planner

| Weekly Schedule | | | Sunday | Monday | Tuesday | Wednesday | Thursday | Friday | Saturday |
|---|---|---|---|---|---|---|---|---|---|
| Musts | Goals | | Day's Priorities | | | | | | |
| Academics: | | | | Run laps | | Run laps | | Finish CS | Run laps |
| Math | Finish Prob | | | | | | | Finish Math | |
| Chem lab | Anal. data | | | | | | | Movie | |
| Comp Sci | Debug Prog | | | | | | | | |
| English | Essay | | | | | | | | |
| | | | Appointments/Commitments | | | | | | |
| | | 8:00 | | English | | English | | English | |
| | | 9:00 | Chapel | Chem | | Chem | | Chem | |
| Hall Dance | | | Synagogue | | | | | | |
| | | 10:00 | | | Chem Lab | | CS Lab | | |
| | | 11:00 | | CS | | CS | | | |
| | | 12:00 | | | | | | | |
| | | 1:00 | | | | | Chem Rec | | |
| | | 2:00 | | Calculus | Calculus | Calculus | | Calculus | |
| | | 3:00 | | Meet prof | | Engr 101 | | Engr 101 | |
| | | 4:00 | | | | | | | |
| | | 5:00 | | | | | | | |
| Physical | Run and Ride | Evening | | | | | | | |
| Mental | Read a novel | | | | | | | | |
| Spiritual | Devotions | | | | | | | | |
| Social | Hall dance | | | | | | | | |
| Emotional | Lunch w/Dave | | | | | | | | |

# 5.11 Overcoming Challenges

We all at some point in our lives encounter adversity. As a student, adversity may take the form of a bad grade on a test or for an entire course, a professor you feel is treating you unfairly, not getting into the academic program of your choice, or any of a myriad other college-related obstacles. Since everyone encounters adversity and failure at some point in their life, your personal success is not always dependent on how well you avoid failing, but rather on how you learn from it, move on and improve. This is especially true as a student.

Almost every student will run into a course or a test where they struggle. The key is to avoid focusing on the negative and continuing to be productive. Every academic advisor has stories about good students who tripped up on a test or in one class and let that affect their performance in other classes.

**Focus on "Controllables"**

If something bad happens, figure out what can be done. This may mean writing down a list of actions you can take. Don't focus on things you can't change. If you receive a low grade on a test, you can't change the grade. You may, however, be able to talk to the professor about the test. You can study the material you missed and be ready for the next one.

**Keep It in Perspective**

Examine the real impact of the situation. What are the consequences of the situation? One class can affect your graduation GPA by about 0.05. This will not produce life-altering consequences, unless you would have graduated with a 4.0! If you are unsure about the consequences, see your professor or an academic advisor and have them help you put it in perspective.

**Move to the Positive**

It is very important if you hit a bump in the road to move beyond it quickly and remain effective in your studies. It is necessary to examine the problem to learn from it. Perhaps you failed a test. This is a great opportunity to examine how you are studying and take corrective action to improve your effectiveness. You can affect your present and future, but not your past.

**Remember, Everyone Fails**

Again, keep things in perspective. If you run into difficulty, you are not alone. There are numerous examples of failures.

Henry Ford neglected to put a reverse gear in his first automobile. But he recovered and went on to be successful.

Abraham Lincoln, likely our greatest president, provided a tremendous example of overcoming adversity. Here is a brief list of some of his accomplishments and setbacks:

| | |
|---|---|
| 1831 – failed in business | 1843 – defeated for Congress |
| 1832 – defeated for legislature | 1846 – elected to Congress |
| 1833 – failed in business again | 1848 – defeated for Congress |
| 1834 – elected to legislature | 1855 – defeated for Senate |
| 1835 – fiancée died | 1856 – defeated for Vice-President |
| 1838 – defeated for Speaker | 1858 – defeated for Senate |
| 1840 – defeated for Elector | 1860 – elected President |

Lincoln never stopped achieving just because he ran into a setback. You don't have to either.

Another example comes from a grandfather who relayed a story about someone in his family failing. "We were visiting your grandmother's brother and his family. It was the time of year when college was in session, so I was surprised when my nephew, Hughey, opened the door and greeted us. He was in his first year of college at the state university. I jokingly said, "What happened to you, Hughey? Flunk out already?" I sure felt stupid when he replied that in fact he had. After that I felt obligated to try to help him out. That evening, he, his father and I talked about what he was going to do. I encouraged him to enroll in the other big state university. I had heard a lot of good things about it even though their science and engineering programs were not as prestigious. He enrolled and did very well. Not only did he graduate, he even got his master's degree before taking a job with a top company."

As a footnote to the story, Hughey continued to do well. The job he took was with a Fortune 100 company. He married and had four children and five grandchildren at last count. He stayed with the same company for his entire career and retired a senior vice president, one step below CEO. Failing in his first attempt at college didn't stop him. He learned from it, moved on and did very, very well. You can do the same with the right attitude and effort.

## References

Douglass, S. and A. Janssen, *How to Get Better Grades and Have More Fun*, Success Factors, Bloomington, Indiana, 1997.

Maxwell, J. C., *The Winning Attitude, Your Key to Personal Success*, Thomas Nelson Publishers, Nashville, Tennessee, 1993.

Landis, R., *Studying Engineering, A Road Map to a Rewarding Career*, Discovery Press, Burbank, California, 1995.

Tobias, C.U., *The Way They Learn, How to Discover and Teach to Your Child's Strengths*, Focus on the Family, Colorado Springs, Colorado, 1994.

Lumsdaine, E., and M. Lumsdaine, *Creative Problem Solving, Thinking Skills for a Changing World*, 2nd ed., McGraw-Hill, New York, 1993.

Herrmann, N., *The Creative Brain,* Brain Books, Lake Lure, North Carolina, 1988.

Covey, S.R., *The 7 Habits of Highly Effective People*, Simon and Schuster, New York, 1989.

# Exercises and Activities

5.1    Identify three things you can do to keep a positive attitude toward your studies.

5.2    What could you have done if you were in the Physics class as described in section 5.2?

5.3    Actually writing goals as described is an important element in defining your goals.
    a)  Write five goals to achieve by graduation.
    b)  Write five goals to reach within ten years.
    c)  Write ten goals to reach during your lifetime.

5.4    Write some specific goals:
    a)  Set a grade goal for each class this semester.
    b)  Calculate your expected semester GPA.
    c)  Define a stretch goal for your semester GPA.

5.5    Find out what the following grade point averages are:
    a)  Minimum GPA to stay off academic probation
    b)  Minimum GPA to be admitted to Tau Beta Pi
    c)  Minimum GPA to be admitted to the engineering honor society in your engineering discipline (e.g., Pi Tau Sigma for mechanical engineering, Eta Kappa Nu for electrical engineering)
    d)  Minimum GPA to be on the Dean's List or Honors List
    e)  Minimum GPA to graduate with honors

5.6    Identify a particular employer and an entry level position with that employer that you would like to obtain following graduation. Find out what is required to be considered for that position.

5.7    List the barriers you see to achieving your graduation goals.
    a)  Categorize them as those within your control and those beyond your control.
    b)  Develop a plan to address those within your control.
    c)  Develop strategies to cope with those beyond your control.

5.8    For each class you are currently taking:
    a)  Map out where you sit in class.
    b)  Rate your learning effectiveness (A, B, C, D or F) for each class.
    c)  For each class not rated an A, develop a plan to increase your learning effectiveness.

5.9    For one week, make a log of everything you do. Break it up into categories of attending class, studying, working, playing, eating, sleeping, physical activity, intellectual pursuits, emotional health, spiritual well-being, and other.
    a)  Total your time spent on each area. Are you spending the time needed on your studies (2 hours for each hour in class)?

b) Is this the kind of schedule that will make you successful? If not, what will you change?

5.10 Review your current class schedule and identify times and places you can study between classes.

5.11 Develop a "work" schedule. Identify the times you are "on the job," that is, studying and taking classes. Identify a make-up time for weekly catch-up.

5.12 List all your current courses. Put a "+" next to each one in which you are caught up. Put a "–" next to each one in which you are not caught up right now. ("Caught up" is defined as prepared for the next lecture with all assignments completed and all assigned reading done.)

5.13 Ask one of your professors the main objective of his or her course and report the response. Does this information help you in organizing your studying?

5.14 Annotate your notes from a lecture in one of your courses, then take those notes in to your professor to get his or her comments on your effectiveness. Report the professor's comments.

5.15 Make a list of the students you already know who are pursuing the same major as you. Then make a list of the students you know in your science and math classes. Rate each one according to their suitability as a study partner.

5.16 Identify one or more people with whom you might form a study group. Try studying together on a regular basis until the next test. Report on your experience.

5.17 List the group study rules you used in your group studying. Report on the things that went well and those that did not.

5.18 Make a list of problems you can work, in addition to the assigned homework problems, in one of your math, science or engineering courses. Report on the method you used to identify these problems.

5.19 Visit one of your professors and report three personal things about that professor. (For example: Where did she go to school? Why did he choose the field he is in? What does she like best about her job? What one piece of advice does he have for freshmen that he didn't share in class?)

5.20 After completing Exercise 19 above, did this change your feelings about your professor? Does he or she seem more approachable now? Did you learn anything else during your visit that could help you succeed in that class?

5.21 Identify at least one opportunity for you as an undergraduate to become involved in engineering research. What advantages could you gain from this experience?

5.22    Identify the type of environment in which you study the best. Describe it and include a sketch of the location.

5.23    Test your conclusion in Exercise 22 with an experiment. Do one assignment in one environment (perhaps in your room, alone and totally quiet) and another in a different environment (perhaps in a more public place, with background noise). Report your findings.

5.24    At the halfway point in the semester, calculate your grades in each class. Compare these to the goals set in Exercise 4. Are you on track? If not, identify the changes you will make to reach these goals.

5.25    Define at least one long-term goal in each of the five wellness areas discussed in this chapter. Next, define short-term goals to achieve these long-term goals.

5.26    List at least two needs you have in each of the five wellness areas. List at least two activities you can plan to meet these needs.

5.27    Rate your roundness in the wellness wheel (on next page), using 5 for very well, 4 for well, 3 for okay, 2 for some improvement needed, and 1 for much improvement needed. Do you have a well functioning wheel? If not, identify steps to correct the problem areas. List at least two new activities for each problem area.

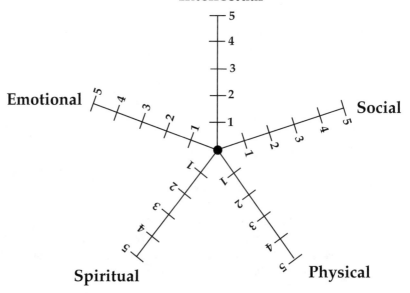

Figure for Exercise 5.27   Wellness assessment.

5.28    Write a paragraph on how each wellness area can help you succeed:
a) as an engineering student.
b) as a practicing engineer.

5.29    Make a weekly schedule with two hours of studying for every hour spent in class. Identify activities which will help develop the different areas in your life. Is this a sustainable schedule? If the answer is no, identify things which can be dropped.

5.30    Identify on your weekly schedule the times of each day when you study most efficiently. Are these the times you are actually scheduled to study? If not, explain why you've scheduled other items instead of studying at these times.

5.31    Adopt one of the weekly schedule instruments discussed in this chapter. Try using it for two weeks and report the results. Was this the right instrument for you?

5.32    Make a "To Do" list. Prioritize the list using A, B and C categories, and then number the items in each category.

5.33    Place the items of Exercise 5.32 into the four quadrants of Covey's matrix shown in Table 5.1. How many A priorities are in quadrant II? Based on this ranking of your "To Do" list, would you change any of your priorities?

5.34    List the top three things you are failing to do that would most help you succeed in your career. Briefly explain why you are currently failing to do each item.

5.35    List three people you would consider for an accountability partner.

5.36    Make a list of at least five questions you want your accountability partner to ask you at each check-up.

5.37    Meet with an accountability partner for one month. Report back on your experience. Did it help? Is this something you will continue?

5.38    Call home and tell your parents that you are developing a success strategy for college. Explain the strategy to them.

5.39    Assume that your grades will be lower than expected at the end of the semester, and write out a strategy for recovering from these low grades.

5.40    Make a list of at least three university people who can help you deal with a problem that is preventing you from succeeding. List each by name and position.

5.41    Make a list of your activities over the last 24 hours. Compare the list and the hours spent on each with your planned schedule.

5.42    List your long-term goals, including every area of your life. Fill out a schedule of what you actually did during the last seven days. Put each activity into categories corresponding to one of your long-term goals, or an extra category titled "not related." Grade yourself (using A, B, C, D or F), indicating how well you did the past week working toward your long-term goals.

5.43    Take the inventory test discussed in Section 5.7 on memory languages. Which is your dominant mode? List three actions you can take to improve your learning based on this dominant mode.

5.44    Which of Herrmann's quadrants best describes your dominant thinking style? Which one is the weakest?

5.45    Which of Herrmann's four quadrants is best suited for engineering? Are there any dominant modes of thinking that are not ideal for engineering? Explain your answers.

5.46    List all your current professors and identify their probable Herrmann quadrant. Compare this with your own style. For the professors with styles that differ from yours, develop an action plan to deal with these differences.

5.47    Explore the availability of electronic organizers and prepare a brief report on the options for students.

5.48    Compare and contrast electronic organizers versus paper ones. What are the benefits of each and which would you recommend to your classmates?

5.49    Interview two professors and report what they use to organize and prioritize their time.

# 6. Problem Solving

## 6.1 Introduction

Seven-year-old Dan could build all kinds of things: toy cars, ramps, and simple mechanical gadgets. Even more amazing was that he could do all this with four simple tools—scissors, hammer, screwdriver, and duct tape. One Christmas, his father asked if he'd like more tools. "No thanks, Dad" was his response. "If I can't make it with these, it's not worth making." It wasn't long, however, before Dan realized that the cool stuff he could create or the problems he could solve were dependent on the tools he had. So he began gathering more tools and learning how to use them. Today, he still solves problems and makes a lot of cool stuff—like real cars, furniture, playgrounds, and even houses. His tools fill a van, a workshop and a garage. Having the right tools and knowing how to use them are critical components of his success.

*The goal of this chapter is to increase the number of problem-solving strategies and techniques you commonly use, and to enhance your ability to apply them creatively in various problem-solving settings.*

Likewise, most people, when they are children, learn a few reliable ways to solve problems. Most fifth graders in math class know common problem-solving methods like "draw a picture," "work backwards," or the ever-popular "guess and check" (check in the back of the textbook for the correct answer, that is). Unfortunately, many people choose to stop adding new problem-solving tools to their mental toolbox. "If I don't know how to solve it, it's probably not worth solving anyway," they may think. Or, "I'll pass this problem over for now and get back to my usual tasks." This severely limits them. Engineers, by definition, need to be good at solving problems and making things. Therefore, filling your toolbox with the right tools and knowing how to use them are critical components to becoming a successful engineer.

The goal of this chapter is to increase the number of problem-solving strategies and techniques you commonly use, and to enhance your ability to apply them creatively in various problem-solving settings. Hopefully, this will be only the beginning of a lifelong process of gaining new mental tools and learning how to use them.

# 6.2 Analytic and Creative Problem Solving

The toolbox analogy is very appropriate for engineering students. Practicing engineers are employed to solve a very wide variety of problems. In the Engineering Majors and the Profiles chapters in this textbook we discuss how engineers can end up filling a number of different roles in daily practice. These roles can require very different problem-solving abilities. Problem solving can be broken into Analytic and Creative. Most students are more familiar with Analytic problem solving where there is one correct answer. In Creative problem solving there is no single right answer. Your analytic tools represent what's in your toolbox. Your creative skills represent how you handle your tools.

*In creative problem solving, there is no single right answer.*

To better understand the difference between these two kinds of problem solving, let's examine the one function that is common to all engineers: design. In the Design chapter, we detail a design process with ten steps. Those ten steps are:

1. Identify the problem
2. Define the working criteria/goal(s)
3. Research and gather data
4. Brainstorm for creative ideas
5. Analyze
6. Develop models and test
7. Make the decision
8. Communicate and specify
9. Implement and commercialize
10. Prepare post-implementation review and assessment

By definition, design is open-ended and has many different solutions. The automobile is a good example. Sometime when you are riding in a car, count the number of different designs you see while on the road. All of these designs were the result of engineers solving a series of problems involved in producing an automobile. The design process as a whole is a **creative problem-solving process.**

*Engineering curricula provide many opportunities to develop analytic problem-solving abilities.*

The other method, **analytic problem solving,** is also part of the design process. Step 5, Analyze, requires analytic problem solving. When analyzing a design concept, there is only one answer to the question "Will it fail?" A civil engineer designing a bridge might employ a host of design concepts, though at some point he or she must accurately assess the loads the bridge can support.

Most engineering curricula provide many opportunities to develop analytic problem-solving abilities, especially in the first few years. This is an important skill. Improper engineering decisions can put the public at risk. For this reason, it is imperative that proper analytical skills be developed.

As an engineering student you will be learning math, science and computer skills that will allow you to tackle very complex problems later in your career. These are critical, and are part of the problem-solving skill-set for analysis. However, other tools are equally necessary.

Engineers also can find that technical solutions applied correctly can become outmoded, or prove dangerous, if they aren't applied with the right judgment. Our creative skills can help greatly in the big-picture evaluation, which can determine the long-term success of a solution. The solution of one problem can cause others if foresight is neglected, and this is not easily avoided in all cases. For instance, the guilt that the developers of dynamite and atomic fission have publicly expressed relates to a regret for lack of foresight. Early Greek engineers invented many devices and processes which they refused to disclose because they knew they would be misused. The foresight to see how technically correct solutions can malfunction or be misapplied takes creative skill to develop. What about fertilizers which boost crops but poison water supplies? Modern engineers need to learn how to apply the right solution at the right time in the right way for the right reason with the right customer.

Most people rely on two or three methods to solve problems. If these methods don't yield a successful answer, they become stuck. Truly exceptional problem solvers learn to use multiple problem-solving techniques to find the optimum solution. The following is a list of possible tools or strategies that can help solve simple problems:

*Engineers need to learn how to apply the right solution at the right time in the right way for the right reason with the right customer.*

1. Look for a pattern
2. Construct a table
3. Consider possibilities systematically
4. Act it out
5. Make a model
6. Make a figure, graph or drawing
7. Work backwards
8. Select appropriate notation
9. Restate the problem in your own words
10. Identify necessary, desired and given information
11. Write an open-ended sentence
12. Identify a sub-goal
13. First solve a simpler problem
14. Change your point of view
15. Check for hidden assumptions
16. Use a resource
17. Generalize
18. Check the solution; validate it
19. Find another way to solve the problem
20. Find another solution
21. Study the solution process
22. Discuss limitations
23. Get a bigger hammer
24. Sleep on it
25. Brainstorm
26. Involve others

In both analytic and creative problem solving, there are different methods for tackling problems. Have you ever experienced being stuck on a difficult math or science problem? Developing additional tools or methods will allow you to tackle more of these problems effectively, making you a better engineer.

The rest of this chapter is a presentation of different ways of solving problems. These are some of the techniques that have been shown to be effective. Each person has a unique set of talents and will be drawn naturally to certain problem-solving tools. Others will find a different set useful. The important thing as an engineering student is to experiment and find as many useful tools as you can. This will equip you to tackle the wide range of challenges which tomorrow's engineers will face.

# 6.3 Analytic Problem Solving

Given the importance of proper analysis in engineering and the design process, it is important to develop a disciplined way of approaching engineering problems. Solving analytic problems has been the subject of a great deal of research, which has resulted in several models. One of the most important analytic problem-solving methods that students are exposed to is the Scientific Method. The steps in the Scientific Method are as follows:

1. Define the problem
2. Gather the facts
3. Develop a hypothesis
4. Perform a test
5. Evaluate the results

In the Scientific Method, the steps can be repeated if the desired results are not achieved. The process ends when an acceptable understanding of the phenomenon under study is achieved.

In the analysis of engineering applications, a similar process can be developed to answer problems. The advantage of developing a set method for solving analytic problems is that it provides a discipline to help young engineers when they are presented with larger and more complex problems. Just like a musician practices basic scales in order to set the foundation for complex pieces to be played later, an engineer should develop a sound fundamental way to approach problems. Fortunately, early in your engineering studies you will be taking many science and math courses that will be suited to this methodology.

The Analytic Method we will discuss has six steps:

1. Define the problem and make a problem statement
2. Diagram and describe
3. Apply theory and equations
4. Simplify the assumptions
5. Solve the necessary problems
6. Verify accuracy to required level

Following these steps will help you better understand the problem you are solving and allow you to identify areas where inaccuracies might occur.

**Step 1: Problem Statement**

It is important to restate the problem you are solving in your own words. In textbook problems, this helps you understand what you need to solve. In real life situations, this helps to ensure that you are solving the correct problem. Write down your summary, then double-check that your impression of the problem is the one that actually matches the original problem. Putting the problem in your own words is also an excellent way to focus on the part of the problem you need to solve. Often, engineering challenges are large and complex, and the critical task is to understand what part of the problem you need to solve.

The next step is to describe the problem and list all that is known. In addition to restating the problem, list the information given and what needs to be found. This is shown in Example 6.2, which follows this section. Typically, in textbook problems, all the information given is actually needed for the problem. In real problems, more information is typically available than is needed to do the calculations. In other cases, information may be missing. Formally writing out what you need and what is required helps you to clearly sort this out.

It is also helpful to draw a diagram or sketch of the problem you are solving to be able to understand the problem. Pictures help many people to clarify the problem and what is needed. They are also a great aid when explaining the problem to someone else. The old saying could be restated as "A picture is worth a thousand calculations."

**Step 2: Description**

State explicitly the theory or equations needed to solve the problem. It is important that you write this out completely at this step. You will find that most real problems and those you are asked to solve as an undergraduate student will not require exact solutions to complete equations. Understanding the parts of the equations that ought to be neglected is vital to your success.

It is not uncommon to get into a routine of solving a simplified version of equations. These may be fine under certain conditions. An example can be seen in the flow of air over a body, like a car or airplane wing. At low speeds, the density of the air is considered a constant. At higher speeds this is not the case, and if the density were considered a constant, errors would result. Starting with full equations and then simplifying reduces the likelihood that important factors will be overlooked.

**Step 3: Theory**

As mentioned above, engineering and scientific applications often cannot be solved precisely. Even if they are solvable, determining the solution might be cost prohibitive. For instance, an exact solution might require a high-speed computer to calculate for a year to get an answer, and this would not be an effective use of resources. Weather prediction at locations of interest is such an example.

To solve the problem presented in a timely and cost-effective manner, simplifying assumptions are required. Simplifying assumptions can make the problem easier to solve and still provide an accurate result. It is important to write the assumptions down along with how they simplify the problem. This documents them and allows the final result to be interpreted in terms of these assumptions.

While estimation and approximation are useful tools, engineers also are concerned with the accuracy and reliability of their results. Approximations are often possible if assumptions are made to simplify the problem at hand. An important concept for engineers to understand in such situations is the Conservative Assumption.

In engineering problem solving, "conservative" has a non-political meaning. A conservative assumption is one that introduces errors on the safe side. We mentioned that estimations can be used to determine the bounds of a solution. An engineer should be able to look at those bounds and determine which end of the spectrum yields the safer solution. By selecting the safer condition, you are assuring that your calculation will result in a responsible conclusion.

Consider the example of a materials engineer selecting a metal alloy for an engine. Stress level and temperature are two of the parameters the engineer must consider. At the early stages of a design, these may not be known. What he or she could do is estimate the parameters. This could be done with a simplified analysis. It could also be

**Step 4: Simplifying Assumptions**

done using prior experience. Often, products evolve from earlier designs for which there are data. If such data exist, ask how the new design will affect the parameters you are interested in. Will the stress increase? Will it double? Perhaps a conservative assumption would be to double the stress level of a previous design.

After taking the conservative case, a material can be selected. A question the engineer should ask is: "If a more precise answer were known, could I use a cheaper or lighter material?" If the answer is No, then the simple analysis might be sufficient. If Yes, the engineer could build a case for why a more detailed analysis is justified.

Another example of conservatism can be seen in the design of a swing-set (see Fig. 6.1). In most sets, there is a horizontal piece from which the swings hang, and an inverted V-shaped support on each end. If you were the design engineer, you might have to size these supports.

One of the first questions to ask is, "For what weight do we size the swing-set?" One way to answer this question is to do a research project on the weight distribution of children, followed by doing research into the use of swings by children at different ages, and then analyzing those data to determine typical weights of children that would use such swings. You might even need to observe playground activity to see if several children pile on a swing or load the set in different ways—say, by climbing on it. This might take weeks to do, but you would have excellent data.

On the other hand, you could just assume a heavy load that would exceed anything the children would produce. For this case, say, 500 pounds per swing. That would assume the equivalent of two large adults on each swing. Make the calculation to determine the needed supports, and then ask if it makes sense to be more detailed.

To answer this, one has to answer the question "What problem am I solving?" Am I after the most precise answer I can get? Am I after a safe and reliable answer regardless of other concerns?

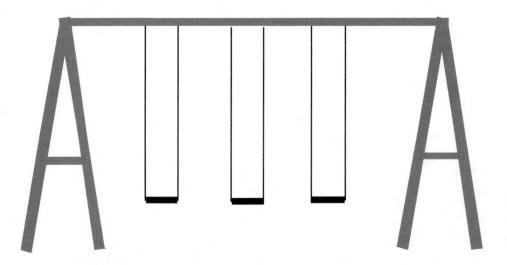

Figure 6.1 Simple swing set.

In engineering, the "other concerns" include such things as cost and availability. In the current example, the supports would most likely be made of common materials. Swing-sets are typically made of wood or steel tubing. So, one way to answer the above questions is to quickly check and see if making a more detailed analysis would be justified. Check the prices of the materials needed for your conservative assumption against potential materials based on a more detailed analysis. Does the price difference justify a more detailed analysis? Another way to ask this is "Does the price difference justify spending time to do the analysis?" Remember, as an engineer, your time costs money.

The answer may be, "It depends." It would depend on how much the difference is and how many you are going to produce. It wouldn't make sense to spend a week analyzing or researching the answer if it saves you or your company only $500. If your company was producing a million sets and you could save $5 per set, it would justify that week's work.

Approximation can improve decision-making indirectly as well. By using it to quickly resolve minor aspects of a problem, you can focus on the more pressing aspects.

Engineers are faced with these types of decisions every day in design and analysis. They must accurately determine what problem is to be solved, and how. Safety of the public is of paramount concern in the engineering profession. You might be faced with a short-term safety emergency where you need to quickly prioritize solutions. Or you might be designing a consumer product where you have to take the long-term view. Swing-sets are often used for generations, after all.

As an engineer, you will need to develop the ability to answer, "What problem am I really solving?" and, "How do I get the solution I need most efficiently?"

## Example 6.1

A manager for an aerospace firm had a method for "breaking in" young engineers. Shortly after the new engineer came to work in his group, the manager would look for an appropriate analysis for the young engineer. When one arose, he would ask the young engineer to perform the analysis. The engineer would typically embark on a path to construct a huge computer model requiring lengthy input and output files that would take a few weeks to complete. The manager would return to the engineer's desk the following morning, asking for the results. After hearing the response that all that was really accomplished was to map out the next three weeks worth of work. The manager would erupt and explain at a very high volume that he or she had been hired as an engineer and what they were supposed to be doing was an engineering analysis, not a "science project". He would show the young engineer how doing a simplified analysis that would take an afternoon to perform and would answer the question given the appropriate conservative assumptions. He would then leave the engineer to the new analysis method.

While his methods were questionable, his point is well taken. There are always time and cost concerns as well as accuracy. If conservative assumptions are made, safe and reliable design and analysis decision can often be made.

Now the problem is set up for you actually to perform the calculations. This might be done by hand or by using the computer. It is important to learn how to perform simple hand calculations; however, computer applications make complex and repetitive calculations much easier. When using computer simulations, develop a means to document what you have done in deriving the solution. This will allow you to find errors faster, as well as to show others what you have done.

**Step 5: Problem Solution**

**Step 6:
Accuracy
Verification**

Engineers work on solutions that can affect the livelihood and safety of people. It is important that the solution an engineer develops is accurate.

I had a student once whom I challenged on this very topic. He seemed to take a cavalier attitude about the accuracy of his work. Getting all the problems mostly correct would get him a B or a C in the class. I tried to explain the importance of accuracy and how people's safety might depend on his work. He told me that in class I caught his mistakes, and on the job someone would check his work. This is not necessarily the case! Companies do not hire some engineers to check other engineers' work. In most systems, there are checks in place. However, engineers are responsible for verifying that their own solutions are accurate. Therefore, it is important for the student to develop skills to meet any required standard.

A fascinating aspect of engineering is that degree of accuracy is a variable under constant consideration. A problem or project may be properly solved within a tenth of an inch, but accuracy to a hundredth might be unneeded and difficult to control, rendering that kind of accuracy incorrect. But the next step of the solution might even require accuracy to the thousandth. Be sure of standards!

There are many different ways to verify a result. Some of these ways are:

- Estimate the answer.
- Simplify the problem and solve the simpler problem. Are the answers consistent?
- Compare with similar solutions. In many cases, other problems were solved similarly to the one currently in question.
- Compare to previous work.
- Ask a more experienced engineer to review the result.
- Compare to published literature on similar problems.
- Ask yourself if it makes sense.
- Compare to your own experience.
- Repeat the calculation.
- Run a computer simulation or model.
- Redo the calculation backwards.

It may be difficult to tell if your answer is exactly correct, but you should be able to tell if it is close or within a factor of 10. This will often flag any systematic error. Being able to back up your confidence in your answer will help you make better decisions on how your results will be used.

The following example will illustrate the analytical method.

## Example 6.2

Problem:  A ball is projected from the top of a 35 m tower at an angle 25° from horizontal with an initial velocity $v_0 = 80$ m/s. What is the time it takes to reach the ground, and what is the horizontal distance from the tower to the point of impact?

**Problem Statement:**  Given:  Initial velocity $v_0 = 80$ m/s

Initial trajectory $= 25°$ from horizontal

Ball is launched from a height of 35 m

**Find:**  1) Time to impact
2) Distance of impact from the tower

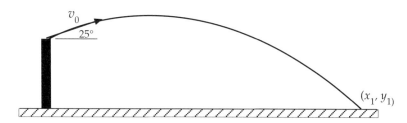

**Equations:**    Found in a physics text:

(a)  $x_1 = x_0 + v_0 t \cos\theta + \dfrac{1}{2} a_x t^2$

(b)  $y_1 = y_0 + v_0 t \sin\theta + \dfrac{1}{2} a_y t^2$

**Assumptions:**  Neglect air resistance

Acceleration $a_y = -9.81 \text{ m}/\text{s}^2$ is constant in the vertical direction

Acceleration is zero in the horizontal direction, $a_x = 0$

Ground is level and impact occurs at $y = -35$ m

**Solution:**  Begin with equation (b):

$$-35 = 0 + 80t \sin 25° - \frac{1}{2} 9.8t^2 \quad \text{or} \quad t^2 - 6.90t - 7.14 = 0$$

$$\therefore t = \frac{-b \pm \sqrt{b^2 - 4ac}}{2a} = \frac{6.9 \pm \sqrt{6.9^2 + 4 \times 1 \times 7.14}}{2} = 7.81 \text{ sec}$$

The solution must be 7.81 sec, since time must be positive.

The distance is found using equation (a):

$$x_1 = 0 + 80 \times 7.81 \times \cos 25° = 566 \text{ m}$$

**Estimation**

Estimation is a problem solving tool that engineers need to develop. It can provide quick answers to problems and to verify complicated analyses. Young engineers are invariably amazed when they begin their careers at how the older, more experienced engineers can estimate so closely to problems before the analysis is even done.

I was amazed at one such case that I experienced first hand. I was responsible for doing a detailed temperature analysis on a gas turbine engine that was to go on a new commercial aircraft. We presented our plan to some managers and people from the Chief Engineers Office. The plan called for two engineers working full time for a month on the analysis.

One of the more senior engineers who had attended the presentation sent me some numbers the next day. He had gone back to his office and made some approximate calculations and sent me the results. His note told me to check my results against his numbers.

After the month long analysis, which involved making a detailed computer model of the components, I checked our results with his "back of the envelope" calculations. He was very close to our result! His answers were not exactly the same as ours, but they were very close. The detailed analysis did provide a higher degree of precision. However, the senior engineer was very close.

What he had done was simply use suggestion #13 from the problem solving methods. He solved a simpler problem. He used simple shapes with known solutions to approximate the behavior of the complex shapes of the jet engine's components.

While estimation or approximation may not yield the precision you require for an engineering analysis, it is a very useful tool. One thing that is can do is to  help you check you analysis. Estimation can provide bounds for potential answers. This is especially critical in today's dependence on computer solutions. Your method may be correct, but one character mistyped will throw your results off. It is important to be able to have confidence in your results by developing tools to verify accuracy.

Estimation can also be used to help decide if a detailed analysis is needed. Estimating using a best case and worst case scenario can yield an upper an lower bound on the problem. if the entire range of potential solutions is acceptable, why do the detailed analysis?  What if for our case, the whole range of temperatures met our design constraints?  It would not have justified a month of our work. In this case, the added precision was needed and did lead to a design benefit. As engineers, you will be asked to decide when a detailed analysis is required and when it is not.

Estimation can be a very powerful tool to check accuracy and make analysis decision. Senior engineers in industry often comment that current graduates lack the ability to do an approximation. This has become more important with the dependence of computer tools and solutions. Computer analyses lend themselves to typos on inputs or arithmetic errors. Being able to predict the ballpark of the expected results can head off potentially disastrous results.

## Example 6.3

With all of today's sophisticated analytical tools, it may be hard to imagine that accurate measurements could be made with out them. Take the size of a molecule. Today's technology makes this measurement possible. Could you have made this measurement in the 18th century? Benjamin Franklin did.

He found the size of an oil molecule. He took a known volume of oil and poured it slowly into a still pool of water. The oil spread out in a circular pattern. He measured the circle and calculated the surface area. Since the volume of the oil remained the same, he calculated the height of the oil slick and assumed that was the molecular thickness, which turned out to be true. His result was close to the correct value.

Do we have more accurate measurements now? Certainly we do. However, Benjamin Franklin's measurement was certainly better than no value. As engineers, there will be times when no data exists and you will need to make a judgment. Having an approximate result is much better than no data.

# 6.4 Creative Problem Solving

Many engineering problems are open-ended and complex. Such problems require creative problem solving. To maximize the creative problem-solving process, a systematic approach is recommended. Just as with analytic problem solving, developing a systematic approach to using creativity will pay dividends in better solutions.

Revisiting our toolbox analogy, the creative method provides the solid judgment needed to use your analytic tools. Individual thinking skills are your basics here. The method we'll outline now will help you to apply those skills and choose solution strategies effectively.

There are numerous ways to look at the creative problem-solving process. We will present a method which focuses on answering these five questions:

1. What is wrong?
2. What do we know?
3. What is the real problem?
4. What is the best solution?
5. How do we implement the solution?

*Dividing the process into steps allows you to break a large, complex problem into simpler problems where your various skills can be used.*

By dividing the process into steps, you are more likely to follow a complete and careful problem-solving procedure, and more effective solutions are going to result. This also allows you to break a large, complex problem into simpler problems where your various skills can be used. Using such a strategy, the sky is the limit on the complexity of projects which an engineer can complete. It's astonishing and satisfying to see what can be done.

At each phase of the process, there is both a divergent and a convergent part of the process, as shown in Figure 6.2. In the divergent process, you start at one point and reach for as many ideas as possible. Quantity is important. Identifying possibilities is the goal of the divergent phase of each step. In this portion of the process (indicated by the outward pointing arrows, ◄——  ——►) use the brainstorming and idea-generating techniques described in subsequent sections. Look for as many possibilities as you can. Often, the best solutions come from different ideas.

**Divergence and Convergence**

In the convergence phase (indicated by inward pointing arrows, ——► ◄——) use analytical and evaluative tools to narrow the possibilities to the one(s) most likely to yield results. Quality is most important, and finding the best possibility to move the process to the next phase is the goal. (One common method is to use a matrix to rate ideas based on defined criteria.) If one choice fails to produce satisfactory results, go back to the idea lists for another method to tackle the problem.

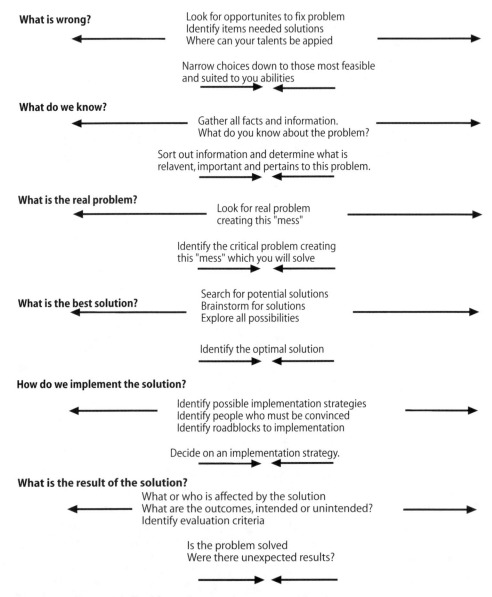

**What is wrong?**

Look for opportunites to fix problem
Identify items needed solutions
Where can your talents be appied

Narrow choices down to those most feasible
and suited to you abilities

**What do we know?**

Gather all facts and information.
What do you know about the problem?

Sort out information and determine what is
relavent, important and pertains to this problem.

**What is the real problem?**

Look for real problem
creating this "mess"

Identify the critical problem creating
this "mess" which you will solve

**What is the best solution?**

Search for potential solutions
Brainstorm for solutions
Explore all possibilities

Identify the optimal solution

**How do we implement the solution?**

Identify possible implementation strategies
Identify people who must be convinced
Identify roadblocks to implementation

Decide on an implementation strategy.

**What is the result of the solution?**

What or who is affected by the solution
What are the outcomes, intended or unintended?
Identify evaluation criteria

Is the problem solved
Were there unexpected results?

Figure 6.2  Problem-solving process.

**Sample Problem: Let's Raise Those Low Grades!**

In order to illustrate the problem-solving process, let's consider an example of a problem and work through the process. The problem we will use for illustration is a student who is getting low grades. This certainly is a problem that needs a solution. Let's see how the process works.

In the first step of the problem-solving process, an issue is identified. This can be something stated for you by a supervisor or a professor, or something you determine on your own. This is the stage where entrepreneurs thrive—looking for an opportunity to meet a need. Similarly, engineers look to find solutions to meet a need. This may involve optimizing a process, improving customer satisfaction, or addressing reliability issues.

**What's Wrong?**

To illustrate the process, let's take our example of the initial difficulty that beginning engineering students can have with grades. Most students were proficient in high school and didn't need to study much. The demands of engineering programs can catch some students off guard. Improving grades is the problem we will tackle.

At this stage, we need to identify that there is a problem that is worth our effort in solving. A good start would be to identify whether it truly is a problem. How could you do this? Total your scores thus far in your classes and compare them with the possible scores listed in your syllabus. Talk with your professor(s) to find out where you are in the classes and if grades will be given on a curve. Talk to an advisor about realistic grade expectations. At the end of this process, you should know if you have a grade problem.

## Example 6.4

Engineers are trained problem solvers. Many engineers move into other careers where their problem-solving abilities are valued. One such case is Greg Smith, who was called into the ministry after becoming a mechanical engineer.

"As soon as people discover that I have a degree in mechanical engineering and am in the ministry, they always ask me the same question, 'So, do you still plan on using your engineering degree some day?' It is difficult for people to understand that I use my engineering training every day as a full time minister.

"One of the greatest skills I learned at Purdue was problem solving. Professors in mechanical engineering were notorious for giving us more information than we would need to solve the problem. Our job was to sift through all the facts and only use what was needed to reach a solution. I find myself using this very skill whenever I am involved in personal counseling. For example, just this week I was listening to someone unload about all the stress he was experiencing from life. He listed his job, a second business at home, family and finances as the primary sources of stress. After further conversation, I was able to "solve" his problem by showing that most of his stress was related to only one of the four sources he listed. The root issue was that this person was basing much of his self-image on his success in this one area of life. I helped him regain his sense of balance in seeing his self-worth not in job performance, but in who he was as a whole person."

The second step in problem solving is the gathering of facts. All facts and information related to the problem identified in the first step are gathered. In this initial information gathering stage, do not try to evaluate whether the data are central to the problem. As will be explained in the brainstorming and idea generating sections to follow, premature evaluation can be a hindrance to generating adequate information.

**What Do We Know?**

With our example of the student questioning his grades, the information we generate might include:

1. Current test grades in each course
2. Current homework and quiz grades in each course

3.  Percent of the semester's grade already determined in each course
4.  Class average in each course
5.  Professors' grading policies
6.  Homework assignments behind
7.  Current study times
8.  Current study places
9.  Effective time spent on each course
10.  Time spent doing homework in each course
11.  Performance of your friends
12.  Your high school grades
13.  Your high school study routine

Part of this process is to list the facts that you know. To be thorough, you should request assistance from someone with a different point of view. This could be a friend, an advisor, a parent or a professor. They might come up with a critical factor you have overlooked.

## What Is the Real Problem?

This stage is one that is often skipped, but it is critical to effective solutions. The difference between this stage and the first is that this step of the process answers the question *why*. Identifying the initial problem answers the question of *what*, or what is wrong. To effectively fix the problem, a problem solver needs to understand *why* the problem exists. The danger is that only symptoms of the problem will be addressed, rather than root causes.

In our example, we are problem solving low class grades. That is the *what* question. To really understand a problem, the *why* must be answered. Why are the grades low? Answering this question will identify the cause of the problem. The cause is what must be dealt with for a problem to be effectively addressed.

In our case, we must understand why the class grades are low. Let us assume that poor test scores result in the low grades. But this doesn't tell us why the test scores are low. This is a great opportunity for brainstorming potential causes. In this divergent phase of problem definition, don't worry about evaluating the potential causes. Wait until the list is generated.

Possible causes may include:

- Poor test-taking skills
- Incomplete or insufficient notes
- Poor class attendance
- Not understanding the required reading
- Not spending enough time studying
- Studying the wrong material
- Attending the wrong class
- Studying ineffectively (e.g., trying to cram the night before tests)
- Failing to understand the material (*note*: might reveal need for tutor or study group)
- Using solution manuals or friends as a crutch to do homework
- Not at physical peak at test time (i.e., up all night before the test)
- Didn't work enough problems from the book

After you have created the list of potential causes, evaluate each as to its validity. In our example, there may be more than one cause contributing to the low grades. If so, make a rank-ordered list. Rank the causes in order of their impact on class performance.

Assume the original list is reduced to:
- Poor class attendance
- Studying ineffectively
- Not spending enough time studying
- Not understanding the material

Of these things, rank them in order of their impact on grades. You may be able to do this yourself or you may need help. An effective problem solver will seek input from others when it is appropriate. In this case, a professor or academic advisor might be able to help you rank the causes and help determine which ones will provide the greatest impact. For this example, say the rank order was:

1. Not understanding the material
2. Studying ineffectively
3. Not spending enough time studying
4. Poor class attendance

We determined that while class attendance was not perfect, it was not the key factor in the poor performance. The key item was not understanding the material. This goes with ineffective studying and insufficient time.

Identifying key causes of problems is vital, so let's look at another example.

## Example 6.5

**"What was the confusion?"** Defining the real problem is often the most critical stage of the problem-solving process and the one most often skipped. In engineering, this can be the difference between a successful design and a failure.

An example of this occurred with an engine maker. A new model engine was introduced and began selling quickly. A problem developed, however, and failures started to happen. Cracks initiating from a series of bolt holes were causing the engines to fail. A team of engineers was assembled and the part was redesigned. They determined that the cause was high stress in the area of the bolt hole. Embossments were added to strengthen the area. It was a success.

Until, that is, the redesigned engines accumulated enough time in the field to crack again. A new team of engineers looked at the problem and quickly determined that the cause was a three-dimensional coupling of stress concentrations created by the embossments, resulting in too high a stress near the hole where the crack started. The part was redesigned and introduced into the field. It was a success.

Well, it was a success, until the redesigned engines accumulated enough time in the field to crack yet again. A third team of engineers was assembled. This time, they stepped back and asked what the real cause was. Part of the team ran sophisticated stress models of the part, just like the other two teams had done. The new team members looked for other causes. What was found was that the machine creating the hole introduced microcracks into the surface of the bolt holes. The stress models predicted that further stresses would start a crack. However, if a crack were intentionally introduced, the allowable stresses were much lower and the part would not fail. The cause wasn't with cracking but with the wrong kind of cracking. The third team was a success because it had looked for the real problem. Identifying the real cause initially, which in this case involved the machining process, would have saved the company millions of dollars, and spared customers the grief of the engine failures.

## What Is the Best Solution?

Once the problem has been defined, potential solutions need to be generated. This can be done by yourself or with the help of friends. In an engineering application, it is wise to confer with experienced experts about the problem's solution. This may be most productive after you have begun a list of causes. The expert can comment on your list and offer his own as well. This is a great way to get your ideas critiqued. After you gain more experience, you will find that technical experts help you narrow down the choices rather than providing more. Also, go to more than one source. This may provide more ideas as well as help with the next step.

In our example, the technical expert may be a professor or advisor who is knowledgeable about studying. Let's assume that the following list of potential solutions has been generated.

- Get a tutor
- Visit the professors during office hours
- Make outside appointments with professors
- Visit help rooms
- Form study group
- Outline books
- Get old exams
- Get old sets of notes
- Outline lecture notes
- Review notes with professors
- Do extra problems
- Get additional references
- Make a time schedule
- Drop classes
- Retake the classes in the summer
- Go to review sessions

The list now must be evaluated and the best solution decided upon. This convergent phase of the problem-solving process is best done with the input of others. It is especially helpful to get the opinions of those who have experience and/or expertise in the area you are investigating. In an engineering application, this may be a lead engineer. In our case, it could be an academic advisor, a teaching assistant or a professor. These experts can be consulted individually or asked to participate as a group.

When refining solutions, an effective tool is to ask yourself which of them will make the biggest impact and which will require the most effort. Each solution gets ranked high or low as to impact and effort. Ideal solutions are those that have a high impact yet require a low level of effort. A visual tool which helps make this process easier is the problem-solving matrix shown in Figure 6.3. Solutions that fall in quadrant B should be avoided. These solutions require a great deal of effort for marginal payoff. Solutions that are the most effective are those in quadrant C. These are the ones that are easy to implement and will produce significant results. It may be necessary to include solutions in quadrant D, requiring high effort and yielding high impact. These should be carefully evaluated to ensure that the investment is worth the reward.

**Effort**

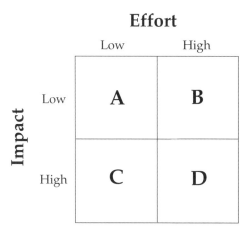

Figure 6.3  A Problem-Solving Matrix.

Implementing the solution is a step that may seem trivial and not worth discussion. This step, however, is a critical phase of the problem solving process. In our scenario, the solution selected may have been:

### Implementing the Solution

- Do extra problems

To accomplish this solution, appropriate additional problems must be selected, done and corrected. Implementation probably requires the assistance of the instructor to help select appropriate problems. To randomly pick additional practice problems could move the effort from quadrant C to D and even B. To have the greatest chances of success, you would want to gain the assistance of your instructor. Other sources of assistance could come from a study group which poses the challenges of getting them to do extra problems with you.

As with all the other steps, a divergent phase begins the step including such activities as brainstorming. The convergent phase of the step culminates with the selection of the implementation plan.

In engineering applications, implementation can be a very critical phase of the process. Most of the solutions to problems either require additional resources including money or the cooperation of other groups that are either not directly affected by the problem or not under your control. Especially early in your career, you will not have control over the people who need to implement solutions so you will need to sell your ideas to the managers who do. An effective implementation plan will be critical to getting your solutions accomplished.

President Jimmy Carter was educated as an engineer and was one of the most respected presidents as a strong moral person. During his presidency, the country was in an energy crisis and action was needed to address the problems. The story goes that he summoned Tip O'Neal, the speaker of the House of Representatives, to the White House to explain his plan. Mr. O'Neal responded by giving the President a list of people who would need to be sold on the plan before it went public. President Carter who was new to Washington was going to skip the development of an implementation plan and enact the plan. He believed that with the country in such an energy crisis that it was obvious what was needed and cooperation was a given. Tip O'Neal was very experienced in the Washington political system and understood that they needed to spend time gaining acceptance to their plan before it could be implemented.

Engineering situations can be very similar. While most are not as political as Washington D.C., most require gaining acceptance of others to implement a solution. In all of these cases, gaining the recognition of those involved is a critical part of the problem solving process.

**Evaluating the Solution**

Implementation does not necessarily end the problem solving process. Just as the design process is a circular process with each design leading to possible new designs, problem solving can also be cyclic. Once a solution has been found and implemented, an evaluation should be performed. As with the other steps in the problems solving process, this step begins with a divergence phase as the problem solver asks what makes a successful solution and how it will be evaluated. To effectively evaluate a solution, a criterion for success must be established. Sometimes there are objective criteria for evaluation but in many circumstances the criterion for success is more subjective.

Once the criterion has been established, the evaluation process should be defined including who will evaluate the solution. Often it is desirable to get a neutral view of someone who was not involved in the formulation of the solution process to be the evaluator.

If the solution is a success, the process may be done. Often in engineering applications, solutions are intermediate and lead to other opportunities. The software industry is a prime example. Software is written to utilize the current computer technologies to address issues. Almost as soon as it is completed, new technologies are available which open up new opportunities or require new solutions.

Some times a "success" does not address the true problem as was the case of the turbine disk failures described earlier. In these cases, the evaluation process may need time to deem it a true success. In other cases, solutions have unintended outcomes that require a new solution even if the initial solution was deemed a success.

Whether a solution is effective or not, there is value in taking time to reflect on the solution and its implications. This allows you as a problem solver to learn both from the process and the solution. Critically evaluating solutions and opportunities is a skill that is extremely valuable as an engineer and is discussed further in the critical thinking section of this chapter.

## 6.5 Personal Problem Solving Styles

The creative problem solving model presented previously is one of the models that can be used to solve the kind of open-ended problems engineers face in every day situations. You may find yourself in an organization that uses another model. There are many models for problem solving and each breaks the process up into slightly different steps. Isaken and Treffinger, for example, break the creative problem solving process into six linear steps [Isaken and Treffinger]. These six steps are:

| | |
|---|---|
| 1. Mess Finding | 4. Idea Finding |
| 2. Data Finding | 5. Solution Finding |
| 3. Problem Finding | 6. Acceptance Finding |

Dr. Min Basadur of McMaster University developed another model, a circular one, which he calls Simplex, based on his experience as a product development engineer with the Procter and Gamble Company. This model separates the problem solving process into eight different steps that are listed below [Basadur]

1. Problem finding
2. Fact finding
3. Problem defining
4. Idea finding

5. Evaluating and selecting
6. Action planning
7. Gaining acceptance
8. Taking action

All these problem solving processes provide a systematic approach to problem solving. Each process has divergent phases where options need to be generated along with convergent phases when best options need to be selected.

Basadur has created and patented a unique method of helping individuals participating in the creative problem solving process identify which parts of the process they are more comfortable in than others. This method, called the Basadur Simplex Creative Problem Solving Profile, reflects your personal creative problem solving style. Everyone has a different creative problem solving style. Your particular style reflects your relative preferences for the different parts of the problem solving process.

Basadur's Creative Problem Solving Profile method identifies four styles and each style correlates with two of the eight problem solving steps in the eight step circular model of creative problem solving above. Thus the eight problem solving steps are grouped into four quadrants or stages of the complete problem solving process.

These four steps are shown as the quadrants in Figure 6.4. The Basadur Creative Problem Solving Process begins with quadrant one, the generation of new problems and opportunities. It cycles through quadrant two, the conceptualization of the problem or opportunity and of new, potentially useful ideas and quadrant three, the optimization of new solutions. It ends with quadrant four, the implementation of new solutions. Each quadrant requires different kinds of thinking and problem solving skills.

Basadur describes the four different quadrants or stages as follows.

## Generating

*Generating* involves getting the problem solving process rolling. Generative thinking involves gathering information through direct experience, questioning, imagining possibilities, sensing new problems and opportunities, and viewing situations from different perspectives. People and organizations strong in generating skills prefer to come up with options or diverge rather than evaluate and select, or converge. They see relevance in almost everything and think of good and bad sides to almost any fact, idea or issue. They dislike becoming too organized or delegating the complete problem, but are willing to let others take care of the details. They enjoy ambiguity and are hard to pin down. They delight in juggling many new projects simultaneously. Every solution they explore suggests several new problems to be solved. Thinking in this quadrant includes problem finding and fact finding.

## Conceptualizing

*Conceptualizing* keeps the innovation process going. Like generating, it involves divergence. But rather than gaining understanding by direct experience, it favors gaining understanding by abstract thinking. It results in putting new ideas together, discovering insights that help define problems, and creating theoretical models to explain things. People and organizations strong in conceptualizing skills enjoy taking information scattered all over the map from the generator phase and making sense of it. Conceptualizers need to 'understand': to them, a theory must be logically sound and precise. They prefer to proceed only with a clear grasp of a situation and when the problem or main idea is well defined. They dislike having to prioritize, implement or agonize over poorly understood alternatives. They like to play with ideas and are not overly concerned with moving to action. Thinking in this quadrant includes problem defining and idea finding.

## Optimizing

*Optimizing* moves the innovation process further. Like conceptualizing, it favors gaining understanding by abstract thinking. But rather than diverge, an individual with this thinking style prefers to converge. This results in converting abstract ideas and alternatives into practical solutions and plans. Individuals rely on mentally testing ideas rather than on trying things out. People who favor the optimizing style prefer to create optimal solutions to a few well-defined problems or issues. They prefer to focus on specific problems and sort through large amounts of information to pinpoint 'what's wrong' in a given situation. They are usually confident in their ability to make a sound logical evaluation and to select the best option or solution to a problem. They often lack patience with ambiguity and dislike 'dreaming' about additional ideas, points of view, or relations among problems. They believe they 'know' what the problem is. Thinking in this quadrant includes idea evaluation and selection and action planning.

## Implementing

*Implementing* completes the innovation process. Like optimizing, it favors converging. However, it favors learning by direct experience rather than by abstract thinking. This results in getting things done. Individuals rely on trying things out rather than mentally testing them. People and organizations strong in implementing prefer situations in which they must somehow make things work. They do not need complete understanding in order to proceed and adapt quickly to immediate changing circumstances. When a theory does not appear to fit the facts they will readily discard it. Others perceive them as enthusiastic about getting the job done but also as impatient or even pushy as they try to turn plans and ideas into action. They will try as many different approaches as necessary and follow up or 'bird dog' as needed to ensure that the new procedure will stick. Thinking in this quadrant includes gaining acceptance and implementing.

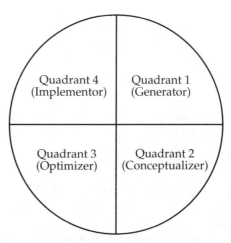

Figure 6.4 The four quadrants of Basadur's Simplex Creative Problem Solving Process (with permission from the Center for Research in Applied Creativity).

Basadur's research with thousands of engineers, managers and others has shown that everyone has a different creative problem solving style. Your particular style reflects your relative preferences for each of the four quadrants of the creative problem solving process: generating/initiating, conceptualizing, optimizing and implementing. Your behavior and thinking processes can not be pigeonholed in any single quadrant. Rather they're a combination or blend of quadrants: you prefer one quadrant in particular, but you also have secondary preferences for one or two adjacent quadrants. Your blend of styles is called the creative problem solving style. Stated another way, your creative problem solving style shows which particular steps of the process you gravitate toward.

**Your Creative Problem Solving Style**

Typically as an engineer, you will be working as part of a team or larger organization. Basadur's research has shown that entire organizations also have their own problem solving process profiles. An organization's profile reflects such things as the kinds of people it hires, its culture and its values. For example, if an organization focuses almost entirely on short-term results, it may be overloaded with implementers but have no conceptualizers or generators. The organization will show strengths in processes that deliver its current products and services efficiently. But it will show weaknesses in processes of long-term planning and product development that would help it stay ahead of change. Rushing to solve problems, this organization will continually find itself reworking failed solutions without pausing to conduct adequate fact finding and problem definition. By contrast, an organization with too many generators or conceptualizers and no implementers will continually find good problems to solve and great ideas for products and processes to develop. But it will never carry them to their conclusion. You can likely think of many examples of companies showing this imbalances in innovation process profiles.

Basadur suggests that in order to succeed in creative problem solving, a team requires strengths in all four quadrants. Teams must appreciate the importance of all four quadrants and find ways to fit together their members' styles. Team members must learn to use their differing styles in complementary ways. For example, generating ideas for new products and methods must start somewhere with some individuals scanning the environment picking up data and cues from customers and suggesting possible opportunities for changes and improvement. Thus, the generator raises new information and possibilities—usually not fully developed, but in the form of starting points for new projects. Then the conceptualizer pulls together the facts and idea fragments from the generator phase into well-defined problems and challenges and more clearly developed ideas worth further evaluation. Good conceptualizers give sound structure to fledgling ideas and opportunities. The optimizer then takes these well-defined ideas and finds a practical best solution and well-detailed efficient plans for proceeding. Finally, implementers must carry forward the practical solutions and plans to make them fit real-life situations and conditions.

Skills in all four quadrants are equally valuable. As an engineering student, you will find yourself working on group projects. Often, these projects are designs that are open-ended and are well suited for creative problem solving. Use these opportunities to practice utilizing your problem solving skills and preferences as well as those of your team members.

### Example 6.6

What is your own preferred problem solving style? You can find out what your preferred problem solving style is by taking the Basadur Simplex Creative Problem Solving Profile. This profile and additional information about the Simplex model can be obtained from the Center for Research in Applied Creativity at:

Center for Research in Applied Creativity, 184 Lovers Lane, Ancaster, Ontario, Canada, L9G 1G8 or http://basadursimplex.com.

## Why Are We Different?

What causes the main differences in people's approaches to problem-solving? Basadur's research suggests that they usually stem from inevitable differences in how knowledge and understanding—"learning"—are gained and used. No two individuals, teams or organizations learn in the same way. Nor do they use what they learn in the same why. As shown in Figure 6.5, some individuals and organizations prefer to learn through direct, concrete experiencing (doing). They gain understanding by physical processing. Others prefer to learn through more detached abstract thinking (analyzing). They gain understanding by mental processing. All individuals and organizations gain knowledge and understanding in both ways but to differing degrees. Similarly, while some individual and organizations prefer to use their knowledge for ideation, others prefer to use their knowledge for evaluation. Again, all individuals and organizations use their knowledge in both ways but to differing degrees.

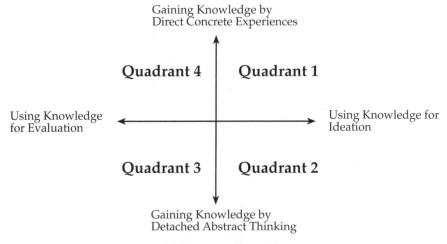

Figure 6.5  The two dimensions of the Basadur Simplex Creative Problem Solving Profile. (The Basadur Problem Solving Model and Profile are copyrighted and are used with the permission of Dr. Min Basadur.)

How an individual or organization combines these different ways of gaining and using learning determines their problem solving process profile. When you understand these differences, you can shift your own orientation in order to complement the problem solving process preferences of others. Equally important you can take various approaches to working with people. You can decide on the optimum strategy for helping someone else to learn something. And you can decide whom to turn to for help in ideation or evaluation. Understanding these differences also helps you interact with other people to help them make the best use of the creative problem solving process. For example, you can help strong optimizers discover new problems and facts or present new problems and facts to them. You can help strong implementers better define challenges or present well-defined challenges to them. You can help strong generators/initiators evaluate and select from among solutions and make plans, or present to them evaluated solutions and ready-made plans. You can help strong conceptualizer to convince others of the values of their ideas and push them to act on them, or push their ideas through to acceptance and implementation for them [Basadur]

# 6.6 Brainstorming Strategies

Alex Osborn, an advertising executive in the 1930s, devised the technique of stimulating ideas known as *brainstorming*. Since that time, countless business and engineering solutions have come about as a result of his method. Basically, brainstorming is a technique used to stimulate as many innovative solutions as possible. These then can be used in both analytic and creative problem-solving practices.

*The goal of brainstorming is to stimulate your mind to trigger concepts or ideas that normal problem solving might miss.*

The goal of brainstorming is to stimulate your mind to trigger concepts or ideas that normal problem solving might miss. This has a physiological basis. The mind stores information in a network. Memories are accessed in the brain by remembering the address of the memory. When asked about ideas for potential solutions, some would come just by thinking about the problem. A portion of your brain will be stimulated, and ideas will be brought to the surface. However, there will be areas of your brain that will not be searched because the pathways to get there are not stimulated. In brainstorming, we get the brain to search as many regions of the brain as possible.

At times it may seem difficult to generate ideas. Isakssen and Treffinger suggest using the method of idea lists. The concept is simple. Write down the ideas you have currently. Then, use these ideas to branch out into new ideas. To help stimulate your thinking in new directions, a set of key words was developed by Alex Osborn. These words include:

*Adapt* – What else is like this? What other ideas does this suggest? Are there any ideas from the past that could be copied or adapted? Whom could I emulate?

*Let your mind wander and write down any ideas that come into your head.*

*Put to other uses* – Are there new ways to use the object as is? Are there other uses if modified?

*Modify* – Could you change the meaning, color, motion, sound, odor, taste, form or shape? What other changes could you make? Can you give it a new twist?

*Magnify* – What can you add to it? Can you give it a greater frequency? Make it stronger, larger, higher, longer, or thicker? Can you add extra value? Could you add another ingredient? Could you multiply or exaggerate it?

*Minify* – What could you subtract or eliminate? Could you make it smaller, lighter, or slower? Could you split it up? Could you condense it? Could you reduce the frequency, miniaturize it, or streamline it?

*Substitute* – What if I used a different ingredient, material, person, process, power source, place, approach or tone of voice? Who else instead? What else instead? What if you put it in another place or in another time?

*Rearrange* – Could you interchange the pieces? Could you use an alternate layout, sequence, or pattern? Could you change the pace or schedule? What if you transpose the cause and effect?

*Reverse* – What if you tried opposite roles? Could you turn it backwards, upside down, or inside out? Could you reverse it or see it through a mirror? What if you transpose the positive and negative?

*Combine* – Could you use a blend, an assortment, an alloy, or an ensemble? Could you combine purposes, units, ideas, functions, or appeals?

Creativity expert Bob Eberle took Osborn's list and created an easy-to-remember word by adding "eliminate" to the list. Eberle's word is SCAMPER. The words used to form SCAMPER are:

**S**ubstitute?
**C**ombine?
**A**dapt?
**M**odify? Minify? Magnify?
**P**ut to other uses?
**E**liminate?
**R**everse? Rearrange?

Using the SCAMPER list may help to generate new ideas.

## Individual Brainstorming

Brainstorming can be done either individually or in a group. Brainstorming individually has the advantage of privacy. There may be times when you deal with a professional or personal problem that you don't want to share with a group.

Some research even shows that individuals may generate more ideas by working by themselves. Edward de Bono attributes this to the concept that "individuals on their own can pursue many different directions. There is no need to talk and no need to listen. An individual on his or her own can pursue an idea that seems 'mad' at first and can stay with the idea until it makes sense. This is almost impossible with a group."

When brainstorming individually, select a place that is free from distraction and interruption. Begin by writing down your initial idea. Now you will want to generate as many other ideas as possible. Let your mind wander and write down any ideas that come into your head. Don't evaluate your ideas; just keep writing them down. If you start to lose momentum, refer to the SCAMPER questions and see if these lead to any other ideas.

There are different formats for recording your ideas when brainstorming. Some find generating in ordered lists works best for them. Others will write all over a sheet of paper, or in patterns. Kinesthetic learners often find that putting ideas on small pieces of paper, which allows them to move them around, helps the creative process. Auditory learners might say the ideas out loud as they are generated.

## Group Brainstorming

The goal of group brainstorming is the same as with individual brainstorming—to generate as many potential solutions as possible without judging any of them. The power of the group comes when each member of the group is involved in a way that uses their creativity to good advantage.

There are a few advantages to group brainstorming. The first is simply that the additional people will look at a problem differently, and bring fresh perspectives. When brainstorming is done correctly, these different views trigger ideas that you would not have come up with on your own.

Another advantage to group brainstorming is that it gets others involved in the problem-solving process early. As you will see in the following section, a key component in problem solving is implementing your solution. If people whose cooperation you need are in the brainstorming group, they will be much more agreeable to your final solutions. After all, it was partly their idea.

To run an effective group brainstorming session, it helps to use basic guidelines. Those outlined in this section have been proven effective in engineering applications. They are also very helpful for use in student organizations and the committees that engineering students are encouraged to join. This process not only covers the generation of ideas, but also provides a quick way to evaluate and converge on a solution. In some situations, you may want to stop after the ideas are generated and evaluate them later.

The guidelines for this process include:

1. Pick a facilitator
2. Define the problem
3. Select a small group
4. Explain the process
5. Record ideas
6. Involve everyone
7. No evaluating
8. Eliminate duplicates
9. Pick three

**Pick a Facilitator**

The first step is to select a facilitator who will record the ideas and keep the group focused. The facilitator is also responsible for making sure the group obeys the ground rules of brainstorming.

**Define the Problem**

It is important that all the participants understand the problem you are looking to solve before generating solutions. Once you start generating ideas, distractions can bring the definition process to a grinding halt. Idea generation would be hampered if the group tried to define the problem they are trying to solve at the wrong time. The definition discussion should happen before the solution idea generating phase.

**Small Group**

The group size should be kept manageable. Prof. Goldschmidt from Purdue University recommends groups of six to 12 members. Groups smaller than that can work, but with less benefit from the group process. Groups larger than 12 become unwieldy, and people won't feel as involved. If you need to brainstorm with a larger group, break it into smaller sub-groups and reconvene after the groups have come up with separate ideas.

**Explain the Process**

Providing the details of the process the group will follow is important. It gives the participants a feeling of comfort knowing what they are getting into. Once the process begins, it is counterproductive to go back and discuss ground rules. This can stymie idea generation just like a mis-timed definition of the problem.

### Record Ideas

Record ideas in a way that is visible to the whole group. Preferably, arrange the group in a semicircle around the person recording. A chalkboard, flip chart, whiteboard, or newsprint taped to a wall all work well. It is important to record all ideas, even if they seem silly. Often the best ideas come from a silly suggestion that triggers a good idea. By writing the suggestions in a way that is visible to all, the participants can use their sight as well as hearing to absorb ideas which may trigger additional ideas. The use of multiple senses helps to stimulate more ideas. Also, having all the ideas in front of the participants guarantees recollection and allows for new ideas, triggered by earlier ideas. Using the SCAMPER questions is a great way to keep ideas flowing.

### Involve Everyone

Start with one idea from the facilitator or another volunteer and write it down. It is easier to get going once the paper or writing surface is no longer completely blank. Go around the group, allowing each person to add one idea per round. If a person doesn't have an idea, let them say "pass," and move on. It is more important to keep moving quickly than to have the participants feel like they must provide an idea every time.

*The power of group brainstorming lies in taking advantage of the creative minds of every member of the group, not just the ones who could dominate the discussion.*

By taking turns, all the participants are ensured an equal opportunity to participate. If the suggestions were taken in a free-flowing way, some participants might monopolize the discussion. The power of group brainstorming lies in taking advantage of the creative minds of every member of the group, not just the ones who could dominate the discussion. Often, the best ideas come from the people who are quiet and would get pushed out of the discussion in a free-flowing setting. It matters not when or how the ideas come, as long as they are allowed to surface.

Continue the process until everyone has "passed" at least once. Even if only one person is still generating ideas, keep everyone involved and keep asking everyone in turn. The one person who is generating ideas might stimulate an idea in someone else at any point. Give everyone the opportunity to come up with a last idea. The last idea may very well be the one that you use.

### No Evaluating

This may be hard for the facilitator or some participants. Telling someone that their suggestion is dumb, ridiculous, out-in-left-field, or making any other negative judgment makes that person (and the rest of the group) less likely to speak up. The genius of brainstorming is the free generation of ideas without deciding if they are good until the correct stage. There are countless times when the wackiest ideas are brought up right before the best ideas. Wacky suggestions can trigger ideas that end up being the final solution. Write each idea down, as crazy as some may be. This makes participants feel at ease. They will feel more comfortable making way-out suggestions and will not fear being made fun of or censored.

Avoiding negative comments is one thing, but if you are facilitating, you need to watch for more subtle signals. You might be communicating indirectly to someone that his or her ideas are not good. If you are writing the ideas down, it is natural to say "good idea" or "great idea." What happens, however, if one person gets "great idea" comments and the others don't? Or what if it is just one person who doesn't get the "great idea" comments?

The participants need to feel completely at ease during brainstorming. Providing negative feedback in the form of laughter, sarcastic remarks, or even the absence of praise will work to dampen the comfort level and, therefore, the creativity. It is safest not to make any value judgments about any ideas at this point. If you are facilitating, try to respond to each idea the same way and to keep "order in your court," so to speak. If you are a participant, try not to make value statements or jokes about other ideas.

After generating ideas, you may want to study the suggestions. In some cases, it is to your advantage to sort out a solution or the top several solutions. Doing this at the end of group brainstorming can be very easy, and also serves to involve the group members in the decision process.

**Eliminate Duplicates**

    The first step is to examine the list of ideas and eliminate duplicates. There is a fine line between eliminating duplicates and creating limiting categories. This step is designed to eliminate repeated ideas. Ask if they really are the same. If they are similar, leave both. Creating categories only reduces the possible solutions and defeats the purpose of brainstorming.

    The next step is to allow the group to ask clarifying questions about suggestions. This is still not the time for evaluation—only clarification. Don't let this become a discussion of merits. Clarification may help in eliminating and identifying duplicates.

Once everyone understands all ideas as best they can, have them evaluate the suggestions. This can be done quickly and simply. First, ask everyone to pick their three top choices. (Each person can select more if you need more possibilities or solutions.) After each member has determined these in their own minds, let each member vote "out loud" by marking their choices. Do this with a dot or other mark, but not a 1, 2 and 3. The top ideas are selected based on the total number of marked votes they receive. The winners then can be forwarded to the problem-solving phases to see if any survive as the final solution.

**Pick Three**

    Note that with this method of voting, the group's valuable time is optimized. Time is not wasted determining whether an idea deserves a "1" or a "3." Also, no time is spent on the ideas that no one liked. If all ideas are discussed throughout the process, the ideas that are not chosen by anyone or which are least preferred often occupy discussion for quite a while. It's best to eliminate these likely time sinkholes as soon as possible.

    Another benefit of not ranking choices at first is that each member could think that their ideas were the fourth on everyone's list, which might frustrate them. Also, how might it affect someone if their ideas were ranked low by the other members? Such a discouraged person could be one who would have very valuable input the next time it is needed, or it could be that this person will be one whom you need as an ally to implement the group's decision. Either way, there is no need to cast any aspersions on ideas which weren't adopted.

# 6.7 Critical Thinking

Since engineering is so dominated by systematic problem solving, it is easy to get caught up in the methodologies and look at everything as a problem with a solution. Problem-solving methods teach a systematic manner for analyzing problems which is necessary to tackle engineering problems, but many of life's issues are much more complex and have no one solution.

I once assigned a project in which students were to design a program using MATLAB® which would predict their own financial future and that of a potential spouse. I jokingly suggested that the class use the program when dating to evaluate the impact of a potential spouse on their financial future. ('Sorry, MATLAB says we wouldn't be compatible…") The class laughed at the suggestion. Clearly, whom you will marry, or even if you will marry, is a complex question and not one that can be solved using some problem-solving methodology or computer algorithm.

Many issues and challenges you will face will have similar complexity and will require a critical perspective. Engineering issues with human, environmental, or ethical impact are among the issues with complex ramifications and require a critical perspective. The issue of poverty is a good example. There have been many programs that have been put into place, and many involving engineering-based solutions, but poverty has remained as a pressing social issue. Solutions to the issue of poverty are complex and varying. Though engineers continue to work in this area, programs or solutions are not being brought to bear on an issue that can simply be "solved."

It is easy to get caught up in the processes and methodologies and move quickly to solutions and implementation plans. There will be many times in your career when you will need to critically evaluate what you are doing or what you are working on. It is an important skill to learn to be able to critically evaluate why you are doing something and how you are going about it by taking a step back to gain perspective. At each step, one of the divergent questions that could be asked is "why?"

Sometimes when you critically evaluate what you are doing, you discover deeper meanings or larger issues. An example is the design of toys for physically challenged children. In the EPICS program (Engineering Projects in Community Service http://epics.ecn.purdue.edu) students develop engineering-based solutions to community problems. As the students in this program have worked with the children, families, and therapists, they have gained an appreciation for the many complex issues facing the children. When they design toys, they must design them so they can be used within the physical limitations of the children. This is fairly straightforward.

However, what the students have also learned is that there are many other issues, such as socialization issues, they can help to address. Some of the toys are now designed so that the physically challenged children can play right along with those who are not challenged. They have also learned to design them so that the physically challenged children can control the play. Normally, the children who are not physically challenged control the games and "allow" the other children to participate. The engineering students have learned about the socialization of the children and its impact on self-esteem.

When working with issues such as disabled and physically challenged children or adults there are often no simple or straightforward answers. They are not simple problems that must be solved, but rather complex issues that need to be recognized and addressed. By working through these issues and the multiple dimensions of the issues, the students have become aware of the social and emotional aspects of their work. They have become better designers, but more importantly they have become better educated about relevant and current issues of the day. When there are discussions about new facilities for the disabled or about mainstreaming children, they are able to participate.

As engineers, you are entrusted with significant responsibility, as discussed in the chapter on ethics. The fruits of our work have great potential for good, and also potential to harm. While it is vitally important to maintain the highest ethical standards, it is also vital to focus a critical eye on the context and direction of your work. Taking opportunities to step back and evaluate what you are doing and why is as important as the problem-solving methodology we have discussed.

At the beginning of the creative problem-solving process, we discussed how important getting the question right was. Similarly, when critically evaluating an issue, the question is vitally important. When evaluating an issue, it is important that you become aware of the slant you bring based on your own background.

Thomas Edison wanted people around him who would not limit their creative thinking by jumping to conclusions. He would take every prospective engineering employee to lunch at his favorite restaurant and recommend the soup. Of course they all ordered the soup unaware that they were being tested. Edison wanted to see if they would salt their soup before tasting it. Salting without tasting, he reasoned, was an indication of someone who jumps to conclusions. He wanted people who were open to questioning and not limited by their preconceptions.

Our own preconceptions make us vulnerable to salting before tasting in many situations. Periodically, ask yourself if you are jumping to conclusions based on your background or personality or thinking style?

Critical thinkers ask the Why and What questions. Why are we approaching the problems this way? What are the implications of the work we are doing? What are the deeper issues? Asking these questions about yourself is a very healthy part of being a good critical thinker. Why are you studying to become an engineer? What will you do as an engineer to help with your personal fulfillment? Becoming a critical thinker will make you a better problem solver and engineer, and one who is more engaged in and aware of society and its issues.

# References

Basadur, M., *Simplex: A Flight To Creativity*, Creative Education Foundation, Inc., 1994.

Beakley, G. C., Leach, H. W., Hedrick, J. K., *Engineering: An Introduction to a Creative Profession*, 3rd Edition, Macmillan Publishing Co., New York, 1977.

Burghardt, M.D., *Introduction to Engineering Design and Problem Solving*, McGraw-Hill Co., 1999.

Eide, A.R., Jenison, R. D., Mashaw, L. H., Northup, L. L., *Introduction to Engineering Problem Solving*, McGraw-Hill Co., 1998.

Fogler, H. S. and LeBlanc, S. E., *Strategies for Creative Problem Solving*, Prentice Hall, Inc., Englewood Cliffs, New Jersey, 1995.

Gibney, Kate, *Awakening Creativity*, Prism, March 1998, pp. 18-23.

Isaksen, Scott G. and Treffinger, Donald J., *Creative Problem Solving: The Basic Course*, Bearly Limited, Buffalo, New York, 1985.

Lumsdaine, Edward and Lumsdaine, Monika, *Creative Problem Solving—Thinking Skills for a Changing World*, 2nd Edition, McGraw-Hill Co, 1990.

Osborn, A. F., *Applied Imagination: Principles and Procedures of Creative Problem-Solving*, Charles Scriber's Sons, New York, 1963

Panitx, Beth, B*rain Storms*, Prism, March 1998, pp. 24-29.

# Exercises and Activities

6.1    A new school has exactly 1000 lockers and exactly 1000 students. On the first day of school, the students meet outside the building and agree on the following plan: the first student will enter the school and open all the lockers. The second student will then enter the school and close every locker with an even number (2, 4, 6, 8, etc.). The third student will then 'reverse' every third locker (3, 6, 9, 12, etc.). That is if the locker is closed, he or she will open it; if it is open, he or she will close it. The fourth student will then reverse every fourth locker, and so on until all 1000 students in turn have entered the building and reversed the proper lockers. Which lockers will finally remain open?

6.2    You are stalled in a long line of snarled traffic that hasn't moved at all in twenty minutes. You're idly drumming your fingers on the steering wheel, when you accidentally started tapping the horn. Several sharp blasts escaped before you realize it. The driver of the pickup truck in front of you opens his door, gets out and starts to walk menacingly toward your car. He looks big, mean and unhappy. Your car is a convertible and the top is down. What do you do?

6.3    Your school's team has reached the national championship. Tickets are very difficult to get but you and some friends have managed to get some. You all travel a great distance to the site of the game. The excitement builds on the day of the game until you discover that your tickets are still back in your room on your desk. What do you do? What is the problem that you need to solve?

6.4    You are a manufacturing engineer working for an aircraft company. The production lines you are responsible for are running at full capacity to keep up with the high demand. One day, you are called into a meeting where you learn that one of your suppliers of bolts has been forging testing results. The tests were never done, so no data exists to tell if the bolts in question meet your standards. You do not know how long the forging of the test results has been going on. You are only told that it has been a 'while.' This means that your whole production line including the planes ready to be delivered may be affected. You are asked to develop a plan to manage this crisis. What do you do?

*Some background information:*
   •Bolts are tested as batches or lots. A specified number are taken out of every batch and tested. The testing typically involves loading the bolts until they fail.
   •The bolts do not have serial numbers so it is impossible to identify the lot number from which they came.
   •The supplier who forged the tests supplies 30% of your inventory of all sizes of bolts.

•Your entire inventory is stored in bins by size. Each bin has a mixture of manufacturers. (To reduce dependence on any one company, you buy bolts from three companies and mix the bolts in the bins sorted by size.)
•Bolts have their manufacturer's symbol stamped on the head of the bolt.
•It takes weeks to assemble or disassemble an aircraft.
•Stopping your assembly line completely to disassemble all aircraft could put your company in financial risk.
•Your customers are waiting on new planes and any delays could cause orders to be lost.
•The FBI has arrested those who forged the test and their business has been closed, at least temporarily.

6.5    How much of an automobile's tire wears off in one tire rotation?

6.6    A farmer going on a trip with a squirrel, acorns and a fox had to cross a river in a boat in which he couldn't take more that one of them with him each time he crossed. Since he often had to leave two of them together on one side of the river or the other, how could he plan the crossings so that nothing gets eaten, and they all get across the river safely?

6.7    Measure the height of your class building using two different methods. Compare your answers and indicate which is more accurate.

6.8    Pick a grassy area near your class building. Estimate how many blades of grass are in that area. Discuss your methodology.

6.9    How high do the letters on an expressway sign need to be to be readable?

6.10   A company has contacted you to design a new kind of amusement park ride. Generate ideas for the new ride. Which idea would you recommend?

6.11   Estimate the speed of a horse.

6.12   Estimate the maximum speed of a dog.

6.13   Assume that an explosion has occurred in your building. Your room is in tact but sustained damage and is in immediate danger of collapsing on you and your classmates. All exits but one (as specified by your instructor) are blocked. The one remaining is nearly blocked. Quickly develop a plan to get your classmates out. Who goes first and who goes last? Why?

6.14   A child's pool is eight feet in diameter and two feet high. It is filled by a garden hose up to a level of one foot. The children complain that it is too cold. Can you heat it up to an acceptable temperature using hot water from the house? How much hot water would you need to add?

6.15   How else could you heat the water for the children in problem 6.14?

6.16    Without referring to a table or book, estimate the melting temperature of aluminum. What are the bounds of potential melting temperatures? Why? What is the actual melting temperature?

6.17    Select a problem from your calculus class and use the analytic problem solving steps to solve it.

6.18    Select a problem from your chemistry class and use the analytic problem solving steps to solve it.

6.19    Select a problem from your physics class and use the analytic problem solving steps to solve it.

6.20    Write a one to two page paper on a historical figure that had to solve a difficult problem. Provide a description of the problem, solution and the person.

6.21    Aliens have landed on earth and the come to your class first. They are very frustrated because they can not communicate with you. Develop a plan to break this communication barrier and show them that we are friendly people.

6.22    You are working on a project that has had many difficulties that are not your fault. However, you are the project leader and your manager has become increasingly frustrated with you and even made a comment that you are the kind of irresponsible person who locks your keys in your car. Your manager is coming into town to see first hand why the project is having problems and you are going to pick her up. You leave early to take time to relax. With the extra time, you decide to take the scenic route and stop the car next to a bubbling stream. Leaving your car, you go over to the stream to just sit and listen. The quiet of the surroundings is just what you needed to calm down before leaving for the airport. Glancing at your watch, you realize that it is time to leave to pick up your manager. Unfortunately, you discover that you indeed have locked your keys in the car. A car passes on the road about every 15 minutes. The nearest house is probably a mile from your location and the nearest town is 15 miles away. The airport is 20 miles away and the flight is due to arrive in 30 minutes. What will you do?

6.23    You are sitting at your desk in the morning reading your e-mail when your boss bursts into your office. It seems that a local farmer was not happy with your company and has dropped a truckload of potatoes in your plant's entrance. The potatoes are blocking the entrance and must be moved. You have been selected as the lucky engineer to fix this problem. What would you do to move the potatoes and how would you get rid of them? Brainstorm ideas and select the best solution. (Note: You work at a manufacturing plant of some kind that does not use potatoes in its process. The potatoes also appear to be perfectly fine, just in the way.)

6.24    You work in a rural area that is known for chickens. An epidemic has swept through the area killing thousands of chickens. The EPA has mandated that the chickens may not be put into the landfill nor can they be buried. Develop a plan to get rid of the chickens.

6.25    A fire has destroyed thousands of acres of forest. Seedlings have been planted in the area but are being destroyed by the local wildlife. Develop a way to reforest the region by preventing the wildlife from eating the seedlings.

6.26    A survey shows that none of the current highschool students in your state are planning to major in engineering when they get to college. Your dean has asked your class to fix this dilemma (with no students, she would be out of a job!). Develop a plan to convince the highschool students that engineering is worthwhile.

6.27    Select a problem from your math book. How many ways can you solve the problem? Demonstrate each method.

6.28    You and seven of your friends have ordered a round pizza. You need to cut it into eight pieces and are only allowed to make three straight cuts. How can you do this?

6.29    Select an innovative product and research the story of its inventors. Prepare a presentation on how the ideas were formed and the concept generated.

6.30    With which part of the Basadur problem solving process are you most comfortable? Write a one page paper on how your preference affects your problem solving.

6.31    Take the Basadur problem solving profile test. Report on your scores. How will this information affect your ability to solve problems.

6.32    Organizations have problem solving styles. Select an organization and prepare a brief report on their own problem solving styles.

6.33    You and your classmates have discovered that your funding for college will be withdrawn after this semester. Develop ideas on how you could obtain the money for the rest of your college expenses.

6.34    Your class has been hired by a new automotive company that wants to produce a brand new car. Their first entry into the market must be truly innovative. Do a preliminary design of this car. What features would it have to set it apart from current vehicles?

6.35    An alumnus of your engineering program has donated money to build a jogging track that would encircle your campus. The track will be made of a recycled rubberized material. How much of this rubberized material will be needed?

6.36    Every Christmas season it seems that one toy becomes a "I have to have one" craze. You and some classmates have been hired to come up with next year's toy of the season. What is it?

6.37    A professional sports team has decided to locate in your town. A local businessperson is going to build a stadium for the team. Unfortunately, the site is contaminated. The project will fall through unless you can devise a way to get rid of the contaminated soil under the new arena. What do you do with the soil?

6.38    NASA has decided to send astronauts to the other planets in the solar system. Psychologists, however, have determined that the astronauts will need something to keep them occupied during the months of travel or they will develop severe mental problems that would threaten the mission. Your job is to devise ways to keep the astronauts occupied for the long journey. Remember that you are limited to what will fit in a space capsule.

6.39    Landfill space is rapidly running out. Develop a plan to eliminate your city's dependence on the local landfill. The city population is 100,000.

6.40    An appliance manufacturer has hired you to expand their market. Your job is to develop a new household appliance. What is it and how will it work?

6.41    College students often have trouble waking up for early classes. Develop a system that will guarantee that the students wake up and attend classes.

6.42    Select one of your classes. How could that class be more effective in helping your learn?

6.43    Prepare a one-page report on a significant engineering solution developed in the past five years. Evaluate its effectiveness and report (both intended and unintended) on the outcomes of this solution.

6.44    Habitat for Humanity International spearheaded an initiative to eliminate substandard housing within the Georgia county in which it is headquartered. One of the first steps was to develop a set of standards to determine if housing was substandard or not. Develop criteria for your own community for determining whether a housing unit is substandard.

6.45    If your own community were to undertake an initiative like the one presented in the previous question, what would need to be done  (both from an engineering and community perspective)?

6.46    Identify one current topic related to engineering or technology that has no single simple solution and write a brief paper discussing issues related to the topic.

6.47    Brainstorm ideas for toys for physically disabled children. Narrow your choices to the top three and identify the benefits of each design. What are the important issues that you would need to consider in your designs?

6.48    In the critical thinking section in this chapter, engineers had to pass Edison's "salt test." Identify areas in your own thinking where you might tend to "salt before tasting" (where you might jump to conclusions or use your own biases to overlook potential options).

6.49 Critically evaluate why you are majoring in engineering. What do you hope to gain by studying engineering?

6.50 Write a one-page paper on your responsibilities to society as a citizen in your community and as an engineer.

6.51 Early one morning is starts to snow at a constant rate. Later, at 6:00 a.m., a snow plow sets out to clear a straight street. The plow can remove a fixed volume of snow per unit time. In other words, its speed it inversely proportional to the depth of the snow. If the plow covered twice as much distance in the first hour as the second hour, what time did it start snowing?

6.52 While three *wise* men are asleep under a tree a mischievous boy paints their foreheads red. Later they all wake up at the same time and all three start laughing. After several minutes suddenly one stops. Why did he stop?

6.53 Using just a 5-gallon bucket and a 3-gallon bucket, can you put four gallons of water in the 5-gallon bucket? (Assume that you have an unlimited supply of water and that there are no measurement markings of any kind on the buckets.)

6.54 A bartender has a three-pint glass and a five-pint glass. A customer walks in and orders four pints of beer. Without a measuring cup but with an unlimited supply of beer how does he get a single pint in either glass?

6.55 You are traveling down a path and come to a fork in the road. A sign lays fallen at the fork indicating that one path leads to a village where everyone tells the truth and the other to a village where everyone tells lies. The sign has been knocked down so you do not know which path leads to which village. Then someone from one of the villages (you don't know which one) comes down the path from which you came. You may ask him one question to determine which path goes to which village. What question do you ask?

6.56 Four mathematicians have the following conversation:

> Alice: I am insane.
> Bob: I am pure.
> Charlie: I am applied.
> Dorothy: I am sane.
> Alice: Charlie is pure.
> Bob: Dorothy is insane.
> Charlie: Bob is applied.
> Dorothy: Charlie is sane.

You are also given that:
   o Pure mathematicians tell the truth about their beliefs.
   o Applied mathematicians lie about their beliefs.
   o Sane mathematician's beliefs are correct.
   o Insane mathematicians beliefs are incorrect.

Describe the four mathematicians.

6.57    Of three men, one man always tells the truth, one always tells lies, and one answers yes or no randomly. Each man knows which man is which. You may ask three yes/no questions to determine who is who. If you ask the same question to more than one person you must count it as a question used for each person asked. What three questions should you ask?

6.58    Tom is from the census bureau and greets Mary at her door. They have the following conversation:

| | |
|---|---|
| Tom: | I need to know how old your three kids are. |
| Mary: | The product of their ages is 36. |
| Tom: | I still don't know their ages. |
| Mary: | The sum of their ages is the same as my house number. |
| Tom: | I still don't know their ages. |
| Mary: | The younger two are twins. |
| Tom: | Now I know their ages! Thanks! |

How old are Mary's kids and what is Mary's house number?

6.59    Reverse the direction of the fish image below by moving only three sticks:

# 7. Computer Tools for Engineers

## 7.1 Introduction

Assume that you are a food process engineer, and a dairy company has hired you for a project:

*"Design a process to manufacture high-quality powdered milk for a profit."*

This is no small undertaking! Table 7.1 shows a list of the steps that you must complete.

Table 7.1 Steps Required to Complete the Powdered-milk Process Design

| Step | Computer Tool |
| --- | --- |
| Research the dairy industry | Internet |
| Design the process | Mathematical Software<br>Design Software<br>Programming Languages |
| Select equipment | Internet |
| Perform economic analyses | Spreadsheet<br>Mathematical Software |
| Analyze experimental data | Spreadsheet<br>Mathematical Software |
| Document the design process | Word Processor |
| Present findings to management | Presentation Software |

With this huge project ahead of you, you realize that you need help. So you turn to the computer. Computer tools can facilitate nearly every task you need to perform. These tools include operating systems, programming languages, and computer applications. For example, word processing applications enable computer users to write papers. To be able to use the computer to solve an engineering problem, you need strong computer skills and a working familiarity with the types of computer tools and applications that are available. Table 7.1 indicates the types of computer tools you would use to complete this project.

The following sections of this chapter will highlight some of the computer tools that an engineer might use to solve problems. Many of the tools will be discussed in the context of solving the dairy design problem.

# 7.2 The Internet

### Introduction

The first step required to solve the dairy design problem is to fill gaps in your knowledge base. That is, you need to do research so that you have a better understanding of the design problem. Some things you should learn are:

*The Internet is a worldwide network of computers that allows electronic communication with each other by sending data packets back and forth.*

• How is powdered milk manufactured?
• What are the consumer safety and health issues involved?
• Who manufactures equipment for dairy processing?
• What other dairy companies manufacture powdered milk?
• Who does dairy processing research?

Research, under the best conditions, can take a significant amount of time. So your aim is to avoid excessive trips to the library and the use of bound indexes. It is preferable to access as much information as possible via your computer. The Internet provides such capabilities.

### History of the Internet

The Internet is a worldwide network of computers that allows electronic communication with each other by sending data packets back and forth. The Internet has been under development since the 1960s. At that time, the Department of Defense was interested in testing out a concept that would connect government computers. It was hoped that such a network would improve military and defense capabilities (Glass and Ables, 1999). Before the end of 1969, machines at the University of Southern California, Stanford Research Institute, the University of California at Santa Barbara, and the University of Utah were connected.

The linking of those four original sites generated much interest in the Internet. However, extensive customization was required to get those original sites linked, because each site used its own operating system. It was apparent that the addition of more sites would require a set of standard protocols that would allow data sharing between different computers. And so the 1970s saw the birth of such protocols.

*Protocols* are used to move data from one location to another across the network, and verify that the data transfer was successful. Data are sent along the Internet in *packets*. Each packet may take a different route to get to its final destination. The *Internet Protocol* (IP) guarantees that if a packet arrives at its destination, the packet is identical to the one originally sent. A common companion protocol to the IP is the *Transmission Control Protocol* (TCP). A sending TCP and a receiving TCP ensure that all the packets that make up a data transfer arrive at the receiving location.

Internet applications were also developed in the 1970s and are still used today. The Telnet program connects one computer to another through the network. *Telnet* allows a user on a local computer to remotely log into and use a second computer at another location. The *ftp* program allows users to transfer files on the network. Both applications were originally non-graphical; they require that text commands be entered at a prompt.

*Internet IP addresses* were established in the 1970s. All organizations that become part of the Internet are assigned a unique 32-bit address by the Network Information Center. IP addresses are a series of four eight-bit numbers. For instance, IP addresses at Purdue University take the form 128.46.XXX.YYY, where XXX and YYY are numbers between zero and 255.

In the 1980s, corporations began connecting to the Internet. Figure 7.1 shows the growing number of sites connected to the Internet. With so many sites joining, a new naming system was needed. In came the *Domain Naming Service*. This hierarchical naming system allowed easier naming of individual hosts (servers) at a particular site. Each organization connected to the Internet belongs to one of the top-level domains (Table 7.1). Purdue University falls under the U.S. educational category; therefore, it has the domain name purdue.edu. Once an organization has both an IP address and a domain name, it can assign addresses to individual hosts. This author's host name is pasture.ecn.purdue.edu. "Pasture" is a large computer (server) on the Purdue University Engineering Computer Network (ecn) , a network of engineering department servers.

As *Internet Service Providers* (ISP), such as America On-Line and CompuServe, came into being in the 1990s, the Internet became more mainstream. The development of the browser named Mosaic made accessing information on the Internet easier for the general public. A *browser* is software that allows you to access and navigate the Internet. Mosaic displays information (data) using both text and graphics. Each page of information is written using the *HyperText Markup Language* (HTML). The most useful aspect of HTML is the hyperlink. A *hyperlink* connects two pages of information. The beauty of the hyperlink is that it allows an Internet user to move through information in a non-linear fashion. That is, the user can "surf" the *World Wide Web* (WWW). The Web, not to be confused with the Internet, refers to the interconnection of HTML pages. Today, the most common browsers are Netscape Navigator and Microsoft Internet Explorer. Updates to these browsers have allowed for the construction of more complicated web pages, such as multiple-frame pages. Sound and video can now be played through the use of helper applications. Greater user interaction with the Web has been made possible through HTML coded references to Java applets (defined later).

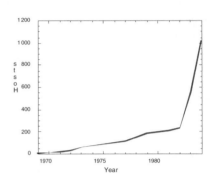

Figure 7.1 Number of Internet hosts by year.

Table 7.1 Selection of Top-Level Domain Names

| Name | Category |
|------|----------|
| edu | U.S. Educational |
| gov | U.S. Government |
| com | U.S. Commercial |
| mil | U.S. Military |
| org | Non-profit Organization |
| net | Network |
| XX<br>(e.g., fr, uk) | Two-Letter Country Code<br>(e.g., France, United Kingdom) |

## Searching the Web

Let's return to the powdered-milk problem. You'd like to use the Internet to answer some of your research questions. First focus on searching for equipment vendors. Assume that you already know about one company that supplies food processing equipment, called APV. Check to see what kinds of equipment they have available for dairy processing. Using your Web browser, you access APV's web site by providing the URL (*Universal Resource Locator*) for APV's home page. In other words, you tell the Web browser what Web address to use to access APV's information. The URL is "http://www.apv.com." This URL consists of the two primary components of a Web address: the protocol (means for transferring information) and the Internet address. The protocol being used here is http (Hypertext Transport Protocol) which allows access to HTML. The Internet address consists of the host name (apv) and the domain (com).

APV's URL take you to APV's introductory (home) page. Once there, you can browse their Web site for information that pertains to your specific research problem. At one point, you find yourself on a page that describes a one-stage spray dryer designed specifically for temperature-sensitive materials such as milk. A spray dryer is a piece of equipment that converts liquid into a powder, or as in your application, milk to pow-dered milk. The URL for this page is http://www.apv.com/anhydro/spray/onestage.htm. This Web address is longer than that for APV's home page. The address contains a path name (/anhydro/spray/) and a file name (onestage.htm). The path name tells you the location within APV's directory structure of the file that you are accessing with your Web browser, just as you store files in directories on your hard drive.

APV is just one equipment vendor. You'd like to investigate others, but you do not know any other Web addresses. To further your research, you use a Web search engine. A *search engine* matches user-specified key words with entries in the search engine's database. The engine then provides a list of links to the search results. There are a number of search engines such as Yahoo! (www.yahoo.com) and Excite (www.excite.com). A keyword search performed using one search engine will yield dif-ferent results than the same keyword search on another search. The reason for the dif-ferent results lies in the type of information that a particular search engine tends to have in its database.

You decide to use the search engine called Lycos (www.lycos.com) to do a keyword search for "spray dryer." When your search results come up, you immediately notice that your query was not specific enough. You get back links to sites on laundry dryers! There are so many sites on the Internet that well-constructed queries are a must. Each search engine provides links to searching tips. In your case, you find that a keyword search for "spray dryer and food" improves your search results.

*Home pages* are Web pages developed by users that contain information about topics of interest to the user, and often include links to the user's favorite Web sites. The lan-guage of all Web pages is HTML. An *HTML document* is a text file containing information the user wants posted to the Web, and instructions on how that information should look when the Web page is accessed by a browser. HTML code can be written from scratch using a text editor, or a composer can be used to generate the code. Many word processors and desktop publishers now have the capability of saving files in HTML format so that the document is displayed by the Web browser exactly as it was in the word processing program. The HTML code consists of instructional tags as seen in Figure 7.2. This code generates the Web page shown in Figure 7.3.

Your employer may ask you to design a Web page discussing the process you de-veloped through the completion of the powdered-milk process design project. A per-sonal reason for creating your own home page is to make available your resume and samples of your work to all Web users. This sort of visibility may result in employment opportunities such as internships and post graduation positions.

More information on creating home pages can be found on the Web or in any book on HTML or the Internet. You may find "HTML: The Complete Reference" by T. A. Powell (1998) to be helpful when writing HTML code.

## Home Pages

*A personal reason for creating your own home page is to make available your resume and samples of your work to all Web users.*

```
<HTML>
<HEAD>
<TITLE>FOODS BLOCK LIBRARY – BASIC ABSTRACT</TITLE>
</HEAD>

<BODY BGCOLOR="#FFFFFF">
<H1>A Computer-Aided Food Process Design Tool for
Industry, Research, & Education</H1>
<H2>Heidi A. Diefes, Ph.D.</H2>
<H2><I>Presented at the Food Science Graduate Seminar<BR>
Cornell University, Fall 1998</I></H2>
<HR ALIGN=LEFT NOSHADE SIZE=5 WIDTH=50%>
<P>
There is a great need in the food industry to link food science research with food engi-
neering and process technologies in such a way as to facilitate an increase in product
development efficiency and product quality. Chemical engineers achieve this through
industry standard computer-aided flowsheeting and design packages. In contrast, ex-
isting programs for food processing applications are limited in their ability to handle
the wide variety of processes common to the food industry. This research entails the
development of a generalized flowsheeting and design program for steady-state food
processes which utilizes the design strategies employed by food engineers.
<P>
</BODY>
</HTML>
```

Figure 7.2 HTML code for Web page shown in Figure 7.3.

Figure 7.3 Web page generated using HTML code in Figure 7.2.

Not all information is currently made available through the Internet, especially highly technical information. Although the Internet may not make such information available in every instance, it can be used to locate the information. Most universities and many public libraries maintain on-line catalogs and links to electronic databases. For instance, Purdue University maintains Thor: The Online Resource (http://thorplus.lib.pur-due.edu/index.html) which provides access to the school library catalog, indexes, and electronic databases. One of the many database links to which Thor provides access is ArticleFirst (Copyright © 1992-1998 OCLC Online Computer Library Center, Inc.), a database for journals in science, technology, medicine, social science, business, the humanities, and popular culture.

**Libraries & Databases on the Internet**

The Internet can also be used to send *electronic mail* (e-mail) to individuals, groups, or mailing lists. E-mail is currently exchanged via the Internet using the *Simple Mail Transport Protocol* (SMTP). However, a new standard is currently under development called *Multipurpose Internet Mail Extension* (MIME). MIME will enable users to send formatted text messages, programs, audio files, encapsulated messages, multipart messages, videos, and image data via e-mail as easily as plain text (James, 1999). MIME is a mail system that identifies file contents by the file extension. For example, MIME can identify HTML files from the .html or .htm extension. Sending and receiving MIME files requires that both the sender and receiver have MIME-compliant mail tools.

**Contacting People through the Internet via E-mail**

*MIME will enable users to send formatted text messages, programs, audio files, encapsulated messages, multipart messages, videos, and image data via e-mail.*

To send e-mail, two things are required: an e-mail package and e-mail addresses. There are many e-mail packages from which to choose. PC and MacIntosh users might select PEGASUS or Eudora; UNIX users might select PINE or ELM. *Graphical user interfaces* (GUIs) and Web *browsers* provide their own e-mail tools. Regardless of the brand, e-mail tools have certain common functions including an ability to send mail, receive mail, reply to mail, forward mail, save mail to a file, and delete mail.

To communicate via the Internet, an *e-mail address* is required. All e-mail users are assigned a unique e-mail address that takes the form username@domain.name. However, it can be difficult to locate a person's e-mail address. One method of locating an e-mail address is to use one of the many people-finding Web sites. For example, the Netscape People Finder can be used to find both home and e-mail addresses. To learn more about other options for locating e-mail addresses, see *Introduction to the Internet* by Scott James (1999). A Web site that allows you to search many people-finding sites at once is http://www.theultimates/white. In addition, most colleges and universities have methods for searching for staff and students on the eb page of the school.

Once you have selected an e-mail address and e-mail software, and after you have signed up with an ISP, you can send e-mail. As part of your dairy design problem, you may keep in touch with the dairy company contacts and other engineering consultants via e-mail. A typical e-mail message might read like the example that follows.

```
To:  dairyconsult@purdue.edu
cc:  dairycontacts@purdue.edu
Attachments:  spdry.bmp
Subject:  Spray dryer design model
Message:
I have been working with the spray dryer vendor on
a model of the system we intend to use. Attached is
a current picture of the model. Please advise.
H. A. Diefes
```

The above e-mail message contains both the typical required and optional components. An e-mail is sent:

**To:** username@domain.name or alias. A copy (**cc:**) may also be sent to a username@domain.name or alias.

**Attachments:** are files to be appended to the main e-mail message.

**Subject:** is a brief description of the message contents.

**Message:** is the actual text of the message.

An *alias* is a nickname for an individual or group of individuals, typically selected by the "owner." An e-mail to an alias gets directed to the actual address. Address books, another common feature of e-mail packages, allow the user to assign and store addresses and aliases. For example, in your address book you may group all the e-mail addresses for your dairy company contacts under the alias "dairycontacts." If you put "dairycontacts" in the **To** or **cc** line of your mail tool, the e-mail message will be sent to everyone on the "dairy contacts" list.

**Words of Warning about the Internet**

With millions of people having access to the Internet, one must be very discerning about the information one accesses. Common sense and intuition are needed to determine the validity of material found on the Web.

In addition, care must be taken not to plagiarize materials found on the Web. As with hard copy materials, always cite your sources.

Further, computer viruses can be spread by executable files (.exe) that you have copied. The virus is unleashed when you run the program. Text files, e-mail, and other non-executable files cannot spread such viruses. Guard against viruses attached to executable files by downloading executable files only from sites that you trust.

## 7.3 Word Processing Programs

Throughout the powdered-milk process design project, you will need to do various types of writing. The largest document will be the technical report for the design. You also may have to complete a non-technical report for management. In addition, you may write memorandums and official letters, and you will need to document experimental results. The computer tool that will enable you to write all these papers is a *word processor.*

The most basic function of a word processor is to create and edit text. When word processors first became available in the early 1980s, that was about the extent of their ability. The user could enter, delete, copy, and move text. Text style was limited to bold, italic, and underline, and text formats were few in number.

"Word processing," as understood today, is really a misnomer, since the capabilities of word processors have become indistinguishable from what was once considered desktop publishing. Word processors such as Microsoft Word, Corel WordPerfect, and Adobe FrameMaker have features that are particularly useful to engineers. Equation, drawing, and table editors are features which make the mechanics of technical writing relatively easy. Many programs include spelling and grammar checkers and an on-line thesaurus to help eliminate common writing errors.

Figure 7.4 shows the graphical user interface (GUI) for Microsoft Word. All word processors display approximately the same components. The title bar indicates the application name and the name of the file that is currently open. The menu bar provides access to approximately 10 categories of commands such as file, edit, and help commands. Some word processors have a toolbar that gives direct access to commands that are frequently used such as "save," "bold," and "italic." A typical word processor GUI also has scroll bars and a status bar. The scroll bars allow the user to move quickly through the document. The status bar provides information on current location and mode of operation. The workspace is where the document is actually typed. The workspace in Figure 7.4 shows a snippet of the technical information related to the spray dryer of our example.

*Equation, drawing, and table editors are features which make the mechanics of technical writing relatively easy.*

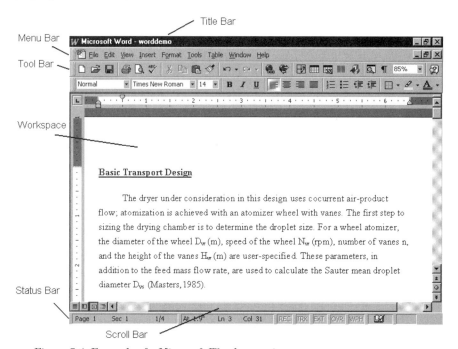

Figure 7.4 Example of a Microsoft Word paper in progress.

This text you are reading was written using word processors. Throughout these pages you can find examples of tables, equations, and figures created using a word processor. A wide variety of typefaces, paragraph styles and formats are also in evidence. You also will see images and pictures that have been imported into this document. A user may import materials that are created using other computer applications. For instance, the photos of practicing engineers in the Profiles Chapter were image files sent to the author for inclusion in his chapter.

Documenting is an ongoing process. you will need to return time and again to your word processor to record all subsequent steps of the dryer design process. Periodically, you will use the word processor to write memos and official letters and to prepare experimental reports, technical documents, and non-technical papers.

# 7.4 Spreadsheets

## Introduction

Much data collection is performed during the research and design phases of an engineering project. Data come from a variety of sources including experiments, design calculations, product specifications, and product statistics. While designing the powdered-milk process, you have gathered cost information from your equipment vendors and collected experimental data on powder solubility. you may like to use the cost information in an economic analysis of your process; (recall that your process is supposed to make a profit). you may also attempt to analyze your laboratory data. Due to the tabular nature of these data sets, this sort of information is best recorded with a spreadsheet.

*The first spreadsheet came in 1978 from a student who was frustrated by the repetitive nature of solving financial-planning problems.*

*Spreadsheet* packages have their origins in finance. The idea for the first spreadsheet application, called VisiCalc, came in 1978 from a Harvard Business School student who was frustrated by the repetitive nature of solving financial-planning problems (Etter, 1995). VisiCalc was originally written for Apple II computers. IBM came out with its own version of VisiCalc in 1981. In 1983, a second spreadsheet package was released by Lotus Development Corporation, called Lotus 1-2-3. While Lotus 1-2-3 remains popular today, there are a number of other spreadsheet packages from which to choose, including Microsoft Excel, Corel Quattro Pro for PC, and Xess for UNIX. One thing to note about the evolution of the spreadsheet is the increase in functionality. VisiCalc centered on accounting math: adding, subtracting, multiplying, dividing, and percentages. Spreadsheets today have more than 200 different functions.

## Common Features

A spreadsheet or worksheet is a rectangular grid which may be composed of thousands of columns and rows. Typically, columns are labeled alphabetically (A, B, C, ..., AA, AB, ...) while rows are labeled numerically (1, 2, 3, ...). At the intersection of a given column and row is a cell. Each cell has a column-row address. For example, the cell at the intersection of the first column and first row has the address A1. Each cell can behave like a word processor or calculator in that it can contain formattable text, numbers, formulas, and macros (programs). Cells can perform simple tasks independently of each other, or they can reference each other and work as a group to perform more difficult tasks.

Figure 7.5 shows the GUI for Microsoft Excel. Notice that many of the interface features are similar to those of a word processor, such as the title bar, menu bar, toolbar, scroll bars, and status bar. Many menu selections are similar as well, particularly the "File" and "Edit" menus. The standard features of all spreadsheets are the edit line and workspace. The edit line includes the name box and formula bar. The name box contains the active cell's address or user-specified name. Within the worksheet in Figure 7.5 is the beginning of the equipment costing information and economic analysis of our example.

Figure 7.5  Example of a Microsoft Excel spreadsheet.

Today's spreadsheets have many features in common including formatting features, editing features, built-in functions, data manipulation capabilities, and graphing features. Let's look at each one of these areas individually.

As with word processing, the user has control over the format of the entire worksheet. Text written in a cell can be formatted. The user has control over the text's font, style, color, and justification. The user can also manipulate the cell width and height as well as border style and background color.

Editing features are also similar to a word processor. Individual cell contents can be cleared or edited. The contents of one or more cells can be cleared, copied, or moved. Excel even has a spell checker.

Spreadsheets have categories of built-in functions. At a minimum, spreadsheets have arithmetic functions, trig functions, and logic functions. Arithmetic functions include a function to sum the values in a set (range) of cells (SUM) and a function to take an average of the values in a range of cells (AVERAGE). Trig functions include SIN, COS, and TAN, as well as inverse and hyperbolic trig functions. Logic (Boolean) functions are ones that return either a true (1) or false (0) based on prescribed conditions. These functions include IF, AND, OR, and NOT. You might use a logic function to help you mark all equipment with a cost greater than $100,000 for further design consideration. The formula to check the heat exchanger cost would be IF(G6>100000," re-design," okay"). This formula would be placed in cell H6. If the value in cell G6 is greater than 100,000, the word *true* is printed in cell H6. Otherwise, the word *false* is placed in cell H6.

Spreadsheet developers continually add built-in, special purpose functions. For instance, financial functions are a part of most spreadsheets. An example of a financial function is one that computes the future value of an investment. In addition, most spreadsheets have a selection of built-in statistical functions.

When a user wants to perform repeatedly a task for which there is not a built-in function, the user can create a new function called a macro. A *macro* is essentially a series of statements that perform a desired task. Each macro is assigned a unique name that can be called whenever the user has need of it.

Spreadsheets also provide a sort function to arrange tabular data. For instance, vendor names can be sorted alphabetically, while prices can be sorted in order of increasing or decreasing numeric value.

*A database is a collection of data that is stored in a structured manner that makes retrieval and manipulation of information easy.*

Graphing is another common spreadsheet feature. Typically, the user can select from a number of graph or chart styles. Suppose that you are interested in studying ways to reduce the energy it takes to operate the spray dryer. Figure 7.6 is a Microsoft Excel-generated *x–y* plot of such an analysis. Here you are investigating the impact that recycling hot air back into the dryer has on steam demand.

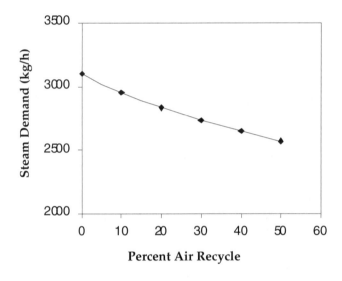

Figure 7.6 An *x-y* plot generated using Microsoft Excel's Chart Wizard.

## Advanced Spreadsheet Features

Database construction, in a limited fashion, is possible in some spreadsheet applications such as Excel. A database is a collection of data that is stored in a structured manner that makes retrieval and manipulation of information easy. A database consists of records, each having a specified number of data fields. Typically, the user constructs a data entry form to facilitate entry of records. Once a database has been constructed, the user can sort and extract information. you could use the database feature to catalog all expenses during implementation of your powdered-milk process. This would require a record for each piece of equipment purchased. Data fields would include size, quantity, list price, discounts, and actual cost of each item.

Real-time data collection is another advanced feature of spreadsheets. This feature allows data and commands to be the received from, and sent to, other applications or computers. During the design phase of the powdered-milk process, you might use this feature to log data during pilot plant experiments. Later, you might use this feature to create a spreadsheet that organizes and displays data coming from on-line sensors in your powdered-milk process. This would enable line workers to detect processing problems quickly.

# 7.5 Mathematics Software

Mathematical modeling is a large component of the design process. For the powdered-milk process, you may need to develop a mathematical model of parts of the process. Problems of this nature are best solved using mathematical or computational software.

MATLAB is a scientific and technical computing environment. It was developed in the 1970s by the University of New Mexico and Stanford University to be used in courses on matrix theory, linear algebra, and numerical analysis (MathWorks, 1995). While matrix math is the basis for MATLAB ("MATrix LABoratory"), the functionality of MATLAB extends much farther. The MATLAB environment enables the user to "manage variables, import and export data, perform calculations, generate plots, and develop and manage files for use with MATLAB" (Palm, 1999). Simply put, MATLAB can be used like a calculator; it can be used like a programming language; it can be used as a data analysis tool.

**MATLAB**

*MATLAB enables the user to manage variables, import and export data, perform calculations, generate plots, and develop and manage files.*

MATLAB is best explained through an example. Consider again the powdered-milk design. As with any food process, you must be concerned about consumer safety. The best way to ensure that the dairy company's customers do not get food poisoning from this product is to pasteurize the milk before drying. The first step in pasteurization is to heat milk to a specified temperature. Milk is typically heated in a plate heat exchanger where a heating medium (e.g., steam) is passed on one side of a metal plate, and milk is passed on the other side.

The required total surface area of the plates depends on how fast the heat will be transferred from the heating medium to the milk, and how much resistance there is to that heat transfer. Mathematically, this is represented by

$$A = Q \times R$$

where $A$ is the surface area of the plates ($m^2$), $Q$ is the rate of heat transfer (W), and $R$ is the resistance to heat transfer ($m^2/W$).

From your previous design calculations, you know that the rate of heat transfer is $1.5 \times 10^6$ W. The resistance to heat transfer range can be any value from $4 \times 10^{-6}$ to $7 \times 10^{-6}$ $m^2/W$. You may wish to study how the required surface area changes over the range of the resistance.

A typical MATLAB session would resemble that in Figure 7.7. The plot generated by these commands is shown in Figure 7.8.

```
>   Q = 1.5e6;
>   R = 4e-6:1e-7:6e-6;
>   A = Q.*R;
>   plot(R,A,'s-');
>   xlabel('Area m^2');
>   ylabel('Resistance m^2/W');
```

Figure 7.7 **MATLAB code for determining heat transfer surface area.**

Figure 7.8 MATLAB plot as seen on a UNIX platform.

Note that MATLAB has a text-based interface. On each line, one task is completed before going on to the next task. For instance, on the first line, the value of $Q$ is set. On the second line, the variable $R$ is set equal to a range of values from $4 \times 10^{-6}$ to $7 \times 10^{-6}$ in steps of $1 \times 10^{-7}$. On the third line, the arithmetic operation $Q \times R$ is performed. On command lines 4 through 6, MATLAB's built-in plotting commands are used to generate a plot of the results.

The commands shown in Figure 7.7 could be saved as a file with a ".m" extension (MATLAB file) and run again and again, or edited at a future date. MATLAB has a built-in editor/debugger interface that resembles a very simple word processor (Figure 7.9). The debugger helps the user by detecting syntax errors that would prevent the proper execution of the code. The editor shown in Figure 7.9 is displaying the code for the heat transfer area calculation.

MATLAB maintains, and periodically adds, toolboxes that are groups of specialized built-in functions. For example, one toolbox contains functions that enable image and signal processing. There is also a Symbolic Math Toolbox that enables symbolic computations based on Maple (see next paragraph). SIMULINK is another toolbox, a graphical environment for simulation and dynamic modeling. The user can assemble a personal toolbox using MATLAB script and function files that the user creates.

## MathCad and Maple

MathCad and Maple are graphical user interface-based mathematics packages. Each is 100% WYSIWY*G*—computer jargon for "What You See Is What You Get." That is, what the user sees on the screen is what will be printed. There is no language to learn since mathematical equations are symbolically typed into the workspace. The workspace is unrestrictive—text, mathematics, and graphs can be placed throughout a document. These packages are similar to a spreadsheet in that updates are made immediately after an edit occurs. MathCad and Maple feature built-in math functions, built-in operators, symbolic capabilities, and graphing capabilities. Perhaps the greatest advantage of MathCad is its ability to handle units. In MATLAB, the user is charged with keeping track of the units associated with variables. In MathCad, units are carried as part of the assigned value.

You could work the same heat transfer problem in MathCad that you did using MATLAB. Figure 7.10 shows the GUI for MathCad and the text used to solve the problem.

Figure 7.9 The MATLAB editor as seen operating on a UNIX platform.

Figure 7.10 Example of a MathCad window and problem solving session.

## Example 7.1

The key to being able to use MATLAB to solve engineering problems effectively is to know how to get help with MATLAB's many commands, features, and toolboxes. There are three easy ways to access help.

**Method 1.** Access the hypertext documentation for MATLAB. At the MATLAB prompt type:

```
>> helpdesk
```

**Method 2.** Use the help command. The help command can be used in three ways. Typing help at the MATLAB prompt will generate a list of all available help topics. Try it out.

```
>> help
```

Near the top of this list, is a topic called matlab\specfun. To see the commands included in a given topic the general format is to type `>> help topic`. To see the commands included in specfun, type:

```
>> help specfun
```

This will generate a list of commands under that topic heading. To get help on a specific command, type `>> help command`. Let's see how the MATLAB command factor is used:

```
>> help factor
```

The following statement should appear:

```
FACTOR Prime factors.
FACTOR(N) returns a vector
containing the prime factors of
N.
See also PRIMES, ISPRIME.
```

Even though MATLAB displays the description in capital letters. MATLAB is case sensitive and will only recognize MATLAB commands in lower case.

**Method 3.** MATLAB has a means of looking for key words in the command descriptions. This enables you to discover command names when you only know what type of operation you wish to perform. For instance, I may wish to learn how to plot. So I would type:

```
>> lookfor plot
```

A list of all the available commands that have the word plot in their descriptions will be generated. How many different commands have the word plot in their description?

## Example 7.2

MATLAB is designed to work with matrices. Therefore, by default, matrix math is performed. Given the following two matrices:

$$\mathbf{A} = \begin{bmatrix} 1 & 0 & 4 \\ 7 & 9 & 3 \\ 3 & 6 & 9 \end{bmatrix} \qquad \mathbf{B} = \begin{bmatrix} 9 & 0 & 1 \\ 3 & 4 & 5 \\ 5 & 9 & 7 \end{bmatrix}$$

**Problem 1.** Find the product of A and B (Section 7.5):

```
% Assign each matrix to a variable name
A_matrix = [1 0 4; 7 9 3; 3 6 9]
B_matrix = [9 0 1; 3 4 5; 5 9 7]
% Perform matrix multiplication to find C
C_matrix = A_matrix * B_matrix
```

The result in MATLAB will be:

```
C_matrix  =
     29      36      29
    105      63      73
     90     105      96
```

**Problem 2.** Find the element-by-element product of **A** and **B** and set it equal to **D** (e.g., $a11 * b11 = d11$):

```
% Perform element-by-element
    multiplication to find D
D_matrix = A_matrix. * B_matrix
```

Note that the notation for element-by-element multiplication is . * . This same notation is used to do any element-by-element operations (e.g., . * , ./, .\, .^). The resulting matrix **D** is:

```
D_matrix  =
      9       0       4
     21      36      15
     15      54      63
```

## Example 7.3

MATLAB can be used to plot data points or functions. We will use both methods to plot a circle of unit radius.

### Method 1: Plotting Data Points

First generate a vector of 100 equally spaced points between 0 and $2 * \pi$ using linspace. Note that the semicolon is used after each command line. This suppresses the echo feature on MATLAB so the contents of the vectors are not displayed on the screen:

```
>> angle=linspace(0,2*pi,100);
```

Define vector of *x* and *y* to plot later:

```
>> x=sin(angle);
>> y=cos(angle);
```

Use the plot function to generate the plot:

```
>> plot(x,y);
```

Use the axis command to set the scales equal on the *x*- and *y*-axes. Otherwise you might produce an oval.

```
>> axis('equal');
```

Label the *x*-axis and *y*-axis and title, respectively.

```
>> xlabel('x-axis');
>> ylabel('y-axis');
>> title('Circle of Radius One');
```

**Method 2:  Plotting With a Function**

First define the function. The name for this function is func_circle. You may choose any name, but be sure to avoid names that MATLAB already uses. Also note that the .^ is used for the exponent. Remember that MATLAB assumes everything is a matrix. The .^ function indicates that each element in the data should be raised to an exponent individually rather that using a matrix operation:

```
>> func_circle=('sqrt(1-x.^2)')
```

Plot the function using the command *fplot* instead of *plot* over the range −1 to 1:

```
>> fplot(func_circle,[-1,1])
```

This function is only positive and gives a semi-circle. To get the bottom half of the circle, define a second function that is negative:

```
>> func_neg=('-sqrt(1-x.^2)')
```

Activate the hold feature to plot the second function on the same plot:

```
>> hold on
```

Plot the negative function over the same range:

```
>> fplot(func_neg,[-1,1])
```

Make the axes equal in length to get a circle:

```
>> axis square
```

Label the x-axis and y-axis and title, respectively:

```
>> xlabel('x-axis');
>> ylabel('y-axis');
>> title('Circle of Radius One');
```

## Example 7.4

MATLAB can be used to perform functional analysis. For examples, find the two local minimums of the function $y = x\cos(x + 0.3)$ between $x = -4$ and $x=4$.

**Method**: First, you need to create a function assignment. Then it is best to plot the function over the specified range because MATLAB's command for finding minimums (fmin) requires that you provide a range over which to look for the minimums:

```
% Assign the function to a variable name as a string
func_y = ëx.*cos(x + 0.3)í


% Plot the function over the entire range
```

```
fplot(func_y,[-4 4])

% Find each local minimum over the specified
sub-ranges
x_min1 = fmin(func_y,-2,0)
x_min2 = fmin(func_y,2,4)

% Find the y values at the minimum locations.
Note that MATLAB can only evaluate functions in
terms
% of x.
x = [x_min1 x_min2]
y_mins = eval(func_y)
```

The MATLAB solution will yield:

```
            x_min1 =
            -1.0575
            x_min2 =
            3.1491
            y_mins =
                -0.7683   -3.0014
```

## Example 7.5

MATLAB can also be used to find the roots or zeroes of a function. As with locating the minimums in the previous example, it is best to plot the function first.

Define the function of interest. In this case, we will find the zero of sin($x$) near 3:

```
>> func_zero=('sin(x)')
```

Plot the function over the interval –5 to 5:

```
>> fplot(func_zero,[-5,5])
```

Use the *fzero* command to find the exact location of the zero. An initial guess must be provided (in this case it is 3.0):

```
>> fzero(func_zero,3)
```

MATLAB returns the answer:

```
ans =
    3.1416
```

Note that *fzero* finds the location where the function crosses the *x*-axis. If you have a function that has a minimum or maximum at zero but does not cross the axis, it will not find it.

## Example 7.6

MATLAB can be used to write scripts to be used later. On a PC or Mac, select "New M-File" from the File menu. A new edit window should appear. On a Unix workstation, open an editor. In either case, type the following commands:

```
disp('This program calculates the square
root of a number')
x=input('Please enter a number: ');
y=sqrt(x);
disp('The square root of your number is')
disp(y)
```

Save this file under the name "calc_sqrt". Note that on a PC or Mac, a ".m" extension will be added to your file name. Using Unix, you will need to add a ".m" to the end of your file name so MATLAB will recognize it as a script file. In MATLAB, type:

```
>> calc_sqrt
```

MATLAB should respond with the first two lines below. Assuming that "3" is entered and the Return key is pressed, MATLAB will display the last two lines:

```
This program calculates the square root of
a number
Please enter a number: 3
The square root of your number is
   1.7321
```

If MATLAB did not recognize "calc_sqrt", check to see that it has a ".m" extension attached to the end of the file name. Alternatively, the file may not be in a directory or folder in which MATLAB is searching.

# 7.6 Presentation Software

At regular intervals during the design process, presentations must be made to research groups and to management. Presentation programs allow a user to communicate ideas and facts through visuals. Presentation programs provide layout and template designs that help the user create slides.

**PowerPoint**

Microsoft PowerPoint is a popular electronic presentation package. Figure 7.11 shows a slide view window of PowerPoint. As you have seen in other Microsoft programs, the GUI for this package has a title bar, a menu bar, a toolbar, and a status bar. The unique feature of this tool is the drawing toolbar.

PowerPoint has five different viewing windows. The "slide view" window displays one slide at a time. Typically, this is the window through which slide editing is done. The "outline view" window lists the text for each slide in a presentation. The "slide sorter view" window graphically displays all the slides. As the name implies, slide sorting is done through this window. The "notes page view" allows the user to attach text with each slide. Since bullets are used on most slides, this feature comes in very handy. It allows the user to come back to a presentation file at a future date, and still remember what the presentation covered. The slides can be viewed sequentially using the "slide show view." This is the view used during a presentation.

Presentation packages have a number of features that help the user put together very professional presentations. PowerPoint has a gallery of templates. Templates specify color scheme and arrangement of elements (e.g., graphics and text) on the slide. Figure 7.11 shows a "slide view" with Clip Art. PowerPoint also has animation capabilities. With the proper computer hardware, sound and video clips also can be incorporated into a presentation.

*With the proper computer hardware, sound and video clips also can be incorporated into a presentation.*

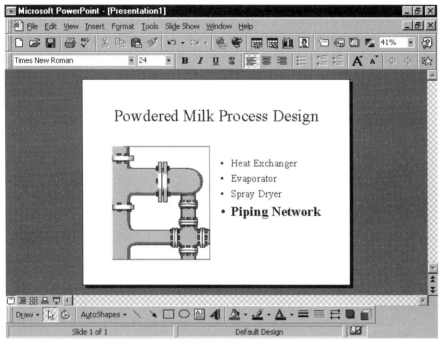

Figure 7.11 Example of a PowerPoint "slide view" window.

# 7.7 Operating Systems

An *operating system* is a group of programs that manage the operation of a computer. It is the interface between the computer hardware, the software, and the user. Printing, memory management, and file access are examples of processes that operating systems control. The most recognized operating systems are UNIX, originally from AT&T Bell Laboratories, and MS-DOS, from Microsoft Corporation. Microsoft's Windows family (95, 98, NT, etc.) are also operating systems, as is the Apple OS for Macintosh Computers.

# 7.8 Programming Languages

**Introduction**

*Programming languages* enable users to communicate with the computer. A computer operates using machine language, which is binary code—a series of 0's and 1's (e.g., 01110101 01010101 01101001). Machine language was used primarily before 1953. Assembly language was developed to overcome the difficulties of writing binary code. It is a very cryptic symbolic code which corresponds one-to-one with machine language. FORTRAN, a higher-level language which more closely resembles English, emerged in 1957. Other higher-level languages quickly followed. However, regardless of the language the programmer uses, the computer only understands binary code. A programming language is an artificial language with its own grammar and syntax that allows the user to "talk" to the computer.

A program is written to tell the computer how to complete a specific task. A program consists of lines of code (instructions) written in a specific programming language. The program is then compiled or converted into machine language. If the compiler detects no syntax or flow control problems in the code, the user may then execute the code. Upon execution of the code, the computer performs the specified task. If the code is well written, the computer completes the task correctly. If you are unsatisfied with the computer's execution of the task, or the computer is unsatisfied with the user's code, you must debug (fix) the program and try again.

There are a number of programming languages in use. The following sections will highlight the most popular languages in the context of their classification.

**Procedural Languages**

Procedural languages use a fetch-decode-execute paradigm (Impagliazzo and Nagin, 1995). Consider the following two lines of programming code:

```
value = 5
new_value = log(value)
```

In the second line of code, the computer "fetches" the variable named **value** from memory, the **log** function is learned or "decoded", and then the function `log(value)` is "executed." FORTRAN, BASIC, Pascal, and C are all procedural languages.

The development of FORTRAN (Formula Translating System) was led by IBM employee John Backus. It quickly became popular with engineers, scientists, and mathematicians due to its execution efficiency. Three versions are in use today: FORTRAN 77, FORTRAN 90, and FORTRAN 95. Each new version has added functionality. The American National Standards Institute (ANSI), which facilitates the establishment of standards for programming languages, recognizes FORTRAN 77 and FORTRAN 90. Thousands of FORTRAN programs and routines written by experts are gathered in standard libraries such as IMSL (International Mathematics and Statistics Library) and NAG (Numerical Algorithm Group).

BASIC (Beginner's All-Purpose Symbolic Instruction Code) was created in 1964 at Dartmouth College by John Kemeny and Thomas Kurtz. Its purpose was to encourage more students to use computers by offering the user-friendly, easy-to-learn computing language.

Pascal is a general-purpose language developed for educational use in 1971. The developer was Prof. Nicklaus Wirth (Deitel and Deitel, 1998). Pascal, named for 17th century mathematician and philosopher Blaise Pascal, was designed for teaching structured programming. This language is not widely used in commercial, industrial, or government applications because it lacks features that would make it useful in those sectors.

Dennis Ritchie of AT&T Bell Laboratories developed the C programming language in 1972. It was an outgrowth of the BCPL (Basic Combined Programming Language) and B programming languages. Martin Richards developed BCPL in 1967 as a language for writing operating system software and compilers (Deitel and Deitel, 1998). Ken Thomas of AT&T Bell Laboratories modeled the B programming language after BCPL. B was used to create the early versions of UNIX. ANSI C programming language is now a common industrial programming language and is the developmental language used in creating operating systems and software such as word processors, database management tools, and graphics packages.

## Object-Oriented Languages

*Object-oriented languages* treat data, and routines that act on data, as single objects or data structures. Referring back to the programming code example, **value** and **log** are treated as one single object that, when called, automatically takes the **log** of **value**. Objects are essentially reusable software components. Objects allow developers to take a modular approach to software development. This means software can be built quickly, correctly, and economically (Deitel and Deitel, 1998).

C++ (C plus plus) is an object-oriented programming language developed by B. Stroustup at AT&T Bell Laboratories in 1985. It is an evolution of the C programming language. C++ added additional operators and functions that support the creation and use of data abstractions as well as object-oriented design and programming (Etter, 1997).

Java's principle designer was James Gosling of Sun Microsystems. The creators of Java were assigned the task of developing a language that would enable the creation of very small, fast, reliable, and transportable programs. The need for such a language arose from a need for software for consumer electronics such as TVs, VCRs, telephones, and pagers (Bell & Par, 1999). Such a language also makes it possible to write programs for the Internet and the Web. Two types of programs can be written in Java—applications and applets. Examples of application programs are word processors and spreadsheets (Deitel and Deitel, 1998). An *applet* is a program that is typically stored on a remote computer that users can connect to through the Web. When a browser sees that a Java applet is referred to in the HTML code for a Webpage, the applet is loaded, interpreted, and executed.

*There are many software tools designed to handle specific types of engineering problems.*

# 7.9 Advanced Engineering Packages

The computer tools introduced so far are useful to all engineers. At one time or another, all engineers need to write papers, work with tabular data, and develop mathematical models. An engineer also needs to be able to solve particular classes of problems. For instance, a food process engineer may need to analyze and optimize processes. A mechanical engineer may need to perform finite element analysis. There are many software tools designed to handle specific types of engineering problems. Described here are just a few key packages that an engineer might encounter.

## Aspen Plus

Aspen Plus is software for the study of processes with continuous flow material and energy. These types of processes are encountered in the chemical, petroleum, food, pulp and paper, pharmaceutical, and biotechnology industries. Chemical engineers, food process engineers, and bioprocess engineers are typical users of Aspen Plus. This tool enables modeling of a process system through the development of a conceptual flowsheet (process diagram). The usual sequence of steps to design and analyze a process in Aspen Plus is:

- Define the problem
- Select a units system (SI or English)
- Specify the stream components (nitrogen, methane, steam, etc.)
- Specify the physical property models (Aspen Plus has an extensive property models library that is user-extendible)
- Select unit operation blocks (heat exchanger, distillation column, etc.)
- Define the composition and flow rates of feed streams
- Specify operating conditions (e.g., temperatures and pressures)
- Impose design specifications
- Perform a case study or sensitivity analysis on the entire process (i.e., see how varying the operating conditions affects the design)

*Software tools allow a single computer to take on the function of dozens of pieces of instrumentation with relative ease.*

The output of Aspen Plus is a report that describes the entire process flowsheet. Overall material and energy balances, steam consumption, and unit operation design specifications are just a sampling of what might be included in the final output report.

## GIS Products

Civil engineers, agricultural engineers, and environmental and natural resource engineers work with geographic-referenced data. The Environmental Systems Research Institute, Inc. (ESRI) is the leader in software tool development for *geographical information systems* (GIS). The purpose of these tools is to link geographic information with spatial data. For example, GIS products can be used to study water and waste water management of a particular landmass (watershed). Specifically, ARC/INFO or ArcView (GIS products) could be used to link data sets concerning drainage and weather with a geographic location as specified by latitude and longitude. GIS products also enable visualization of the data through mapping capabilities.

## Virtual Instrumentation

Engineers in all disciplines of engineering have occasions that require them to acquire and analyze data. Modern software tools make it possible to transform a computer into almost any type of instrumentation required to perform these tasks. One such software tool is LabVIEW, created by National Instruments Corporation. This software allows a user to create a virtual instrument on their personal computer. LabVIEW is a graphical programming development environment based on the C programming language for data acquisition and control, data analysis, and data presentation. The engineer creates programs using a graphical interface. LabVIEW then assembles the computer instructions (program) that accomplishes the tasks. This gives the user the flexibility of a powerful programming language without the associated difficulty and complexity.

The software can be used to control data acquisition boards installed in a computer. These boards are used to collect the data the engineer would need. Other types of boards are used to send signals from the computer to other systems. An example of this would be in a control systems application. LabVIEW allows the engineer to quickly configure the data acquisition and control boards to perform the desired tasks.

The tools also can be used to make the computer function as a piece of instrumentation. The signals from the data acquisition boards can be displayed and analyzed like an oscilloscope or a filter. A sample window of a LabVIEW window performing this function is shown in Figure 7.12. In this example, data were collected and processed through a filter to yield a sine wave.

Software tools such as LabVIEW allow a single computer to take on the function of dozens of pieces of instrumentation with relative ease. This reduces cost and improves the speed of data acquisition and analysis.

Engineers who work with mechanical or structural components include aeronautical, agricultural, biomedical, civil, and mechanical engineers. All must verify the integrity of the systems they design and maintain. Testing these systems on a computer is the most cost-effective and efficient method. Modern finite element tools allow engineers to do just that. One such commonly used tool is ANSYS, created by ANSYS, Inc.

Finite element tools take the geometry of a system and break it into small elements or pieces. Each element is given properties and allowed to interact with the other elements in the model. When the elements are made small enough, very accurate simulations of the actual systems can be made.

The first step in using such a tool is to separate the physical component into a mesh. A sample mesh of centrifugal compressor vanes is shown in Figure 7.13. This mesh defines the physical boundaries of the elements. ANSYS contains aids to allow the user to create the mesh from a CAD drawing of the system. Other tools are also available to create the mesh to be used by ANSYS.

After the physical mesh is generated, the properties of the elements must be defined. The properties of the elements represent the material that the engineer is modeling. The elements can be solids or fluids and can change throughout the

model. Part of the system may be one material, and another part may be different. Initial conditions such as temperatures and loads also must be defined. The engineer can use tools such as ANSYS to perform steady-state as well as transient analyses.

## Finite Element Tools

*Finite element tools take the geometry of a system and break it into small elements or pieces.*

Figure 7.12  Sample LabVIEW window.

A final useful aspect of these tools is the post-processing capability. Tools such as ANSYS allow the engineer to examine the results graphically. The finite element model can be colored to show temperature, stress, or displacements. The engineer can then visually interpret the results, rather than searching through lists of numbers, which can highlight problems with a design.

## Other Tools

We would need an entire book, or more likely a series of books, to touch on all the computer tools engineers use today. There are different sets of computer tools within each discipline and each subspecialty of each discipline. Every day more tools are produced, and those existing are improved. New versions are constantly being developed to add features to help engineers perform their jobs more efficiently and more accurately.

To a large degree, the development of new software tools is capitalizing on rapidly changing computer hardware capabilities. Microprocessors increase in speed according to Moore's Law, which states that microprocessor speed doubles every 18 months. That means that in the four and a half years from your senior year in high school to your first engineering job, processors will go through three doublings —they will be eight times faster.

*The 1972 computer would be the snail and the computer of 2002 would be the jet.*

This may not sound impressive, but consider what this has meant since the first microprocessor was developed in 1972. In 2002, thirty years after the first microprocessor was invented, 20 doublings will have taken place.  The processor speed has increased over a million times. An analogy of how much faster this is would be to compare a snail inching along the ground with a commercial jet flying overhead. The 1972 computer would be the snail and the computer of 2002 would be the jet.

Figure 7.13  A sample mesh of a centrifugal compressor and diffuser vanes.

Considering that you will be entering a career that will last over 30 years, the computers you will be using as a senior engineer could be a million times faster than the machines you use now. Will microprocessors continue to advance at this pace? We don't know. We do know, however, that the computers of tomorrow will be much faster than the snails you are using now.

What this means in terms of computer tools is that they will continue to change. Experiment with as many tools as you can. Just as mechanics learn how to use an assortment of tools, so should engineers learn to use an assortment of computer tools. They also should always be ready to learn how to use new tools, because there always will be new tools to learn.

# References

Andersen, P.K., et. al., *Essential C: an Introduction for Scientists & Engineers*, Saunders College Publishing, Fort Worth, 1995.

Bell, D. and M. Parr, *Java for Students*, 2nd Ed., Prentice-Hall Europe, London, 1998.

Chapman, S.J. , *Introduction to FORTRAN 90/95*, WCB/McGraw-Hill, Boston, 1998.

Dean, C.L., *Teaching Materials to accompany PowerPoint 4.0 for Windows*," McGraw-Hill, New York, 1995.

Deiter, H.M., and P.J. Deiter, *Java: How to Program*, 2nd Ed., Prentice-Hall, Upper Saddle River, 1998.

Etter, D.M., *Microsoft Excel for Engineers*, Addison-Wesley, Menlo Park, CA, 1995.

Etter, D.M., *Introduction to C++ for Engineers and Scientists*, Prentice-Hall, Upper Saddle River, 1997.

Glass, G. and K. Ables, *UNIX for Programmers and Users*, 2nd Ed., Prentice-Hall, Upper Saddle River, 1999.

Gottfried, B.S., *Spreadsheet Tools for Engineers: Excel 97 Version*, WCB/McGraw-Hill, Boston, 1998.

Grauer, R.T. and M. Barber, *Exploring Microsoft PowerPoint 97*, Prentice-Hall, Upper Saddle River, 1998.

Greenlaw, R. and E. Hepp, *Introduction to the Internet for Engineers*, WCB McGraw-Hill, Boston, 1999.

Hanselman, D. and B. Littlefield, *Mastering MATLAB 5: A Comprehensive Tutorial and Reference*, Prentice-Hall, Upper Saddle River, 1998.

Heal, K.M, Hansen, et. al., *Maple V Learning Guide*, Springer-Verlag, New York, 1998.

Impagliazzo, J. and P. Nagin, *Computer Science: A Breadth First Approach with C*, John Wiley & Sons, New York, 1995.

James, S.D., *Introduction to the Internet*, 2nd Ed., Prentice-Hall, Upper Saddle River, 1999.

Kincicky, D.C., *Introduction to Excel*, Prentice-Hall, Upper Saddle River, 1999.

Kincicky, D.C., *Introduction to Word*, Prentice-Hall, Upper Saddle River, 1999

King, J., *MathCad for Engineers*, Addison-Wesley, Menlo Park, 1995.

MathWorks, Inc., *The Student Edition of MATLAB, Version 4*, Prentice-Hall, Upper Saddle River, 1995.

Nyhoff, L.R. and S.C. Leestma, *FORTRAN 90 for Engineers and Scientists*, Prentice-Hall, Upper Saddle River, 1997.

Nyhoff, L.R. and S.C. Leestma, *Introduction to FORTRAN 90 for Engineers and Scientists*, Prentice-Hall, Upper Saddle River, 1997.

Palm, William J., *MATLAB for Engineering Applications*, WCB/McGraw-Hill, Boston, 1999.

Powell, T.A., *HTML: The Complete Reference*, Osbourne/McGraw-Hill, Berkeley, 1998.

Pritchard, P.J., *Mathcad: A Tool for Engineering Problem Solving*, WCB/McGraw-Hill, Boston, 1998.

Rathswohl, E.J., et. al., *Microcomputer Applications – Selected Edition – WordPerfect 6.1*, The Benjamin/Cummings Publishing Company, Redwood City, 1996.

Sorby, S.A., *Microsoft Word for Engineers*, Addison-Wesley, Menlo Park, 1996.

Wright, P.H., *Introduction to Engineering*, 2nd Ed., John Wiley, 1989.

# Exercises and Activities

7.1    Access your school's home page; surf the site and locate information about your department.

7.2    Perform a keyword search on your engineering major using two different Web search engines.

7.3    Determine if your school has on-line library services. If so, explore the service options.

7.4    Send an e-mail to a fellow-student and your mom or dad.

7.5    Search for the e-mail address of an engineering professional society.

7.6    Find the Web site of an engineering society. Prepare a one-page report on that organization based on the information found at their Web site.

7.7    Select the Web site of the publisher of this text (see the copyright page). Prepare a critique of the website itself. What could be done to improve it? How does it compare with other Web sites you have accessed? If you wish, you may send your suggestions to the publisher via e-mail.

7.8    Prepare an oral presentation on the reliability or unreliability of information obtained from the Internet. Include techniques to give yourself confidence in the information.

7.9    Prepare a written report on the reliability or unreliability of information obtained from the Internet. Include techniques to give yourself confidence in the information.

7.10   Do a case study on a situation where inaccurate information from the Internet caused a problem.

7.11   Write a paper for one of your classes (e.g., Chemistry) using the following word processor features: text and paragraph styles and formats.

7.12   Create a table in a word processor.

7.13   Prepare a resume using a word processor.

7.14   Prepare two customized resumes using a word processor for two different companies you might be interested in working for someday.

7.15   Prepare a two-page report on why you selected engineering, what discipline you wish to enter, the possible career paths you might follow, and why. Create the first page using a word processor. Create the second page using a different word processor.

7.16    Prepare a brief presentation on your successes or frustrations using the two different word processors. Which would you recommend?

7.17    Use a spreadsheet to record your expenses for one month.

7.18    Most spreadsheets have financial functions built in. Use these to answer the following question: If you now started saving $200 per month until you were 65, how much money would you have saved? Assume an annual interest rate of a) 4%, b) 6%, and c) 8%.

7.19    Assume that you wish to accumulate the same amount of money on your 65th birthday as you had in the previous problem, but you don't start saving until your 40th birthday. How much would you need to save each month?

7.20    A car you wish to purchase costs $25,000; assume that you make a 10% down payment. Prepare a graph of monthly payments versus interest rate for a 48-month loan. Use a) 6%, b) 8%, c) 10%, and d) 12%.

7.21    Repeat the previous problem, but include plots for a 36- and 60-month loan.

7.22    Prepare a brief report on the software tools used in an engineering discipline.

7.23    Interview a practicing engineer, and prepare an *oral* presentation on the computer tools he or she uses to solve problems.

7.24    Interview a practicing engineer, and prepare a *written* report on the computer tools he or she uses to solve problems on the job.

7.25    Interview a senior or graduate student in engineering. Prepare an oral presentation on the computer tools he or she uses to solve problems.

7.26    Interview a senior or graduate student in engineering. Prepare a written report on the computer tools he or she uses to solve problems.

7.27    Select an engineering problem and prepare a brief report on which computer tools could be used in solving the problem. Which are the optimum tools, and why? What other tools could be used?

7.28    Prepare a brief report on the kinds of situations for which engineers find the need to write original programs to solve problems in an engineering discipline. Include the languages most commonly used.

7.29    Prepare a presentation using PowerPoint which discusses an engineering major.

7.30    Prepare a presentation which speculates on what computer tools you will have available to you in five, 10, and 20 years.

7.31    Use MATLAB help to learn about the command linspace.

7.32   What MATLAB command will put grid lines on a plot?

7.33   Perform the following operations using the matrices in Example 7.2.
   a.   A/B and A./B (right division)
   b.   A\B and A.\B (left division)
   c.   A^3 and A.^3
   d.   A^B and A.^B
   Can all of these operations be performed? If not, why not?

# 8. Teamwork Skills

## 8.1 Introduction

More than ever these days, engineering schools are requiring their students to work in teams. These teams take various forms: collaborative study groups; laboratory groups; design groups as part of individual classes; and design groups participating in extracurricular competitions such as the solar car, human powered vehicles, SAE formula cars, concrete canoes, bridge designs, Rube Goldberg projects, and many others. This team emphasis in engineering schools mirrors a management philosophy which capitalizes on the use of teams which has swept through the corporate world over the past few decades. Engineering employers, in fact, are the principal drivers for the use of teams in engineering schools. They seek engineering graduates who have team skills as well as technical skills. The purpose of this chapter is to impart a team vision, to show why teams and collaboration are so important to organizational and technical success, and to give practical advice for how to organize and function as a team. If you study this chapter, work through the exercises, and examine some of the references, you should have most of the tools you need to be a successful team member and team leader while in engineering school. Apply the principles in this chapter to your team experiences while in college, and learn from and document each. By so doing, you will have a portfolio of team experiences that will be a valuable part of your resume. Above all, you will have the satisfaction that comes from being a part of great teams, and great teams can be fun.

A full-page advertisement in *USA Today* in 1991 stated clearly the highly competitive climate in industry and what Chrysler (now DaimlerChrysler), for example, was doing about it. The use of interdisciplinary teams was an important part of the solution:

**Why Do Corporations Focus on Teams?**

> *"No more piece-by-piece, step-by-step production. Now it's teams. Teams of product and manufacturing engineers, designers, planners, financing and marketing people—together from the start….It's how we built Dodge Viper…from dream to showroom in three years, a record for U.S. car makers. From now on, all our cars and trucks will be higher quality, built at lower cost and delivered to the market faster. That's what competition is all about."*

The greater the problem, the greater the need for the use of teams. Individuals acting alone can solve simple problems, but tough problems require teams. This section lists and then discusses the reasons many organizations have embraced the use of teams. Why is the use of teams so popular today?

- Engineers are asked to solve extremely complex problems.
- More factors must be considered in design than ever before.
- Teams provide for greater understanding through the power of collaboration.
- Many corporations are international in scope, with design and manufacturing engineering operations spread across the globe.
- Since "Time to Market" is extremely important to competitive advantage, "concurrent engineering," inherently a team activity, is widely employed.
- Corporations increasingly are using project management principles.

**Quick Profiles of Great Teams**

There is a greater understanding today than ever before of the importance of and power of creative collaboration. One recent book on the subject is *Organizing Genius: The Secrets of Creative Collaboration*, by Warren Bennis and Patricia Ward Biederman. Their first chapter is titled "The End of the Great Man," in which they argue that great groups are able to accomplish more than talented individuals acting alone, stating that "None of us is as smart as all of us." They cite a survey by Korn-Ferry (the world's largest executive search firm) and *The Economist* in which respondents were asked who will have the most influence on global organizations in the next ten years. Sixty-one percent of the respondents answered "teams of leaders," whereas 14 percent said "one leader." Individual chapters of their book focus on "great groups." A chapter on the Manhattan Project discusses the development of the atomic bomb during the second world war by a team of great scientists. One chapter discusses Lockheed's "Skunk Works," developers of the U-2 spy plane, the SR-71 Blackbird, and the F117-A Stealth fighter-bomber used in the Persian Gulf war. Another example is the development of the graphical user interface at Xerox's PARC (Palo Alto Research Center) and its adoption into the Macintosh computer by Steve Jobs and Steve Wozniak. Steve Jobs' goal was to lead his team to create something not just great, but "insanely great," to "put a dent in the universe."

Another recent book on collaboration is *Shared Minds: The New Technologies of Collaboration*, by Michael Schrage. He defines collaboration as "the process of *shared creation*: two or more individuals with complementary skills interacting to create a shared understanding that none had previously possessed or could have come to on their own." He cites examples of collaboration in science (Watson and Crick; Heisenberg, Bohr, Fermi, Pauli, and Schrodinger; Einstein and mathematician Marcel Grossman), music (Gilbert and Sullivan; Rodgers and Hart; McCartney and Lennon), art (Monet and Renoir; Van Gogh and Gauguin; Picasso and Braque), and literature (editor Maxwell Perkins and writers F. Scott Fitzgerald and Thomas Wolfe; Ezra Pound and T. S. Eliot).

James Watson and Stanley Crick, winners of the Nobel prize for discovering the double helix structure of DNA credit their collaboration for their success: "Both of us admit we couldn't have done it without the other—we were interested in what the answer is rather than doing it ourselves." Crick wrote that "our...advantage was that we had evolved...fruitful methods of collaboration." Sir Isaac Newton, father of classical physics, in a letter to his contemporary (and rival) Robert Hooke, recognizing his debt to fellow scientists, wrote: "If I have seen further (than you and Descartes) it is by standing on the shoulders of Giants." Modern physics also found its genesis in collaboration. Heisenberg, Bohr, Fermi, Pauli, and Schrodinger, all Nobel laureates for their roles in founding quantum physics, worked, played and vacationed together as they developed together their revolutionary new ideas.

**Increasing Complexity of Projects**

Engineers are asked to solve increasingly complex problems. The complexity of mechanical devices has grown rapidly over the past two hundred years. David Ullman, in his text *The Mechanical Design Process*, gives the following examples. In the early 1800's, a musket had 51 parts. The Civil War era Springfield rifle had 140 parts. The bicycle, first developed in the late 1800's, has over two hundred parts. An automobile has tens of thousands of parts. The Boeing 747 aircraft, with over 5 million components, required over 10,000 person-years of design time. Thousands of designers worked over a three-year period on the project. Most modern design problems involve not only many individual parts, but also many subsystems—mechanical, electrical, controls, thermal, and others—each requiring specialists acting in teams.

Engineering designers must consider more factors than ever before, including initial price, life cycle costs, performance, aesthetics, overall quality, ergonomics, reliability, maintainability, manufacturability, environmental factors, safety, liability, and acceptance in world markets. Satisfying these criteria requires the collective teamwork of design and manufacturing engineering, marketing, procurement, business and other personnel. The typical new engineering student, being interested primarily in technical things, tends to believe that engineers work isolated from other factors, always seeking the "best technical solution." This is not true. The very act of engineering involves solving difficult problems and finding technical solutions while considering numerous constraints. Engineers make things happen; they are doers. Engineers often stand astride boundaries in organizations. They are technical experts who also understand economics and the dynamics of the business enterprise.

**The International Factor**

Many corporations are international in scope, with design and manufacturing engineering operations spread across the globe. These operations require teams to work together who may never physically meet. Instead, they meet and share data via electronic means. For example, it is not unusual for design engineers in the United States to collaborate on a design with Japanese engineers for a product that will be manufactured in China. Manufacturing personnel from the Chinese plant may also contribute to the project. As another example, software engineers in the U.S., the U.K. and India may collaborate on a software project literally around the clock and around the world. At any time during a 24-hour period, programmers in some part of the world will be working on the product.

**The Need for Speed**

"Concurrent engineering" is widely employed in order to achieve better designs and to bring products to market more quickly. Time to Market is the total time needed to plan, prototype, procure materials, create marketing strategies, devise tooling, put into production, and bring a new product to market. In the traditional company, each of these steps is done serially, one at a time. It is a relay race, with only one person running at a time, passing the baton to the next runner. The design engineer "tosses a design over the wall" to a manufacturing engineer who may not have seen it yet. She then "tosses it over the wall" to procurement personnel, and so on. Concurrent engineering, on the other hand, is a parallel operation. Like rugby, all of the players run together, side-by-side, tossing the ball back-and-forth. Marketing, manufacturing, and procurement personnel are involved from the beginning of the design phase. The use of teams, as well as new technologies such as CAD/CAM, rapid prototyping, shared data, and advanced communications have radically changed the process of engineering. By using teams and CAD/CAM technologies, in the face of fierce international competition, U.S. auto makers have cut the time needed to bring a new car to market from approximately 5 years to less than 3 years.

Speed—timely delivery of products to the marketplace—is critical for companies to be profitable. Jack Welch, General Electric's widely respected CEO (his biographical sketch is included in Chapter 3), has said "we have seen what wins in our marketplaces around the globe: speed, speed, and more speed." Xilinx Corporation has stated that their research shows that a six-month delay in getting to market reduces a product's profitability by a third over its life cycle. A Business Week article stated: "Reduce product development time to one third, and you will triple profits and triple growth." Figure 8.1 shows graphically the losses incurred over the lifetime of an equivalent product due to delays in bringing product to market. Early product introduction leads to higher profits because early products bring higher profit margins, are more apt to meet customer needs, and are more likely to capture a larger market share. This figure was created by Dr. Charles J. Nuese and included in his wisdom-laden book on product development, *Building the Right Things Right*. The figure is fictitious, but it is based on principles Dr. Nuese has observed during his career as an engineer and a consultant in the semiconductor industry.

Developing new products is a key to growth and profits. During a recent year, Rubbermaid Corporation introduced 400 new products—more than one per calendar day! At least one-third of Rubbermaid's $2 billion in annual sales come from products which have been in existence less than five years. 3M Corporation CEO L. D. DiSimone, facing flat revenues and stiff competition, adopted a business focus of generating 30 percent of revenues from products less than 4 years old. The creative thought processes, the innovation needed to create a steady stream of new ideas, and the expertise to manufacture and bring these to market quickly, are all functions that teams perform well.

It is important for engineering students to understand the importance of speed (without compromising quality). Delays have many negative consequences. Engineering professors who supervise teams should provide incentives to students for early completion of team projects, or at least penalties for failure to meet deadlines. Student teams should recognize the importance of speed, push themselves and their teams toward early completion of projects, and document their successes on their resumes and/or portfolios.

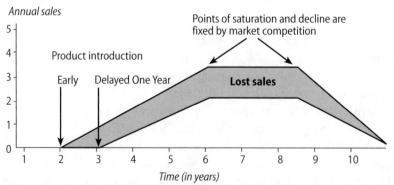

Figure 8.1  The cost of delay in bringing an equivalent new product to market. (Used by permission of Charles J. Nuese.)

Project management is widely practiced in industries and government labs. Many engineers discover on their first job that they will frequently work in project-oriented teams. Unfortunately, most engineering students never take a project management class. Project management principles were developed in the defense industry in the 1950's and 1960's as a way to productively manage department of defense contracts. The typical corporation is organized by vertical divisions or lines, where individuals are clustered by job functions. For example, a corporation may have separate research, manufacturing, engineering, human resources, product development, marketing, and procurement divisions. Project management is a way of organizing individuals not by function but by products or projects. Selected individuals from research, engineering, manufacturing, human resources, procurement, and other divisions are gathered together as a team for a particular purpose—for example, solving a corporate problem, developing a new product, or meeting a crucial deadline. A project management approach, therefore, is inherently a cross-functional team approach and an excellent way to solve problems. Successful project management, however, is not an easy task.

The project manager's job is to complete a project, on time, within budget, with the personnel that he or she is given. Unfortunately, they are rarely able to hand pick their personnel, and their supervisors and/or customers typically set the project completion time and budget. In other words, project managers are virtually never given all the people, time, and money they wish for. These tough constraints are often mirrored in student design teams. Their teacher picks the team members, they don't have much money or resources to work with, and their teacher determines the time frame. These conditions are uncomfortable for students, but they prepare students for the engineering world.

Project managers use tools that any team leader would benefit from learning how to use. Project managers plan work requirements and schedules and direct the use of resources (people, money, materials, equipment). One popular project management tool that student teams should learn to use is the Gantt chart. (A simple Gantt chart is pictured in Fig. 8.2 below.) A Gantt chart organizes and clearly presents information on task division and sequencing. Tasks are divided into sub-tasks. The shading conveys the sequence in which the tasks should be completed and how long a task is expected to take. In the version on the next page, the person(s) responsible for each task is listed.

Project managers monitor projects by tracking and comparing progress to predicted outcomes. They use project management software such as Microsoft Project for planning and monitoring purposes. Project managers carefully guard the scope of the project from excessive mission creep (a never-ending expansion of purpose). A project is successful if it is completed on time, within budget, at the desired quality, and with effective utilization of resources. For more information on project management, consult any text on Project Management and read the publication titled "PMBOK: Project Management Body of Knowledge" on-line at www.pmi.org, the web site of the Project Management Institute.

**Project Management Uses Teamwork**

| Team A-9 ("Semi-Conscious Objectors"): Gantt Chart for Object Relocator: Fall 2000 | | | | | | | | | | |
|---|---|---|---|---|---|---|---|---|---|---|
| Task | Leader(s) | Oct 11-17 | Oct 18-25 | Oct26-31 | Nov1-7 | Nov8-14 | Nov15-21 | Nov22-28 | 11/29-12/5 | Dec6-12 |
| Finalize Design | Team | ▓ | | | | | | | | |
| Procure Materials | Joe | ▓ | | | | | | | | |
| Pick-Up Module | Jaime | | ▓ | | | | | | | |
| Conveyor Assembly | George | | ▓ | | | | | | | |
| Working prototype | Team | | | ▓ | | | | | | |
| Fine Tuning Machine | Jaime/Team | | | | ▓ | | | | | |
| Instruction Manual | George | | | | | ▓ | | | | |
| Professor Tests Projects | Team | | | | | | ▓ | | | |
| Report Rough Draft | Sarah | | | | | | ▓ | | | |
| Finalize CAD Drawings | Ahmad | | | | | | ▓ | ▓ | | |
| Presentation | Ahmad/Joe | | | | | | | ▓ | ▓ | |
| Final Report Due | Team | | | | | | | | | ▓ |
| Judges test best two projects per class | Team | | | | | | | | | ▓ |

Figure 8.2 A Sample Gantt Chart.

## 8.2 What Makes a Successful Team?

Have you ever been a member of a great team? It's a great experience, isn't it? On the other hand, have you ever been part of an unsuccessful team? It was a disappointing experience, wasn't it? How do you measure or evaluate a team's success or lack thereof? What specific factors make a team successful? What factors cause a team to fail?

First, it is important to define what a team is and what it is not. A team is not the same as a group. The term "group" implies little more than several individuals in some proximity to one another. The term "team," on the other hand, implies much more. A team is two or more persons who work together to achieve a common purpose. The two main elements of this definition are *purpose* and *working together*.

All teams have a purpose and a personality. A team's purpose is its task, the reason the team was formed. Its personality is the collective style its people and how the team members work together. Each team has its own style, approach, dynamic, and ways of communicating which are different from those of other teams. For example, some teams are serious, formal, and businesslike, whereas others are more informal, casual, and fun-loving. Some teams are composed of friends, others are not. Friendship, though desirable, is not a necessary requirement for a team to be successful. Commitment to a common purpose and to working together is required. Regardless of style or personality, a team must have professionalism. A team's professionalism ensures that its personality promotes productive progress toward its purpose.

The successful team should have the following attributes:

(1) *A common goal or purpose.* Team members are individually committed to that purpose.

(2) *Leadership.* Though one member may be appointed or voted as the team leader, ideally every team member should contribute to the leadership of the team.

(3) *Each member makes unique contributions to the project.* A climate exists in the team that recognizes and appropriately utilizes the individual talents/abilities of the team members.

(4) *Effective team communication.* Regular, effective meetings; honest, open discussion. Ability to make decisions.

(5) *Creative spark.* There's excitement and creative energy. Team members inspire, energize, and bring out the best in one another. A "can do" attitude. This creative spark fuels collaborative efforts and enables a team to rise above the sum of its individual members.

(6) *Harmonious relationships among team members.* Team members are respectful, encouraging, and positive about one another and the work. If conflicts arise, there are peacemakers on the team. The team's work is both productive and fun.

(7) *Effective planning and use of resources.* This involves an appropriate breakdown of tasks and effective utilization of resources (people, time, money).

Individual team members of a successful team should have the following attributes:

(1) *Attendance*: Attends all team meetings, arriving on time or early. Dependable, like clockwork. Faithful, reliable. Communicates in advance if unable to attend a meeting.

(2) *Responsible*: Accepts responsibility for tasks and completes them on time, needing no reminders nor cajoling. Has a spirit of excellence, yet is not overly perfectionistic.

(3) *Abilities*: Possesses abilities the team needs, and contributes these abilities fully to the team's purpose. Does not withhold self or draw back. Actively communicates at team meetings.

(4) *Creative, Energetic*: Acts as an energy source, not a sink. Brings energy to the team. Conveys a sense of excitement about being part of the team. Has a "can do" attitude about the team's task. Has creative energy and helps spark the creative efforts of everyone else.

(5) *Personality*: Contributes positively to the team environment and personality. Has positive attitudes, encourages others. Acts as a peacemaker if conflict arises. Helps the team reach consensus and make good decisions. Helps create a team environment that is both productive and fun. Brings out the best in the other team members.

# 8.3 Growth Stages of a Team

Simply forming teams to solve problems may seem like a good idea, given the evidence and cases from the previous section, but teams are not a cure-all. Teams require nurturing. Successful team development is a process. Whether in a corporate, academic, or any other setting, all teams must pass through several developmental phases before they truly become productive. These stages are *forming, storming, norming, performing, and adjourning*. Unfortunately, successfully completing the first four stages and becoming a productive, performing team is not guaranteed. *Many teams, in practice, never reach the performing stage*; they remain mired in the forming, storming, or norming stages. Every team's challenge is to grow through these stages and achieve **performance**.

## Stage 1: Forming

The first stage of team development is **forming**. In this stage, typical dialogue may consist of the following: "Nice to meet you. Yeah, I'm not sure why we are here either. I'm afraid this might be a lot of work." In the formation stage, team members get acquainted with one another, with the leader (or they choose a leader if one is not pre-appointed), with the team's purpose, and with the overall level of commitment (work load) required. Team members begin to learn of one another's personalities, abilities, and talents, but also of one another's weaknesses and idiosyncrasies. Individual team members are typically shy, reserved, self-conscious, and uncertain. A key role of the leader in this stage is to lead team ice-breaker activities, facilitate discussion, and encouraging everyone to speak while quieting some who might tend to dominate the conversation. Another role of the leader is to help the team begin to focus on the task at hand.

## Stage 2: Storming

The second stage of team development is **storming**. In this stage, typical dialogue (or private thoughts) may consist of the following: "Do I have to work with *this* team? What did I do to deserve this? There clearly aren't any super-heroes on this team, including that dizzy leader. How are we supposed to solve this messy problem?" During the storming stage, the enormity and complexity of the task begins to sink in, possibly sobering and discouraging the participants. "We are supposed to do what? By when?" Teams are rarely formed to solve easy problems, only very difficult and complex ones. Typically, time schedules are unrealistically short, and budgets are inadequate. Further complicating the issue, teammates have learned enough about their fellow team members to know that there are no super-heroes, no saviors they can count on to do it all. (One person doing all the work is a team failure.) Some team members may not initially "hit it off" well with the others. Cliques or factions may emerge within the team, pitted against other. Since the leader's weaknesses (all leaders have weaknesses) are by now apparent, some individuals or factions may vie for leadership of the team. Though possibly under siege, the leader's role is critical during the storming stage. The leader must help the team to focus on its collective strengths, not its weaknesses, and to direct their energies toward the task. To be a successful team, it is not necessary for team members to like one another or to be friends. A professional knows how to work productively with individuals with widely differing backgrounds and personalities. Everyone must learn the art of constructive dialogue and compromise.

The third stage of team development is **norming**. In this stage, typical dialogue (or private thoughts) may consist of the following: "You know, I think we can do it. True, there are no super-heroes, not by a long shot, but once we stopped fighting and started listening to one another, we discovered that these folks have some good ideas. Now if we can just pull these together…" "Norms" are shared expectations or rules of conduct. All groups have some kinds of norms, though many times unstated. Do you recall a time you joined a group or team and felt subtle influence to act, dress, look, speak, or work in a particular way? The more a team works together, the more they tend to converge toward some common perspectives and behaviors. During the norming stage, team members begin to accept one another instead of complaining and competing. Rather than focusing on weaknesses and personality differences, they acknowledge and utilize one another's strengths. Individual team members find their place in the group and do their part. Instead of directing energies toward fighting itself, the team directs its collective energy toward the task. The key to this shift of focus is a collective decision to behave in a professional way, to agree upon and adhere to norms. Possible norms include working cooperatively as a team rather than individually, agreeing on the level of effort expected of everyone, conducting effective discussions and meetings, making effective team decisions, and learning to criticize one another's ideas without attacking the person. One commanded norm is that all team members are expected to be at all meetings, or to communicate clearly in the event that they can't attend. During the norming stage, feelings of closeness, interdependence, unity, and cooperation develop among the team. The primary role of the leader during the norming phase is to facilitate the cohesion process. Some team members will lag behind the team core in embracing norms. The leader and others on the team (at this stage leadership is beginning to emerge from others on the team as well) must artfully nudge individuals along toward group accountability and a task focus.

**Stage 3: Norming**

The fourth stage of team development is **performing**. In this stage, typical dialogue (or private thoughts) may consist of the following: "This is a fun team. We still have a long way to go, but we have a great plan. Everyone is pulling together and working hard. No super-heroes, but we're a super team." In this stage, teams accomplish a great deal. They have a clear shared vision. Responsibilities are distributed. Individual team members accept and execute their specific tasks in accordance to the planned schedule. They are individually committed, and hold one another accountable. On the other hand, there is also a blurring of roles. Team members "pitch in" to help one another, doing whatever it takes for the team to be successful. In a performing team, so many team members have taken such significant responsibility for the team's success that the spotlight is rarely on a single leader anymore. Typically, whoever initially led the team is almost indistinguishable from the rest of the team.

**Stage 4: Performing**

The fifth and final stage of team development is **adjourning**. Because teams are typically assembled for a specific purpose or project, there is a definite time when the team is disbanded—when that goal is accomplished. If the team was successful, there is a definite feeling of accomplishment, even euphoria, by the team members. On the other hand, an under-performing team will typically feel anger or disappointment upon adjournment.

**Stage 5: Adjourning**

These stages, originally set forth by Bruce W. Tuckman in 1965, appear in some form in most books which discuss teams. Some leave off the final step, adjourning. Others switch the order and present the steps as forming, conforming (instead of "norming"), storming, and performing.

# 8.4 Team Leadership Structures

**Traditional**

Occasionally, engineering design teams are asked to create an organizational chart. Preparing this chart gives the team the opportunity to define roles and to give thought to the relationship between the team and the team leader. Most student teams, in part because they don't know any better, default to a **traditional team structure**. This structure is easy to imagine, but it may or may not accurately represent the best way the team should operate. It implies a strong leader who largely directs the actions of the group, possibly with little participation or discussion from the other team members. It suggests separation between the leader and the other team members.

**Participative**

The second typical structure emphasizes a **participative leadership/team model** in which the leader is positioned in close proximity to all of the members, with short, direct communication paths. The figure implies direct accountability of the leader to all of the members and dependence of the leader on the participation of all the members.

**Flat**

A third structure is similar to the second except it emphasizes the leader's role as a working member of the team. It suggests a **flat structure** where the leader is an equal, rather than a hierarchical structure where the leader is above the team. For this structure it is easy to visualize the leadership function shifting around the ring from member to member as situations arise which require the special expertise of individual members.

**Consultant**

The fourth structure shows the relationship between a **student team and its instructor**. The student team may be a design group, a research team, or a collaborative study unit. The instructor, though not part of the team, will be nearby and will serve as an important resource to the team. He may be asked to advise the team on administrative issues, to act as a technical consultant, or to assist in intervention or disciplinary action for a non-performing team member. All teams—in industry, in the university, and in private organizations—ultimately operate under the authority of a leader or administrator.

There is no one "right" structure. None of these structures is inherently good or bad. Examining these team structure examples gives the team the opportunity to define roles and to give thought to the relationship between the team and the team leader. A team needs to choose a structure that effectively models how they want to work as a team. The ultimate success of the team then hinges upon the team and its leader(s) working together to accomplish the team's purpose. As a team works, they may find that their operational structure may shift periodically in response to different situations and challenges they face.

# 8.5 How Effective Teams Work

Every team makes decisions in order to accomplish its work. The ability to make high quality decisions in a timely manner is a mark of a great team. Not surprisingly, the *process* of how decisions are made greatly affects the *quality* of these decisions. Unfortunately, most teams arrive at decisions without even knowing how they did it. In this section, we present a variety of ways in which decisions are typically made and discuss their advantages and disadvantages. Though it is recommended that a team decides in advance what decision-making process it will use, the best teams use a variety of means of making decisions depending on the circumstances. The following classification of ways teams make decisions is summarized from a section in the book, *Why Teams Don't Wo*rk, by Harvey Robbins and Michael Finley.

**Consensus:** A decision by consensus is a decision in which all the team members find common ground. It does not necessarily mean a unanimous vote, but it does mean that everyone has an opportunity to express their views and to hear the views of others. The process of open sharing of ideas often leads to better, more creative solutions. Unfortunately, achieving consensus can take time and become unwieldy for larger teams.

**Majority:** Another way to make a decision is by majority vote. The option which receives the most votes wins. The advantage of this approach is that it takes less time than reaching consensus. Its disadvantage is that it provides for less creative dialogue than consensus, and there will always be a faction—the minority—who lose the vote and may become alienated.

**Minority:** Sometimes a small subset of the team—a subcommittee, for example—makes the decision. The advantage of this is that it may expedite the decision. The disadvantage is that there is less overall team communication, and some team members may be prevented from making contributions to the decision.

**Averaging:** Averaging is compromise in its worst form. It's the way Congress and some committees arrive at decisions. Averaging is often accomplished with haggling, bargaining, cajoling, and manipulation. Usually no one is happy except the moderates. The advantage of averaging is that the extreme opinions tend to cancel one another out. The disadvantages are that there often is little productive discussion (as in consensus), and that the least informed can cancel the votes of the more knowledgeable.

**Expert:** When facing a difficult decision, there is no substitute for expertise. If there is an expert on the team, he or she may be asked to make a decision. If a team lacks an expert on a particular issue, which is often the case, the best teams recognize this and seek the advice of an expert. The advantage of this is that theoretically, the decision is made with accurate, expert knowledge. The disadvantage of this is that it is possible to locate two experts who, when given the same information, disagree on the best course of action. How is a team assured that the expert gives them the best advice? Team members may be divided on which expert to consult, as well as on their assessment of the expert's credentials.

**Authority Rule without Discussion:** This occurs when there is a strong leader who makes decisions without discussing the details with or seeking advice from the team. This works well with small decisions, particularly of an administrative nature, with decisions that must be made quickly, and with decisions that the team is not well

**Modes of Team Action**

qualified to contribute to, anyway. There are many disadvantages to this approach, however. The greatest is that the team's trust in its leader will be undermined. They will perceive that the leader doesn't trust them and is trying to circumvent them. If one person is continually taking it upon himself to make all the decisions, or if the team is abrogating its responsibility to act together and by default, forcing the decision onto an individual, then this is a group, not a team. It is a loose aggregation of individuals, a dysfunctional team.

**Authority Rule with Discussion:** This can be a very effective way of making decisions. It is understood from the outset that the final decision will be made by one person—the leader or a delegated decision-maker—but the leader first seeks team input. The team meets and discusses the issue. Many, but perhaps not all, viewpoints are heard. Now, more fully informed, the appointed individual makes the decision. The advantage of this method is that the team members, being made part of the process, feel valued. They are more likely to be committed to the result. This type of decision-making process requires a leader and a team with excellent communication skills, and a leader who is willing to make decisions.

## Getting Going In a Team Setting

If you have an opportunity to become part of a team, how should you approach it? What should your attitudes be?

1. *Determine to give your best* to help the team grow and to accomplish its purpose. Don't just tolerate the team experience.

2. *Do not expect to have perfect teammates.* You're not perfect; neither are they. Don't fret that your friends are not on your team. Make friends with those on your team.

3. *Be careful about first impressions* of your team. Some team members may seem to know all the right things to say, but you may never see them again. Others might at first look like losers, but in the long run they are the ones who attend all of the meetings, make steady contributions, and ensure the success of the team. Be careful about selecting a leader at the very first team meeting. Sometimes the person who seems at first to be the most qualified to lead will never make the commitment necessary to be the leader. Look for commitment in a leader.

4. *Be a leader.* If you are not the appointed leader, give your support to the leader and lead from your position on the team. Don't be scared off by team experiences in the past where the leader did all the work. Be one who encourages and grows other leaders.

5. *Help the team achieve its own unique identity and personality.* A team is a kind of corporation, a living entity with special chemistry and personality. No team is just like any other team. Every team is formed at a unique time and setting, with different people and a different purpose. Being part of a productive team that sparkles with energy, personality, and enthusiasm is one of the rewards of teamwork. Being part of a great team is a great memory.

6. *Be patient.* Foster team growth, and give it time to grow. Teammates are at first aliens, unsure of one another and of the team's purpose. Watch for and help the team grow through the stages of Forming, Storming, Norming, and Performing.

7. *Evaluate and grade yourself and your team's performance.* Document its successes and failures. Dedicate yourself to be one who understands teams and who acts as a catalyst—one who helps the team perform at a high level. Note team experiences in your undergraduate portfolio, especially times when you took a leadership role. Prepare to communicate these experiences to prospective employers.

# 8.6 The Character of a Leader

Every successful human endeavor involving collective action requires leadership. Great teams need great leadership. Without leadership, humans tend to drift apart, act alone and lose purpose. They may work on the same project, but their efforts, without synchronization and coordination, interfere with, rather than build on, one another. Without coordinated direction, people become discouraged, frustrations build, and conflicts ensue. Money is wasted, time schedules deteriorate, but the greatest loss is a loss of human potential. A failed team effort leaves a bitter taste not easily forgotten. Incomparably sweet, on the other hand, is the thrill of team success.

**Leader Attributes**

Leaders ensure that the team remains focused on its purpose and that it develops and maintains a positive team personality. They challenge and lead the team to high performance and professionalism. With one hand they build the team, with the other they build the project. In support of these objectives, leaders must do the following:

1. *Focus on the purpose*: Help the team remain focused on its purpose.
2. *Be a team builder:* Leaders may actively work on some project tasks themselves, but their most important task is the team, not the project. They build, equip, and coordinate the efforts of the team so it can accomplish its purpose and succeed as a team.
3. *Plan well and effectively utilize resources* (people, time, money). Leaders effectively assess and use team member abilities.
4. *Run effective meetings*: They ensure that the team meets together regularly and that meetings are productive.
5. *Communicate effectively*: Leaders effectively communicates the team vision and purpose. They also effectively praise good work and give effective guidance for improvement of substandard performance.
6. *Promote team harmony by fostering a positive environment:* If team members focus on one another's strengths instead of weaknesses, conflicts are less likely to arise. Conflict is not necessarily evil, however. The effective leader must not be afraid of conflict. He or she should view conflicts as an opportunity to improve team performance and personality, and to refocus it on its purpose.
7. *Foster high levels of performance, creativity, and professionalism*: Have a high vision. Challenge team members to do the impossible, to think creatively and to stimulate one another to high performance.

Every team needs all the leadership it can get. Most teams have a single individual who is either voted or appointed as its leader. In the best teams, however, many members contribute leadership in ways that support and complement the appointed leader. Like the appointed leader, every member should focus on the team's purpose, build the team, plan, recognize the gifts of others, contribute to effective meetings, communicate well, promote a harmonious team environment, and be creative. Ideally, every member, together with the leader, should build the team with one hand and labor on the project with the other hand. Noted author John Maxwell defines leadership in this way: "Leadership is influence." Every team member, from his or her position on the team, has the opportunity to considerably influence the team's performance. It is possible to have a truly outstanding leader, yet have the team fail—if the team chooses to use its influence to avoid following the leader, to undermine the leader and to promote team disunity. Conversely, a team with a relatively weak appointed leader can be greatly successfully if the team members pitch in and use their influence to build the team. Ultimately, the entire team—both the members and the appointed leader—must work together to build a great team. It takes teamwork to make a great team.

## Leadership Styles

According to John Schermerhorn, author of *Management*, "Leadership style is a recurring pattern of behaviors exhibited by a leader." It is the tendency of a leader to act or to relate to people in a particular way. There are two general categories of leadership style—task-oriented leaders and people-oriented leaders. **Task-oriented leaders** are highly concerned about the team's purpose and task at hand. They like to plan, carefully define the work, assign specific task responsibilities, set clear work standards, urge task completion, and carefully monitor results. **People-oriented leaders** are warm and supportive toward team members, develop rapport with the team, respect the feelings of their followers, are sensitive to followers' needs, and show trust in followers.

You may wonder whether one style of leadership better than another for a team. A successful team needs both styles of leadership in some proportion. Earlier it was stated that a team has a purpose and a personality. The team must be task-oriented, but it also must cultivate a team personality and a positive environment in which team performance can flourish. This is further evidence of the need for individual team members to provide leadership that complements the leadership of the appointed leader. If the appointed leader tends to be more task-oriented, the team can balance this with a people-orientation. If the appointed leader is people-oriented, he will likely need individuals on the team to assist with the details of task management. The best leaders welcome complementary leadership from the teams.

# 8.7 Team Grading and Reports

The primary aim of this chapter has been to give you tools and insights pertaining to teams which will significantly increase the likelihood that you will be part of (and an active leader in) great teams. You likely will be involved in many teams during your college years, in your professional life, and in volunteer capacities. Teams are everywhere, but few know how to make teams work. For teams, Voltaire's words are applicable: "Common sense is not so common." If you work hard to be uncommonly sensible about teams, you will be of great value to an employer. Approach every team experience as an opportunity to learn and to grow as a team member and leader. An excellent way to gain the most benefit is to critically evaluate each team experience. Some suggested dimensions for evaluating teams are as follows:

(1) Did the team accomplish its purpose? Did it get the job done?
(2) Did it do the job well? Were the results of high quality? If not, why?
(3) Did the team grow through all of the developmental stages (Forming, Storming, Norming, Performing)? Were there any detours?
(4) Reflect on the team's personality. Did the team enjoy working together? Did team members inspire one another to greater creativity and energy?
(5) Evaluate team members on a report card like the one shown in Figure 7.2.
(6) Evaluate the team leader(s). Was he/she an effective leader? Why or why not? How could he/she be effective?
(7) Honestly evaluate your contribution to the team.

In addition to learning more about teams, another reason to document team performance is to inform your instructor (when you are in college) or your supervisor (when at work) about the team's work. Virtually all teams have a supervisor to whom the team is accountable. It is important that you report data to them. Some argue that members of student teams should each get the same grade, and in general, this is the best practice. But students should know from the start that if they do not participate in a significant way in the team's work, then their instructor may adjust the individual's grade downward. In order for the instructor to make such determinations, it greatly helps if he or she is given quantitative information on team member contributions (both at midterm and at the end of the semester). Sometimes instructors ask team members to grade one another, but the reason for the grades remains unclear. The act of team members evaluating one another on specific team success criteria gives concrete evidence for team member grades. These criteria are even more valuable to the team if they are held up as a target or goal at the beginning of the team activity, and if it is made clear that they will be the basis for evaluating the team. In addition to the team member report card, the following should be valuable to your instructor:

(1) Attendance records for all team meetings (usually weekly). Who attended? Was anyone habitually late or absent?
(2) Actual contributions of each member. It works well to include in the team's final report a table like the one presented in Figure 8.3 noting the percentage contribution of each individual on each of the team's major projects/ assignments. Add these percentages in order to quan-

tify a team member's total contribution. Variation is expected in the contributions of the various members because it is impossible to balance effort precisely. In the example in Fig. 8.3, four members' contributions varied from 95 to 155 percent. Also included in the table (a sign of a healthy team overall) is the fact that each of the four active team members took a leadership role on at least one aspect of the project effort: Sally took primary lead on the final report, George led in the creation of the Instruction Manual, Ahmad led in the CADKEY Drawings and the PowerPoint presentation, and Jaime led in the project construction. Though their contribution percentages vary, there is little evidence that anyone's grades should be adjusted downward. They each deserve, it appears, the team grade. Joe, on the other hand, contributed nothing, and deserves a grade penalty. Chances are, the 10 percent is a gift from his teammates.

| Team Member Report Card | | | | | |
|---|---|---|---|---|---|
| **Criteria** | **Team Member** | | | | |
| | Ahmad | George | Jaime | Joe | Sarah |
| 1. Attendance: Attends all team meetings, arriving on-time or early. Dependable, faithful, reliable. | | | | | |
| 2. Responsible: Gladly accepts work and gets it done. Spirit of excellence. | | | | | |
| 3. Has abilities the team needs. Makes the most of abilities. Gives fully, doesn't hold back. Communicates. | | | | | |
| 4. Creative. Energetic. Brings energy, excitement to team. "Can do" attitude. Sparks creativity in others. | | | | | |
| 5. Personality: Positive attitudes, encourages others. Seeks consensus. Fun. Brings out best in others. Peace-maker. Water, not gasoline, on fires. | | | | | |
| **Average Grade:** | | | | | |
| **Grading Scale:** 5 - Always; 4 - Mostly; 3 - Sometimes; 2 - Rarely; 1 - Never | | | | | |

Figure 8.3  Sample team member report card.

| Team Members | Team Member Contributions on Primary Semester Projects | | | | | |
|---|---|---|---|---|---|---|
| | CADKEY Drawings | Project Construction | Instruction Manual | Final Report | Power Point Presen-tation | Total Indiv. Contri-bution |
| Sarah | 0% | 15% | 45% | 60% | 15% | **135 %** |
| Joe | 0% | 5% | 0% | 5% | 0% | **10 %** |
| George | 10% | 15% | 55% | 15% | 10% | **105 %** |
| Ahmad | 75% | 10% | 0% | 10% | 60% | **155 %** |
| Jaime | 15% | 55% | 0% | 10% | 15% | **95 %** |
| | **100%** | **100%** | **100%** | **100%** | **100%** | **400 %** |

Figure 8.4  Team member contribution report.

# References

Bennis, Warren, and Biederman, Patricia Ward, *Organizing Genius: The Secrets of Creative Collaboration*, Addison-Wesley, 1997.

Katzenbach, Jon R., and Smith, Douglas K., *The Wisdom of Teams: Creating the High-Performance Organization*, HarperBusiness, 1993.

Nuese, Charles J., *Building the Right Things Right: A New Model for Product and Technology Development*, Quality Resources, 1995.

Robbins, Harvey A., and Finley, Michael, *Why Teams Don't Work: What Went Wrong and How to Make it Right*, Peterson's/Pacesetter Books, Princeton, New Jersey, 1995.

Schermerhorn Jr., John R., *Management,* 5th ed., John Wiley and Sons, 1996.

Scholtes, Peter R., Joiner, Brian L., and Streibel, Barbara J., *The TEAM Handbook,* 2nd ed., Oriel Incorporated, 1996.

Schrage, Michael, *Shared Minds: The New Technologies of Collaboration*, Random House, New York, 1990.

Tuckman, Bruce W., "Developmental Sequence in Small Groups," *Psychological Bulletin*, vol. 63 (1965), pp. 384-389.

Ullman, David G., *The Mechanical Design Process,* 2nd ed., McGraw-Hill, 1997.

Whetten, David A., and Cameron, Kim S., *Developing Management Skills,* 3rd ed., Harper Collins, 1995.

# Exercises and Activities

8.1   Some individuals consider teams to be the answer to all human and organizational problems. Conversely, some individuals asked to participate on teams think that being part of a team is a major frustration. Comment on these propositions.

8.2   From your perspective, is it easy or difficult for humans to work together as teams? Does it come naturally or is a lot of work involved in making a great team?

8.3   Describe in your own words the growth stages of a team. Give at least two examples from your own team development experiences (any will do). Have you been a part of a team that developed to a certain point and stalled? A team that progressed quickly to a productive, performing team? A team that experienced prolonged rocky periods but ultimately made it to the performing stage? Note factors that led to the success and failure of these teams.

8.4    With a team of three to five people, role play a sample of conversation(s) and team behaviors for each of the five team growth stages: Forming, Storming, Norming, Performing, Adjourning.

8.5    With a team of three to five people, role play a sample of conversation(s) and team behaviors for teams making decisions by consensus and three other decision-making methods discussed in the chapter.

8.6    Here are two acronyms, from unknown sources, for the word *team*: "Together Excellence, Apart Mediocrity," and "Together Everyone Achieves Much." These communicate the message that by working together we can achieve more than we could achieve working alone. Other quotes related to this are "The whole is greater than the sum of the parts," and "None of us is as smart as all of us." Do you believe that these statements are true? Why, or why not? Does it depend on the situation? Are some problems better suited for teams than others? Give examples.

8.7    Look up the word "synergy" in a college dictionary and give its basic meaning and its root meaning from the Greek. In an introductory biology textbook look up examples of "synergism." Write and briefly describe three examples of synergism from the world of biology. Discuss lessons that humans can learn from these biologically examples?

8.8    There are a number of historical examples of individuals who collaborated and accomplished what was probably impossible if they acted alone. Have you ever been part of a fruitful collaboration? If so, discuss it? Why was it fruitful? Have you been part of a failed collaboration? If so, why did it fail?

# 9. Engineering Design

## 9.1    What Is Engineering Design?

Engineers create things. Engineers build things. In order to successfully perform these tasks, engineers must be involved in *design* or in a *design process*. So what is engineering design and an engineering design process? Webster's dictionary defines design as "to create, fashion, execute, or construct according to plan." It defines a natural process as "a natural phenomenon marked by gradual changes that lead toward a particular result: a series of actions or operations conducing to an end; a continuous operation or treatment, especially in manufacture."

This chapter may seem closely related to Chapter 8, "Problem Solving." Indeed, there are many similar topics and themes in Chapter 8 that are important in Design. One of the critical areas which impacts Design is the issue of "external constraints" that can influence the outcome of the process. Appropriate references back to Chapter 8 will be noted throughout this chapter.

In the work of engineers, engineering design is an important and ongoing activity. Students who graduate with an engineering degree from an accredited program have had a significant amount of design experience as part of their education. The Accreditation Board for Engineering and Technology (ABET) has traditionally defined engineering design as follows:

> *Engineering design is the process of devising a system, component, or process to meet desired needs. It is a decision-making process ... in which the basic sciences and mathematics and engineering sciences are applied to convert resources optimally to meet a stated objective. Among the fundamental elements of the design process are the establishment of objectives and criteria, synthesis, analysis, construction, testing, and evaluation. ... it is essential to include a variety of realistic constraints, such as economic factors, safety, reliability, aesthetics, ethics, and social impact.*

ABET then describes the basic educational components for a major engineering design experience that must be included in an engineering program:

*Design cannot be taught in one course; it is an experience that must grow with the student's development.*

> Each educational program must include a meaningful, major engineering design experience that builds upon the fundamental concepts of mathematics, basic sciences, the humanities and social sciences, engineering topics, and communication skills. The scope of the design experience within a program should match the requirements of practice within that discipline. … all design work should not be done in isolation by individual students; team efforts are encouraged where appropriate. Design cannot be taught in one course; it is an experience that must grow with the student's development. A meaningful, major design experience means that, at some point when the student's academic development is nearly complete, there should be a design experience that both focuses the student's attention on professional practice and is drawn from past course work. Inevitably, this means a course, or a project, or a thesis that focuses upon design. "Meaningful" implies that the design experience is significant within the student's major and that it draws upon previous course work, but not necessarily upon every course taken by the student.

Many entering engineering students often confuse the "design process" with drafting or art-related work. To clarify this, ABET adds the following:

> Course work devoted to developing computer drafting skills may not be used to satisfy the engineering design requirement.

Most engineering programs concentrate their engineering design coursework in the later part of the student's program, thereby allowing the student to apply much of the prerequisite background in math, science and related fields to the various engineering problems. In most engineering programs, design courses comprise between 20% and 25% of the total curriculum.

## 9.2    The Design Process

It is important to realize that there is not one uniform approach to engineering design that is followed by practicing engineers. Some firms approach engineering design as a short, simple process with only a few steps, while others use a more complex, multi-step method with several stages. No matter what process is used, it is important to realize that engineering design is always continuous. The completion of one design, or the solution of one problem, may serve to open up opportunities for subsequent designs or modifications.

A design process is used whether a product is being developed for an ongoing manufacturing process, where thousands or even millions of a certain item will be produced, or for a one-time design, such as with the construction of a bridge, dam or highway exit ramp.

This chapter will present one design process (the 10-stage process) that is presently in use. This process will then be applied to an existing product to analyze how each of the 10 stages may have been applied in the product's realization.

The 10 stages that make up the process are as follows:

- Stage 1:  Identify the problem/product innovation
- Stage 2:  Define the working criteria/goals
- Stage 3:  Research and gather data
- Stage 4:  Brainstorm/generate creative ideas
- Stage 5:  Analyze potential solutions
- Stage 6:  Develop and test models
- Stage 7:  Make the decision
- Stage 8:  Communicate and specify
- Stage 9:  Implement and commercialize
- Stage 10: Perform post-implementation review and assessment

The actual process begins with the appointment of a project manager or team leader. This individual will be responsible for oversight of the entire process to ensure that certain key elements of each stage have been satisfied before the project moves on to the next stage. This person will also be responsible for recruiting team members of varying backgrounds and expertise for each of the stages. The team will include non-engineers as well as engineers. Some of the team members will be used throughout the process, while others will be needed only for certain parts of the process.

Engineers are problem-solvers; and the problems they solve are often identified as the specific needs and problems of customers. For example, a new prosthesis may be required to overcome a particular handicap, or increased gas mileage standards demand higher-efficiency engines, or a new computer program is needed to monitor a modified manufacturing process, or new safety devices are required to better protect infants in automobiles. Therefore, the first stage to problem solving in engineering design is to establish the actual problem clearly, and to identify sources of information to help understand the scope and nature of the problem.

The project manager will call upon the resources of various individuals to assist with these initial stages in the process. There may be a multitude of sources outside as well as within the organization that can assist with solving the problem. Many firms have a research and development unit made up of scientists and engineers who possess the training and expertise to assist with problem evaluation. In addition, sales engineers, who maintain consistent contact with outside individuals, can provide valuable input on problem identification. If the problem is one of expanding a current product line or modifying an existing system to improve it or make it better fit customer needs, management will likely play a role in the definition of the problem. Each group would be represented on the team at this stage.

External resources may come from trade shows, conferences, technical presentations, patent listings, and publications. Ideas generated from existing or prospective clients may also be valuable. Information gathered from external research agencies, private laboratories and government-funded foundations can be useful as well. In addition, an awareness of the competition and what products and services they are involved with can be beneficial in problem identification. At this early stage in the process, it is helpful to establish a preliminary, formal statement of the problem. For example, the need for a new automobile safety device for infants might evolve into a preliminary statement of the problem as follows: "Develop a better child restraint system that will protect children involved in automobile collisions."

It is important at this point to review the section of Chapter 8 which discusses the "real Problem." Before going ahead with the design process, the team must be certain that the correct issues and background have been thoroughly explored so the most appropriate solutions can be developed.

At this point, if the project manager is satisfied that all necessary issues have been resolved, the project will likely be passed on to the next stage where project criteria and goals will be developed.

## Stage 2: Define the Working Criteria and Goals

Once the problem has been identified, it is important for the team to be able to validate it throughout the design process. This requires the establishment of certain working criteria, or standards, which can be used in each of the 10 stages to measure possible solutions. The ultimate objective in this stage is to be able to establish preliminary goals which will act as the focal point of the team as it works through the process. The development of some working criteria provides a means to compare possible solutions. At this stage of the design process, everything is preliminary, so it is still possible for the team to modify the criteria if necessary.

Examples of working criteria could include answers to the following questions:

1. How much will it cost?
2. Will it be difficult to produce?
3. What will be the size, weight, strength?
4. What will it look like?
5. Will it be easy to use?
6. Will it be safe to use?
7. Are there any legal concerns?
8. Will it be reliable and durable?
9. Can it be recycled?
10. Is this what the customer truly wanted?
11. Will our customers want to purchase it?
12. Will customers want to purchase this version instead of a competitor's product?

Once some preliminary working criteria have been established, it is important to try to develop overall goals for the process. This is basically a statement of objectives which can be evaluated as the design process evolves. Using the example of new standards implemented to increase gas mileage and reduce emissions in automobiles, the goals for the design might be: "To develop an automobile engine which produces 25% less emissions while increasing gas mileage by 10%."

Having overall goals established for the project provides a means of evaluating, monitoring and changing, if necessary, the focus of the process as it evolves through the 10 stages. For the project manager, the criteria and goals become a "checkpoint" for assessing the progress to date, and will help them determine if the project is ready to move to the next stage, or if the process needs to return to Stage 1 for re-evaluation.

This stage is very important to all remaining stages of the design process. Having good, reliable background information is necessary for the team to begin exploring all relevant aspects of the problem. Consistent with the preliminary working criteria and the goals that have been established, the team members selected for this phase of the process must determine what types of information will be needed and the best sources of that information. For example, they may want to know such things as:

**Stage 3: Research and Gather Data**

1. What information has been published about the problem?
2. Is there a solution to the problem that may already be available?
3. If the answer to the above is yes, who is providing it?
4. What are the advantages of their solution?
5. What are the disadvantages of their solution?
6. What is the cost?
7. Is cost a significant issue?
8. What is the ratio of time spent compared to overall costs?
9. Are there legal issues to consider?
10. Are there environmental concerns which must be considered?

There are many possible resources the team members can utilize to assist them in their research. A good starting point may be a simple search using the Internet. This may provide useful sources of material that can serve as the focus for additional research. Other sources of reference information may come from:

1. libraries
2. professional associations (technical and non-technical)
3. trade journals and publications
4. newspapers and magazines
5. market assessment surveys
6. government publications
7. patent searches and listings (U.S. Patent Office: www.uspto.gov.)
8. technical salespersons and their reference catalogs
9. professional experts including engineers, professors and other scientists
10. the competition's product (How do they construct it? Disassemble their product and study it.)

Detailed files, notes, pictures, sketches, and other supporting materials will be maintained by the team to assist them as they proceed through the remaining stages of the design process. As new supporting information is discovered by the team, the material will be added as additional reference resources.

Depending on the type of information that has been collected, at this point it may be appropriate to review the established preliminary working criteria and the overall goals. To assist the project manager throughout the process, some modifications may need to be made. For example, it may be determined that one or more of the criteria may not actually apply to the problem. Likewise, new issues may surface which may necessitate the addition of new criteria or a modification of the goals. It is important to have these issues resolved before moving onto the next stage.

**Stage 4:**
**Brainstorm /**
**Generate Creative**
**Ideas**

The basic concept involved at this stage of the process is to creatively develop as many potential solutions to the problem as possible. The more ideas that can be generated, the better the likelihood of identifying a feasible solution to the problem. The project manager will want to gather a group of individuals from both technical and non-technical backgrounds to provide their unique perspective to the problem. This group may include engineers, scientists, technicians, shop workers, production staff, finance personnel, managers, computer specialists, and perhaps even a few clients.

A major method of generating multiple ideas to a problem is called *Creative Problem Solving*, using a technique called *brainstorming*. (Note: For a thorough discussion of creative problem solving and brainstorming strategies, please refer to Section 8.6 in Chapter 8.) With this method, a large group of individuals with varying backgrounds and training are brought together to attempt to solve a particular problem. Every idea that is spontaneously contributed from the group is recorded. The basic premise is that no idea is deemed too wild or illogical at this stage. No preliminary judgments are made about any member's idea, and no negative comments are allowed. The goal is to develop a long list of possible alternative solutions to the problem at hand. The group leader should be able to encourage participants to suggest random thoughts and ideas.

Some students may have had the opportunity to engage in brainstorming exercises. Brainstorming can be fun and highly stimulating to the creative process. For example: How many ways can you suggest to use a piece of string and a Styrofoam drinking cup? What could be created from a trash bag that contains some old magazines, tape and a ruler? Think of ways your student organization could earn extra funds for a field trip, etc.

It is conceivable that a brainstorming session would be continued on a second occasion to allow members time to consider other possible options. When the group reconvenes, members may have several new ideas or new perspectives for examining the problem.

At the conclusion of this stage, the group should have a long list of potential solutions to the problem. It is important to remember that at this stage of the process no idea has been eliminated from consideration. Each idea will be evaluated eventually, but it is important to keep all options open at this stage.

Once the project manager is satisfied that all possible solutions have been suggested, the project will likely be cleared for the next stage.

**Stage 5:**
**Analyze Potential**
**Solutions**

Note: Chapter 8.3 discussed the Analytic Method in detail. That information is directly related to this material in this stage. In the early part of the analysis stage (Phase I), it is important to try to narrow the ideas generated in the brainstorming stage to a few ideas which can be subjected to more sophisticated analysis techniques. This early narrowing could include:

- Examine the list and eliminate duplicates. As discussed earlier, it is important not to create limited categories, but only to eliminate repeated ideas. If two are similar, both should remain at this point.
- Allow the group to ask clarifying questions. This could help identify duplicate ideas.
- Ask the group to evaluate the ideas. The group members can vote for their top three ideas, and those that gain the most votes will be retained for more detailed analysis.

At this point, there will probably be a small number of ideas remaining. These can now be analyzed using more technical and perhaps time consuming analysis techniques (Phase II).

A variety of individuals should be involved at this stage, but the engineer will be of primary importance. The analysis stage requires the engineer's time and background. It is here that one's training in mathematics, science, and general engineering principles are extensively applied to evaluate the various potential solutions. Some of the techniques in this phase can be time consuming, but a thorough and accurate analysis is important before the project moves to the next stage in the process. For example, if the problem under consideration was the development of an automobile bumper that could withstand a 20-mile-per-hour crash into a fixed object barrier, several forms of analysis could be applied, including:

**Common sense**: Do the results seem reasonable when evaluated in a simple form? Does the solution seem to make sense compared to the goal?

**Economic analysis**: Are cost factors consistent with predicted outcomes?

**Analysis using basic engineering principles and laws**: Do each of the proposed solutions satisfy the laws of thermodynamics? Newton's laws of motion? The basic principles of the resistance of a conductor, as in Ohm's law, etc.?

**Estimation**: How does the performance measure up to the predicted outcomes? If the early prediction was that some of the possible bumper solutions would perform better than others, how did they perform against the estimate? (Review Chapter 8.7 for a more thorough discussion of Estimation.)

**Analysis of compatibility**: Each of the possible solutions and their related mathematical and scientific principles are compared to the working criteria to determine their degree of compatibility. For example: How would each bumper solution meet the criteria of being cost-effective? What would be the size, weight, and strength of each of the proposed solutions? How easy would each one be to produce?

**Computer analysis techniques**: One frequently used method is *finite element analysis*. With this method, a device is programmed on a computer and then numerically analyzed in segments. These segments are then compared mathematically to other segments of the concept. In the bumper crash example, the effects of the impact could be analyzed as a head-on crash and then compared to a 45-degree angle collision or a side-impact crash. As each section is analyzed, the "worst-case scenario" can be evaluated.

**Conservative Assumptions**: As discussed in Chapter 8.8, this technique can be most useful in analysis. It can build safeguards into the analysis until more data is generated.

After each of the working criteria have been examined and compared to the list of possible solutions, a process eliminates those that have not performed well in the various forms of analysis. It is expected at this point that only three to five options from the original list of prospective solutions will remain. These remaining options will then be reviewed by the project manager, who will likely authorize that the project be cleared for the next stage, assuming that these remaining options meet the working criteria and overall goals. If not, the process will need to be terminated, or return to an earlier stage to correct the problem.

## Stage 6: Develop and Test Models

Once each of the prospective solutions has been analyzed and the list of feasible options has been narrowed to a few possibilities, it is time to enter the phase where specific models will be developed and tested. Again, in this stage it is important to have a strong background in engineering coupled with experience and sound judgment. However, this stage will also involve team members who are computer specialists, shop workers, testing technicians, and data analysts.

There are several types of models which are commonly used by engineers and others in this stage. These include:

**Mathematical models**: Various conditions and properties can be mathematically related as functions and compared to one another. Often these models will be computerized to assist in visualizing the changing parameters in each of the models.

**Computer models**: There are various types of computer models which can be used. Typically these models allow the user to create on-screen images which can be analyzed prior to the construction of physical models. The most common computer modeling is referred to as CAD (Computer Aided Design) where models are designed and displayed as three-dimensional wire-frame drawings or as shaded and colored pictures. The CADKEY CD, located inside the back cover of this book, performs such computer modeling. The computer can also be used to control equipment that can generate solid models using techniques such as "stereo-lithography," where quick-hardening liquids are shaped into models or other forms of "rapid prototyping." These on-screen models, or the prototype models they produce, can then be used in the testing process.

**Scale models**: Typically, these smaller models have been built to simulate the proposed design but may not include all of the particular features or functions. These models are often called prototypes or mock-ups and are useful in helping engineers visualize the actual product. Such models may be used to depict dams, highways, bridges, new parts and components, or perhaps the entire body of a prototype automobile.

**Diagrams or graphs**: These models provide a tool for visualizing, on a computer or on paper, the basic functions or features of a particular part or product. These diagrams or graphs could be the electrical circuit components of an operating unit of the product, or a visualization of how the components eventually may be assembled.

Once the models have been developed and created, it is time to test each of them. Performing a variety of tests on each of the models allows for comparison and evaluation against the working criteria and the overall goals that have been established. In actuality, tests are done continually throughout a project including early models, prototypes and the testing of product quality as the product is manufactured or built. However, the results of the testing done in this stage establish the foundation for the decisions that will be made about the future of this project.

Examples of these tests include:

**Durability**: How long will the product run in testing before failure? If the product is a structure, what is its predicted life span?

**Ease of assembly**: How easily can it be constructed? How much labor will be required? What possible ergonomic concerns are there for the person operating the equipment or assembling the product?

**Reliability**: These tests are developed to characterize the reliability of the product over its life cycle, to simulate long-term use by the customer.

**Strength**: Under what forces or loads is a failure likely, and with what frequency is it likely to occur?

**Environmental**: Can the parts be recycled?

**Quality consistency**: Do tests show that product quality is consistent in the various stages? Is the design such that it can be consistently manufactured and assembled? What conditions need to be controlled during manufacture or construction to ensure quality?

**Safety:** Is it safe for consumer use?

**Consistency of testing**: This technique examines each of the testing methodologies against the various results obtained to determine the consistency among the various tests. In the bumper design example, the testing methodology might be different for a head-on impact test than for a 45-degree crash test in order to minimize testing inconsistencies.

All of this information will be evaluated by the project manager. If the manager is satisfied that these results consistently meet the working criteria and overall goal, the project will likely be cleared for the next stage.

At this stage, it is important for the team members to establish a means to compare and evaluate the results from the testing stage to determine which, if any, of the possible solutions will be implemented. The working criteria that have been used throughout the process are the critical factors which will be used to determine the advantages and disadvantages of each of the remaining potential solutions. One of the ways to evaluate the advantages and disadvantages of each of the proposed solutions is to develop a *decision table* to help the team visualize the merits of each. Typically, a decision table lists the working criteria in one column. A second column assigns a *weighted available point total* for each of the criteria. The team will need to determine the order of priority of each of the criteria. The third column then provides *performance scores* for each of the possible solutions. A sample Decision Table might look like the one that follows.

**Stage 7: Make the Decision**

Table 9.1 A Decision Table

| Working Criteria | Points Available | #1 | #2 | #3 |
|---|---|---|---|---|
| Cost | 20 | 10 | 15 | 18 |
| Production Difficulty | 15 | 8 | 12 | 14 |
| Size, Weight, Strength | 5 | 5 | 4 | 4 |
| Appearance | 10 | 7 | 6 | 8 |
| Convenient to use | 5 | 3 | 4 | 4 |
| Safety | 10 | 8 | 7 | 8 |
| Legal issues | 5 | 4 | 4 | 4 |
| Reliability/Durability | 15 | 7 | 9 | 11 |
| Recyclability | 5 | 4 | 3 | 4 |
| Customer Appeal | 10 | 7 | 8 | 9 |
| **Total** | **100** | **63** | **72** | **84** |

Based on this sample decision table, it appears that while none of the proposed solutions have scored near the "ideal" model, Solution #3 did perform better than the others. Using this information, the project manager and team leaders would make the final decision to "go" or "no-go" with this project.. They may decide to pursue Solution

#3, to begin a new process, or even to scrap the entire project. Assuming they decide to pursue Solution #3, the team would prepare the appropriate information for the next stage in the process.

## Stage 8: Communicate and Specify

Before a part, product or structure can be manufactured, there must be complete and thorough communication, reporting, and specification for all aspects of the item. Team member engineers, skilled craft workers, computer designers, production personnel, and other key individuals associated with the proposed project must work together at this stage to develop the appropriate materials. Such materials include detailed written reports, summaries of technical presentations and memos, relevant e-mails, diagrams, drawings and sketches, computer printouts, charts, graphs, and any other relevant and documented material. This information will be critical for those who will be involved in determining final approval for the project, as well as the group involved in the final implementation of the product. They must have total knowledge of all parts, processes, materials, facilities, components, equipment, machinery and systems that will be involved in the manufacturing or production of the product.

Communication is an important tool throughout the design process, but especially in this stage. If team members cannot adequately sell their ideas to the rest of the organization, and be able to appropriately describe the exact details and qualities of the product or process, then many good possible solutions might be ignored. At this stage of the process, it also may be important to create training materials, operating manuals, computer programs, or other relevant resources which can be used by the sales team, the legal staff and prospective clients and customers.

At this point, if the project manager is satisfied that all necessary materials have been adequately prepared and presented, the project will likely be passed on to the next stage, the implementation stage.

## Stage 9: Implement and Commercialize

The next-to-last stage of the design process is critical, as it represents the final opportunity for revision or termination of a project. At this point in the process, costs begin to escalate dramatically, so all serious issues should be resolved by this time.

In addition to the project manager and team leaders, there are a number of other individuals involved at this stage, representing a variety of backgrounds and areas of expertise. While engineers are a part of this stage, many of the activities here are performed by others. Some of those involved in this stage may include:

**Management and key supervisory personnel**. These individuals will make the ultimate decisions concerning the proposed project. They are concerned with the long-term goals and objectives of the organization, determining future policies and programs that support these goals, and making the economic and personnel decisions that affect the overall health of the organization.

**Technical representatives**. These may include skilled craft workers, technicians, drafters, computer designers, machine operators and others involved in manufacturing and production. This group will have primary responsibility for getting the product "out the door."

**Business representatives**. This group may consist of:
- **human resource personnel** if new individuals must be hired
- **financial people** to handle final budget details and financial analysis questions

- **purchasing personnel** who will procure the needed materials and supplies
- **marketing and advertising staff members** who will help promote the product
- **sales people** who will be involved in the actual selling and distribution of the product

**Attorneys and legal support staff**. The legal representatives who will handle a variety of legal issues including patent applications, insurance and risk protection analysis.

If all parties are in agreement that all criteria have been satisfied and the overall goal achieved, the actual production and commercialization will begin. However, there is one remaining stage where the project activities and processes will be evaluated and reviewed. (Note: Some stages of the process may be monitored differently depending on whether the project relates to a one-time product, i.e., a bridge or dam, or an ongoing manufacturing process.)

At this point, it is assumed that the project is in full production. The project manager, key supervisory personnel and team members who had significant input with the project are gathered together for a final project review and assessment. This stage involves the termination of the project team, since the product is now considered to be a regular product offering in the firm's overall product line. The product's performance is reviewed, including the latest data on production efficiency, quality control reports, sales, revenues, costs, expenditures, and profits. An assessment report is prepared which will detail the product's strengths and weaknesses, outline what has been learned from the overall process, and suggest ways that teams can improve the quality of the process in the future. This report will be used as reference for future project managers and teams to consult.

**Stage 10: Perform Post-Implementation Review and Assessment**

## 9.3   A Case Study Using the 10-Stage Process

In this section, a simple engineering problem will be discussed, analyzed and evaluated. The intent is to provide you with a better appreciation for the 10 stages involved in getting a product to market, and the team effort required for implementation.

The following example pertains to a product which is currently being developed by several manufacturers. However, the process described in this section is speculative, intended to illustrate an example of how the 10-stage process may be used in the development of this product.

The ABC Engineering Company is a small to mid-sized engineering subsidiary of the XYZ Research Corporation. XYZ is a world-renowned research and development facility in the areas of computer vision, real-time image processing and design, as well as other advanced video applications.

**Background**

For many years, ABC has been a supplier of products and services for the banking industry. ABC has dominated the field in such areas as security systems, bank cards and PIN identification systems. The parent company, XYZ, an expert in computer vision systems, realizes that the future of face-to-face banking transactions is rapidly changing. The security of electronic transactions is a major problem. The challenge for XYZ and its subsidiary ABC is to develop a new electronic security system to validate banking transactions. They decide to apply this technology to the ever-growing popularity of the automated teller machine (ATM) industry.

## Stage 1: Identify the Problem

A project manager is appointed by ABC to oversee the entire program intended to launch a new product and resolve a particular problem. The team from ABC first must identify the problem and attempt to generate sources of ideas to better understand the scope and nature of the problem. In these beginning stages of the design process, the team probably would include engineers, drafters and computer designers, as well as personnel from manufacturing, production, management, sales and finance. In this case, the group discusses the increasing role of automated teller machines in the banking industry. The current method of conducting transactions relies on an individual personal identification number (PIN) and a security access card. An individual approaches an ATM linked to a compatible financial institution, inserts an access card and enters a PIN number. The machine then allows the user to begin a transaction.

The team realizes that there have been increasing problems with fraudulent use of ATMs. It is possible that ABC's experience with the banking industry and the growth in its field of computer vision may provide an opportunity for them to expand their operations into this area.

Following this premise, the group begins to better understand the overall problem and tries to locate sources that will help them to narrow their focus on the perceived problem. They begin this process by working with sources from within ABC. This includes meetings with members of their research and development staff and input from sales and marketing personnel who are in constant communication with customers. In addition, managers and supervisors may have ideas regarding the expansion of current product offerings through modifications and upgrades. External sources of information include prospective banking industry customers, the Internet, libraries, trade shows, conferences and competition.

This information helps them to more narrowly define the problem. The group may develop a preliminary statement of the problem to guide them at this early stage of the process. Such a problem statement might be: "Eliminate fraudulent access to ATM accounts."

Once the project manager is satisfied that all necessary issues have been addressed at this stage, the project is passed on to the next stage.

## Stage 2: Define the Working Criteria or Goals

Using the preliminary problem statement as a guideline, the team develops a list of working design criteria which will allow them to evaluate various alternative solutions. Based on the information used to identify the problem, the group establishes the following preliminary working criteria:

**Cost**: The machine must be affordable to banks and other buyers of ATM machines.

**Reliability:** The accuracy of the new solution must be near 100%.

**Security**: The security for the cardholder must be fraud-proof.

**Technical feasibility**: The solution must be easily produced with current technology.

**Convenience to the customer**: The solution should be easy to use.

**Acceptance**: The solution must not be deemed too peculiar, or the public may not use it.

**Appearance**: The new solution must have a pleasing appearance.

**Environmental**: The new solution and its parts should be recyclable.

After thorough discussions and meetings, the team decides to re-examine their preliminary problem statement. Following additional discussion and review of the working criteria, the project goal becomes: Design and implement an alternate method of identifying a cardholder at an ATM, while focusing on reducing the acceptance of fraudulent entries.

Establishing this goal for the project provides a means of evaluating, monitoring and possibly changing the focus of the process as it evolves through the 10 stages. For the project manager, these criteria and the goal become a "checkpoint," or a means for assessing the project's progress periodically, and helps them determine when the project is ready to move to the next stage.

| | |
|---|---|
| Now that the goal and working criteria have been developed, it is time to collect all information relevant to the problem. Specifically, the team needs additional information regarding the current PIN method of identification and data on ATM fraud. They will also want to investigate new developments in the field, including information available on current products already on the market. They may look at this information to determine whether there are possibilities for redesign or modification of their current product lines, or to explore the feasibility of developing a new product from scratch. They also may want to explore current public opinion and available market research. | **Stage 3: Research and Gather Data** |

Sources for this information may include the Internet, libraries, newspapers, magazines, journals, trade publications and patent searches. An example of a Web search on ATMs might look like that of Figure 9.1 on the next page.

The team should maintain accurate notes, records and files for the information they collect. Sketches and diagrams of similar products or competitors' models may assist them when they begin to develop their solutions.

The project manager determines that all necessary conditions have been met and gives clearance for the project to move to the next stage.

| | |
|---|---|
| Using the information that has been collected and studied, the team decides to re-evaluate, prior to the brain-storming session, the list of preliminary working criteria and to assign a weighted percentage of importance to each of the factors. The team decides to eliminate the criteria of "appearance" and "environmental" from their list. The resulting working criteria and their respective assigned weights are as follows: | **Stage 4: Brainstorm/ Generate Creative Ideas** |

| | | |
|---|---|---|
| Cost | = | 10% |
| Reliability | = | 25% |
| Security | = | 10% |
| Feasibility | = | 15% |
| Convenience of use | = | 20% |
| Public acceptance | = | 20% |

The team then enters into a series of creative problem-solving/brainstorming sessions. At this stage the group consists of engineers, sales people, managers, supervisors, research and development staff members, production workers, computer specialists, and a few selected clients from the banking industry. Their assignment is to develop as many options as possible for uniquely identifying ATM customers. They are instructed not to make any judgments or negative comments on any of the proposed ideas. No ideas are to be considered ridiculous at this point in the process. The list they develop follows.

## Creative Brainstorming

Ways to uniquely identify ATM customers:

1. Modify the current use of PIN access codes: adjust current access methods to tighten security.

2. Individualized signature verification: An ATM attendant would match the card holder's signature with one written "on the spot" by the person standing at the machine.

3. Machine recognition signature verification: A scanner inside the ATM would match the cardholder's signature with one written "on the spot" by the person standing at the machine.

4. Voice recognition: The ATM would be equipped with a speaker phone so the person could speak their name and an internal computer could match the voice of the card holder to a recording on file.

5. Speech pattern recognition: Similar to above, but the person would speak a complete sentence or longer phrase that could be analyzed by voice characteristics and pattern of speech.

6. Fingerprint match: The ATM would have a fingerprint pad where the customer would press on the pad and a computer would compare the print to one on file.

7. Blood match: The ATM would have some type of device to prick the customer's finger for a small blood sample which could instantly be analyzed by an internal lab testing machine for DNA compatibility.

8. Hair sample match: The individual would submit a hair sample, with root attached, into the machine. As above, it could instantly be analyzed by an internal lab testing machine for DNA compatibility.

9. Eye iris match: The ATM machine would be equipped with a camera which would take a picture of the iris structure, and perform a comparison to a profile on file.

10. Breath analyzer match: The individual would breathe onto a sensitized glass plate and the results would be processed into an internal computer that would analyze a match with that of the card holder.

11. Dental identification match: The ATM machine would be equipped with a sanitized plate that the individual would bite onto and a computer sensor would read the patterns and analyze for a match.

Chapter 9 / Engineering Design—*A Case Study Using the 10-Stage Process*    **261**

At this point, the project manager is satisfied that a sufficient number of ideas have been uncovered. Authorization is given for the project to move to the next stage where all these potential solutions will be analyzed.

Figure 9.1  Sample Internet Web search results for ATM machines.

The team now begins to analyze each of the 11 ideas and compare them to the working criteria. In the early part of analysis (Phase I), it is important to try to narrow the ideas to a few which can be subjected to more sophisticated analysis techniques (Phase II). It is important to remember that there are many different analysis techniques that can be used by the team. The results of this analysis are as follows:

**Stage 5:
Analyze Potential
Solutions**

## Analysis of Alternative Ideas

1. *Modify the current use of PIN access codes:* adjust current access methods to tighten security.

  a) Similar to current solution

  b) Not reliable or secure

  c) Neutral cost

2. *Individualized signature verification:* An ATM attendant would match the card holder's signature with one written "on the spot" by the person standing at the machine.

  a) High cost

  b) Not customer-friendly

  c) Difficult to implement

3. *Machine recognition signature verification:* A scanner inside the ATM would match the cardholder's signature with one written "on the spot" by the person standing at the machine.

a) Technology is currently available
b) Public would probably accept this method
c) Higher cost to implement
d) Long-term reliability is untested

4. *Voice recognition:* The ATM would be equipped with a speaker phone so the person could speak their name and an internal computer could match the voice of the card holder to a recording on file.

a) Technology is currently available
b) Higher cost
c) Easy to use
d) Long-term reliability is untested

5. *Speech pattern recognition:* Similar to above, but the person would speak a complete sentence or longer phrase that could be analyzed by voice characteristics and pattern of speech.

a) Expensive
b) Technology not as well developed

6. *Fingerprint match:* The ATM would have a fingerprint pad where the customer would press on the pad and a computer would compare the print to one on file.

a) Technology is currently available
b) Moderate cost
c) Easy to use
d) Long-term reliability is good

7. *Blood match:* The ATM would have some type of device to prick the customer's finger for a small blood sample which could instantly be analyzed by an internal lab testing machine for DNA compatibility.

a) High cost
b) Good reliability
c) Completely unacceptable to the public

8. *Hair sample match:* The individual would submit a hair sample, with root attached, into the machine. As above, it could instantly be analyzed by an internal lab testing machine for DNA compatibility.

a) High cost
b) Good reliability
c) Public acceptance would be very poor

9. *Eye iris match:* The ATM machine would be equipped with a camera which would take a picture of the iris structure, and perform a comparison to a profile on file.

a) Technology is currently available
b) Moderate cost

c) Very easy to use

d) Long-term reliability is excellent

10. *Breath analyzer match:* The individual would breathe onto a sensitized glass plate and the results would be processed into an internal computer that would analyze a match with that of the card holder.

a) Technology not fully developed

b) Expensive to implement

c) Public acceptance would be poor

11. *Dental identification match:* The ATM machine would be equipped with a sanitized plate that the individual would bite onto and a computer sensor would read the patterns and analyze for a match.

a) Technology not developed

b) High cost

c) Public acceptance would be extremely poor

After each of the working criteria have been examined and compared to the list of possible solutions, those that have not performed well in the various forms of analysis will be eliminated. At this point it appears that only three to five options from the original list of prospective solutions will survive. These options are now reviewed by the project manager, who determines that they satisfy the working criteria and overall goal, so the project is authorized for the next stage.

At this stage, the team will take those proposed solutions that survived the analysis stage and develop models for testing and further evaluation. Based on the results from the analysis stage, the team has selected three proposals for further study: #4 (voice recognition), #6 (fingerprint match), and #9 (eye iris match).

**Stage 6:
Develop and
Test Models**

Initially, several different diagrams and sketches of each idea concept are developed and reviewed. The team has many different styles, options and units to consider. They decide to develop different models of each, including computer-generated models and actual prototypes constructed by craft workers in the model shop. These models are shown in the photographs in Figures 9.2 – 9.4.

Once the various idea concepts have been created into prototypes, the actual testing begins. Based on research and information that was collected in stage #3, and the information needed for comparison to the working criteria, the team decides to concentrate on several specific tests including technical feasibility, quality performance, reliability and ease of use. The results of the tests indicate that proposed solution #9 (eye iris identification) performs very well, as do the other two. Data should be generated for all three proposals. For Solution #9, the following data were generated:

**Technical feasibility test**: It was found that the current technology of specialty cameras and computer systems can identify a human by scanning the iris of the eye for over 400 identifying features and is able to match them to a large database.

**Quality performance test**: Eye scans from 512 individuals who had their iris features stored in the database system were taken. The program correctly identified all those who participated. Those whose records were not on file were correctly rejected.

**Reliability test**: The error rate on the eye scan reliability test was 1 in 131,578 cases. This is by far the lowest error rate among any other form of physical characteristic testing, including fingerprint and voice recognition. In addition, the iris scan provides long-term reliability, since the iris does not significantly change with age as do other body parts.

**Convenience of use test**: The technology is available so that a hidden camera can spot a person approaching and zoom in on the right iris. As the person draws closer, security access control can make a positive identification in a few seconds.

**Cost estimation**: While not a test per se, enough data have been collected to estimate the cost of this system at $5,000 per unit. While the initial estimate is high for machines which typically sell for $25,000 to $30,000, it is predicted that the cost will drop as units are mass produced.

Once all the test results have been gathered, all the information and data is evaluated by the project manager. It is determined that these results consistently meet the working criteria and overall goal, so the project is cleared for the next stage, where a decision will be made.

## Stage 7:
## Make the Decision

A final decision table is developed and the scores from the tests and other sources of information are recorded. In this particular case, the final table could look like this:

Table 9.2  The Decision Table

| Working Criteria | Points Available | #4 | #6 | #9 |
|---|---|---|---|---|
| Cost | 10 | 6 | 8 | 8 |
| Reliability | 25 | 20 | 22 | 24 |
| Security | 10 | 9 | 7 | 9 |
| Technical Feasibility | 15 | 15 | 14 | 15 |
| Convenience of Use | 20 | 15 | 16 | 18 |
| Public Acceptance | 20 | 16 | 17 | 18 |
| **Total** | **100** | **81** | **84** | **96** |

The results confirm that the eye scan identification system performed well. Using this information, the project manager decides to move ahead on solution #9 and the team prepares the appropriate information for the next stage in the process.

## Stage 8:
## Communicate
## and Specify

Team members including engineers, skilled craft workers, computer designers, production personnel, and other key individuals associated with the proposed project work together in this stage to develop the appropriate materials. The team members prepare the necessary detailed written reports, summaries of technical presentations and memos, relevant e-mails, diagrams, drawings and sketches, computer printouts, charts, graphs, and all other relevant and documented material on the eye scan identification system. They provide complete and detailed information on all parts, processes, materials, space, facilities, components, equipment, machinery and systems that will be involved in manufacturing the product. It is also important to create training materials, operating manuals, computer programs, and other relevant resources which can be used by the sales team, the legal staff and prospective clients and customers.

The project manager is now satisfied that all necessary materials have been adequately prepared and presented. Approval is given for the project to be passed on to the implementation stage.

At this point in time, the management of ABC has given approval for the eye scan identification project to move forward into production. Questions concerning overall costs, financial and labor commitments have been resolved. The finance department has developed a project budget. The purchasing department has begun the process of obtaining bids for the needed parts and supplies. The legal staff is resolving final legal issues including patent applications and copyright materials.

**Stage 9: Implement and Commercialize**

A group of engineers has been selected to oversee the production start-up, which includes obtaining necessary production space, facilities, equipment, personnel, and production timetables. Once most of this is in place, a pilot production process begins.

The marketing, advertising, and public relations staffs begin to contact prospective clients and customers to promote the eye scan ATM system. The sales people involved in the actual selling and distribution of the product are trained to work with clients and are given appropriate promotional literature including training and operations material.

All the parties are in agreement that the criteria have been satisfied and the overall goal achieved. The actual production and commercialization now begins. However, there is one remaining stage where the project activities and processes will be evaluated and reviewed.

At this stage, it is assumed that the eye scan ATM system is in full production and commercialization. The project manager, key supervisory personnel and team members who had any significant input to the project gather together for a final project review and assessment. This stage involves the termination of the project team, as the product is now considered to be a regular product offering in the firm's overall product line. The eye scan ATM system's performance is reviewed, including the latest data on production efficiency, quality control reports, sales, revenues, costs, expenditures and profits. An assessment report is prepared which will detail the product's strengths and weaknesses, a report detailing what was learned during the process, and suggestions for ways that future teams can improve the quality of the process. The working criteria are examined and suggestions for future groups will be included in this report. This report is used as another reference resource for future project managers and teams to consult.

**Stage 10: Perform Post-Implementation Review and Assessment**

Note: The Iris Identification System is currently being piloted in several British and Western U.S. locations. The information presented in this case study is a simulated application of the 10-stage design process.

Figure 9.2  Idea Concepts for Fingerprint Identification.

**Idea #1:** Fingerprint Scanner with Monitor, Keypad

**Idea #2:** Fingerprint Scanner, Small

**Idea #3:** Fingerprint Scanner, Larger

**Idea #4:** Fingerprint Scanner System

Figure 9.3  Idea Concepts for Eye Iris Identification.

Internal Eye Scan Unit

**Idea #1:**  ATM Wall Unit
with Internal Eye Scan

**Idea #2:**  ATM Free-Standing Unit
with Internal Eye Scan

**Idea #3:**  Remote Eye-Scan Unit

**Idea #4:**  ATM Top-Mounted Eye Scan Unit

Figure 9.4  Idea Concepts for Voice Recognition Identification.

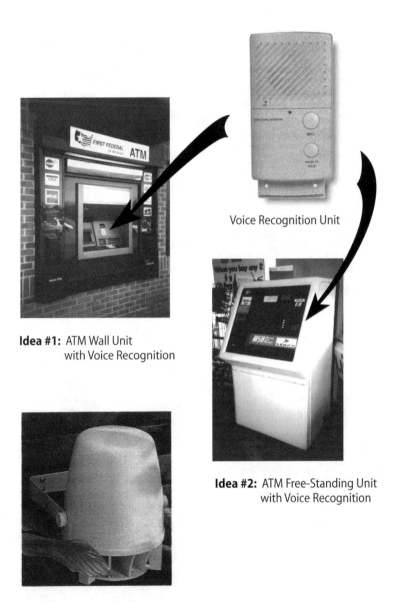

Voice Recognition Unit

**Idea #1:** ATM Wall Unit
with Voice Recognition

**Idea #2:** ATM Free-Standing Unit
with Voice Recognition

**Idea #3:** Ceiling or Wall-Mounted
Voice Recognition Unit

# 9.4 A Student Example of the 10-Stage Design Process
## by Amy Cole

**Background**

The ABC Engineering Company is a mid-sized subsidiary of the XYZ Research Corporation. The XYZ Research Corporation is famous for its progress in the area of athletic training research. ABC is a national supplier for the in-line skating industry. They realize that the industry is quite large, but is stabilizing. In order to keep the industry growing, they need to keep up with current consumer needs. They decide to make teenagers the focus of their studies. College students in particular use in-line skates on a regular basis as one of their primary means of transportation and, therefore, account for a large portion of the buying population.

**Stage 1: Identify the Problem**

A specific problem must be identified for the engineers and designers working on the project to have a clear understanding of what must be accomplished. This must be done in order to begin the design process. Those involved with the design project need to know specifically what to focus on. The team working on the early portion of the design process would probably involve engineers and drafters as well as manufacturers, producers, and marketers. The team must include members with a technical background since they are knowledgeable in the specifications and restrictions of the design. The other members of the team provide an alternate view on the project. They are responsible for answering questions pertaining to the appearance and customer satisfaction of the design. This group will be discussing, specifically, the operation and function of current in-line skate models. Currently, students use in-line skates as a means of getting to and from class, across campus, and even across town. When used in this way, students must deal with carrying their backpacks and school supplies, along with the hassle of carrying their shoes while skating. Students must also deal with the inconvenience of having their skates with them once they get to their destinations.

Realizing this, the team takes on the problem of inconvenience that comes along with the benefits of in-line skating. Considering the team's previous work with the XYZ Corporation, they should be reasonably well-suited to tackle this problem. Team members also should review old products to see if their uses could be modified to fit the project at hand. External sources could include prospective customers' needs, competition advances, the Internet, libraries, and conferences.

This should help the group to sharpen their focus on the problem and hopefully to create a preliminary problem statement. A good statement for such a problem might be: "Eliminate the inconvenience associated with in-line skates."

**Stage 2: Define the Working Criteria or Goals**

The team now must develop basic criteria for the design to follow. In this case, the alternative solutions should be evaluated using criteria similar to the following:

**Cost:** The solution must be affordable to the average customer. This includes college-age and younger students.

**Reliability:** The solution's reliability should be as close to 100% as possible.

**Weight:** The solution should be light since it will involve mobility.

**Customer Convenience:** The solution should be easily operated.

**Technical Feasibility:** The solution needs to be produced easily and quickly.

**Environmental:** The components of the solution should be recyclable.

**Appearance/Acceptance:** The solution should appeal to the customer and make them feel comfortable while using it.

At this point, the team should discuss and re-evaluate their preliminary problem statement. A goal statement should be created in order to keep track of the progress being made and to keep the designers on the right track. A goal statement for the current project might be: "Design and implement a device to eliminate the hassles associated with in-line skating, thereby making them more convenient and increasing their usage."

## Stage 3: Research and Gather Data

At this point in the design process, it is time for the team to collect information related to the problem. They need to research public opinion and market information to make sure they keep the customer in mind. They also need to research information regarding the basic operation of in-line skates and how they can use new developments as part of the solution. Researchers must decide whether to redesign an old project or start over with a completely new one.

Sources for research include libraries, the Internet, journal articles, patent searches, vendor catalogs, and trade publications. The group should produce diagrams of their preliminary ideas, along with notes and sketches of competitors' products. At this stage, it is left up to the project manager to decide when to move on to the next stage.

## Stage 4: Brainstorm/ Generate Creative Ideas

The team must stop now and revise their list of criteria before beginning the brainstorming session. They should review data collected and assign weighted percentages of importance to the factors they keep on the list. Assume that they decide to remove "environmental," but feel that the rest are necessary. Weights are assigned to the remaining criteria as follows:

| | |
|---|---|
| Cost | = 15% |
| Weight | = 15% |
| Feasibility | = 15% |
| Convenience of use | = 25% |
| Appearance/acceptance | = 10% |

The team is now prepared to start the brainstorming process. The group is composed of engineers, marketers, drafters, research and development members, managers, and select others who can offer a different perspective on the problem. They create as many alternatives to the current design as possible. At this stage, no negative comments or judgments should be made on any idea. The following list is produced.

### Creative Brainstorming

Ways to increase the convenience of in-line skates:

1. Make the wheels slide out of the track. This makes storage of the skates easier.
2. Make the wheels flip to the sides. This creates a flat base to walk on with the wheels on the side.
3. Make the whole base of the in-line skate detachable, leaving a shoe suitable for walking.
4. Design a bag especially for carrying the skates and other accessories. Shoes can be carried in the bag during use of the in-line skates.

5. Develop a cover to go over the wheels. This allows the skates to be walked on.
6. Make the wheel assembly strap on to the skater's everyday shoes.
7. Design a handle to make the in-line skates easier to carry.

At this point, it is left again to the project manager's discretion whether there is a sufficient number of ideas. In this case, the manager is comfortable with the results and allows the team to move ahead in the design process.

Having already brainstormed new ideas and revised the list of criteria, the team analyzes each of the ideas generated. While evaluating, the team should keep in mind that they will need to narrow the list for future steps. The analysis proceeds as follows:

**Stage 5: Analyze Potential Solutions**

1. *Removable wheels.* Wheels slide out of the track to make storage easier.
   a) Easy to use
   b) Similar to current design
   c) Convenience level not as high
2. *Flip wheels.* Wheels flip to one side to make a base for walking.
   a) Eliminates need for shoes
   b) Moderate cost
   c) Technology is available
3. *Detachable base.* The base detaches from the skate, leaving a normal shoe.
   a) Very convenient
   b) Moderate cost
   c) Easy to use
4. *Carrying case.* A bag is designed to carry in-line skates, shoes, and accessories.
   a) Low cost
   b) More hassle/inconvenience
   c) Reliable
5. *Wheel cover.* A cover is designed to fit over the wheels of the in-line skate.
   a) Convenient
   b) Low cost
   c) Moderately reliable
6. *Strap on wheels.* The wheels of the skate strap on to the skaters shoe.
   a) Not reliable
   b) Moderate cost
   c) Not as safe
7. *In-line skate handle.* The in-line skates have a handle to make carrying them easier.
   a) Low cost
   b) Reliable
   c) Hassle involved

The analysis is compared to the working criteria. The team eliminates solutions that do not meet the standards and retain three to four solutions to develop further. The project manager reviews the remaining solutions and gives the team permission to continue.

**Stage 6: Develop and Test Models**

At this point in time, the team must develop models for the remaining solutions. The models will undergo testing and more extensive evaluation. The team, after reviewing the analysis and criteria, decides to continue the process with option #2 (wheel cover), #3 (detachable base), #5 (wheel cover), and #7 (in-line skate handle).

Figure 9.5  Option #2: Flip wheels.

Figure 9.6  Option #3: Detachable base.

Figure 9.7  Option #5: Wheel cover.

Figure 9.8: Option #7: Handle.

Mathematical and computer models or diagrams should be produced for each option. The team can choose from a number of programs on the computer which offer many different options for creating the sketches. Actual working prototypes should be developed for each solution.

Once the prototypes have been developed, the team of engineers and technicians can initiate the testing process. The team performs specific tests relating to durability, reliability, and quality. The project manager reviews the data collected from the tests and gives his clearance for the next stage.

**Stage 7:
Make the Decision**

To aid the team in the decision-making process, a final decision table is made. For this case, the table could be as follows:

| Working Criteria | Points Available | #2 | #3 | #5 | #7 |
|---|---|---|---|---|---|
| Cost | 20 | 15 | 16 | 18 | 18 |
| Reliability | 15 | 11 | 15 | 7 | 15 |
| Weight | 15 | 12 | 14 | 14 | 11 |
| Convenience | 25 | 18 | 25 | 15 | 18 |
| Technical Feasibility | 15 | 15 | 15 | 15 | 15 |
| Appearance | 10 | 9 | 10 | 5 | 5 |
| **Totals** | 100 | 80 | 95 | 73 | 82 |

The tests show that solution #3, the detachable base model, is the preferred solution. All previous information confirms this and the project manager agrees to use solution #3 to finish the design process. The team organizes their data and prepares to move on in the design process.

Communication plays a critical role in engineering and really comes into play during this stage of the design process. At this point, the team consists mostly of engineers, design technicians, and production members. This is the time when all documented material is presented. Typical presentations include written reports, relevant e-mails and memos, diagrams, computer printouts, charts, graphs, and summaries of technical presentations. Everything involved in the production of in-line skates including parts, processes, materials, facilities, and machinery is described in detail in the presentations. It also is important that the team communicate how the product works through instruction manuals and videos and training materials. Again, the project manager clears the project to continue to the next phase.

**Stage 8: Communicate and Specify**

The detachable-base in-line skates now have undergone all of the necessary steps to begin production and implementation. At this point, the team usually does not consist of as many engineers as in previous stages. The team usually includes members involved in the managerial, technical, business, and legal aspects of the product. The details, such as the budget and price, costs, and labor commitments, have been decided. The technical members—the craft workers, CAD operators, machinists, and technicians—handle the operations and production of the in-line skates. Select engineers must arrange production space, facilities, and start-up equipment.

**Stage 9: Implement and Commercialize**

The business portion of the team includes human resources personnel, finance, purchasing, marketing, advertising, sales, and technical/customer support. It is their responsibility to use the information collected and to make the in-line skates appealing to the customer. They need to figure out how to sell the product.

Everyone works together to ensure that all of the criteria are met and production runs smoothly. Production now begins, leaving only a final stage remaining in the design process.

At this point, the detachable-base in-line skates are in full production. All of the members of the team must meet one last time to review the data and the performance of the in-line skates. The production efficiency, sales, revenues, costs, expenditures, profits, and quality control reports should be reviewed. This meeting will be the last time the group members meet as a team for this project, as it is now considered a part of the firm's production line. The group produces a final report which can be referenced by future project teams.

**Stage 10: Perform Post-Implementation Review and Assessment**

Internet contact: www.rollerblade.com, Benneton Sports System, Rollerblade, Inc., 2000.

# Exercises and Activities

9.1    Disassemble one of the devices listed below, and put it back together. Sketch a diagram of all the parts and illustrate how they fit together to make the device operate. List at least three ways you think the design could be improved. Choose one of the following devices: a) flashlight, b) lawn sprinkler, c) sink faucet, d) stapler, e) toaster, f) computer mouse.

9.2    Prepare a list of questions that should be resolved in identifying each of the following problems:

   a)  Develop an improved manual gearshift for a mountain bike.
   b)  Develop a better braking system for in-line skates.
   c)  Develop a recliner chair that incorporates six built-in surround sound speakers in the unit.
   d)  Develop a hands-free flashlight.
   e)  Develop a theft-proof bicycle lock.
   f)  Develop a baseball cap with wireless stereo headphones.
   g)  Develop a secure storage area on a bike for a helmet.
   h)  Develop a new idea to reduce overcrowded parking on campus.
   i)  Develop a hands-free cellphone that can be used when doing activities.
   j)  Develop a shockproof cellphone carrying case.

9.3    Develop a list of working criteria that could be used in deciding whether to:

   a)  Accept a co-op job offer from Company A or Company B.
   b)  Study overseas for the Fall semester or remain on campus.
   c)  Buy a new car or repair your old one.
   d)  Purchase a desktop or laptop computer.
   e)  Change your major or remain in engineering.
   f)  Purchase a new computer or upgrade your current model.
   g)  Live on the dormitory or lease an off campus apartment next year.
   h)  Pledge a fraternity/sorority.

9.4    Identify five product, structure, or system designs you think can be improved. Pick one and write a preliminary problem statement for the engineering design process.

9.5    Using an item from your list in Exercise 9.4, develop a list of reference materials that would be used in developing possible solutions to the problem. Provide specific examples.

9.6    Get together with three other classmates and brainstorm at least 30 ways to use one of the following objects:

   a)  Two-foot length of string
   b)  Ping-pong ball
   c)  Three plastic pop bottles
   d)  Page of notebook paper and a 2-inch piece of tape

e) Old telephone directory

f) Trash bag

g) Deck of playing cards

h) Yo-yo

i) Frisbee

j) Metal coat hanger

k) Empty plastic milk container

l) 15 paper clips

m) 2 tubes from paper towel rolls and 2 pages of newspaper

n) deflated balloon

o) bag of marshmallows

p) newspaper and 12″ of masking tape

9.7 Using the information you developed in Exercise 9.4, prepare a decision table for three possible alternative solutions.

9.8 Read a current newspaper or magazine article (possibly on the Web) that describes and discusses a new product, device, or system. Prepare a three- to four-page essay which analyzes and explains, in detail, each of the stages of the design process that were probably involved in the engineering of this product or device. Apply each stage of the 10-stage design process discussed in this chapter to your product or device, and use specific examples to support your statements. Make sure you list your sources in the body of the paper or in a bibliography.

9.9 Put together a team of five students from your class. Using the 10-stage design process, have your team develop a portable, garage-like covering for a bicycle that can be stored somewhere on the bike when not in use.

9.10 Put together a team of four or five students from your class. Using the 10-stage design process, build the tallest possible tower that can support a 12-oz can of soda/pop. **Materials:** one roll of masking tape, one package of straws, and one can of soda/pop. **Constraints:** Your tower must be free standing. The base must fit on a plate. Your tower must stand for 5 seconds under load for measurement. After the design is complete, 30 minutes will be allowed for tower construction.

9.11 Put together a team of four or five students from your class. Using the 10-stage design process, build a bridge that spans three feet between supports that contact the ground. **Materials:** one roll of duct tape, ten feet of rope, a 40 inch by 75 inch piece of cardboard, and 14 cardboard slats. After the design is complete, 40 minutes will be allowed for tower construction and testing. The test: one of your team members must walk heel-to-toe across the bridge.

9.12 Put together a team of five students from your class. Using the 10-Stage design process, have your team develop a carrying mechanism for rollerblades and/or your shoes when one is not in use.

# 10. Technical Communication

## 10.1 Introduction

In this chapter we will discuss the importance of written communication in your professional life, the audiences that will be reading what you write, the methodology you should incorporate, the structure of proper writing, and the grammatical concerns that relate to what you write. We will also touch on verbal communication skills. Though we will merely introduce the subject in this chapter, you should delve into the topic in greater depth on your own in order to master all that is required to be an excellent communicator.

It is always interesting to hear aspiring young engineers say, "I had all that English stuff in high school and I never want to deal with it again. Engineers are math and science types. They're not English majors!" But effective communication is essential in engineering. Information must be presented clearly and concisely on a daily basis.

Maybe you don't think of yourself as a "writer." And perhaps you think that writing engineering-related reports won't require much in the way of literary skills, since they're so straightforward. You'll find that the writing demands for an engineer are more similar to those of a writer than you may think. You'll realize that technical writing can require as much creativity, skill, and discipline as the rest of engineering.

## 10.2 Communication in the Life of the Engineer

To a great extent, your writing will help you make your mark as an engineer, for better or for worse. Your writing will either aid or detract from your success.

As an engineer, you will be required to communicate in a variety of ways. You will need to write, speak, think, listen, use, and understand nonverbal communication. There are many books on the market that deal at length with communication. This chapter is meant to provide you with certain highlights that you should be aware of as you begin your engineering education—information that will lay the groundwork for tasks you will be assigned in the years ahead. So let's get started. The issues that we will address are:

- the reasons for communication
- the audiences you will address
- the power of the language you use
- technical communication
- common grammatical concerns
- formats of written communication
- types of communication

## 10.3    Why Do We Communicate?

What is the purpose of communication? Well, there are many purposes, actually. It might surprise you to consider that we are often the most important readers of our own writing. What we write should be valuable to us as a future resource. If you keep that in mind, written communication will be easier and more rewarding.

Of course there are a number of additional reasons for communicating with excellence: obtaining good grades, creating effective reports, convincing others of your perspective—both on a professional and personal basis. Each will require that you assume a different mindset.

Effective communication:

- transfers vital information
- provides a basis for judging your knowledge
- conveys your interest and competence
- increases the knowledge of others
- identifies to you the gaps in your information
- allows you to carry out desired changes

As you consider your own communication, there are a number of things to keep in mind:

*Is the language you are using suited to your audience?*

- Do you know how to communicate effectively to your audience?
- Do you need to expand your communication skills to communicate more effectively to your audience?

*Is what you are saying clear to your readers?*

- Will your readers draw the same conclusions you draw?
- Do you use words, phrases, or ideas that could lead to multiple interpretations?

*Do you know your audience?*

- Does your audience have any unique preferences?
- Do they have any hidden expectations?
- How do they feel about the things you are communicating?

*Do you know how your audience perceives things?*

- Do all engineers see things the same way?
- If you are addressing an audience of non-engineers, how might they interpret things differently?

# 10.4 The Audience

Before you begin to write, ask yourself, "Who is going to read this, anyway?" You must consider the needs and perspectives of your audience. Their level of education or experience may dictate the language and style of your writing. They may use words or expressions that are unique to their part of the country or their part of the world.

Always remember that technical material needs to be clear and precise. You cannot simply write as you please and hope that your readers will "figure it on their own." It's as if you are the screen writer and science requires a very clear script. You must prepare your readers, give the required information to help them understand, and repeat information as necessary to bring them to full understanding.

Below is a series of fictitious letters. They illustrate the need to consider the audience to whom you are communicating and the necessity of targeting your communication to fit your particular audience's needs. In this humorous instance, the writer has failed to communicate effectively due to an apparent lack of concern for his audience. Let's take a look.

**Miscommunication**

**TO: U.S. BUREAU OF STANDARDS**

DEAR SIRS:

IS IT OKAY TO USE HYDROCHLORIC ACID TO CLEAN DRAINPIPES?

SAM, THE PLUMBER

---

**TO: SAM, THE PLUMBER**

SIR:

THE EFFICACY OF HYDROCHLORIC ACID IS INDISPUTABLE, BUT THE CHLORINE IS INCOMPATIBLE WITH METALLIC PERMANENCE.

U.S. BUREAU OF STANDARDS

---

**TO: U.S. BUREAU OF STANDARDS**

DEAR SIRS:

THANK YOU FOR LETTING ME KNOW THAT IT IS OK TO USE HYDROCHLORIC ACID TO CLEAN DRAINPIPES.

SAM, THE PLUMBER

---

**TO: SAM T. PLUMBER**

SIR:

WE CANNOT ASSUME RESPONSIBILITY FOR THE PRODUCTION OF TOXIC NOXIOUS RESIDUES ARISING FROM THE USE OF HYDROCHLORIC ACID, AND SUGGEST THAT YOU PURSUE AN ALTERNATE PROCEDURE.

U.S. BUREAU OF STANDARDS

---

**TO: U.S. BUREAU OF STANDARDS**

DEAR SIRS:

GLAD TO KNOW THAT YOU AGREE WITH MY IDEA.

SAM, THE PLUMBER

---

**TO: SAM T. PLUMBER**

SAM,

THE SPONTANEOUS AND ENERGY-RELEASING INTERACTION OF ATMOSPHERIC CONSTITUENTS WITH THE MOLECULAR STRUCTURE OF THE MATERIAL CAUSES TO BE PRODUCED A TIME-VARYING CONDITION OF UNSTABLE STRUCTURAL EQUILIBRIUM DISTRIBUTED SPATIALLY OVER THE CYLINDRICAL SURFACE.

**OR, MORE TO THE POINT:**

DON'T USE HYDROCHLORIC ACID. IT EATS THE HECK OUT OF PIPES!

The story of "Sam the Plumber" shows how easy it is to fail to identify properly with your audience. As an engineer, it will be critical that you present your material in a way that your audience will be able to understand clearly. The issue becomes further complicated when multiple, differing audiences must read the same written communication. For instance, a report might be passed laterally from your technical manager over to the Marketing Department and on up to the CEO. So you must know who will be reading your writing and modify it accordingly. The essential information may not change, but the way in which you present it might.

# 10.5  The Power of Language

Keep in mind here the importance of technical terms. Terminology is not the same as jargon. It can be very sharp, precise, understandable and proper. It is nice to simply say "glitches" when there are problems in a process, but a proper technical term would explain clearly what those "glitches" are. Jargon can create severe problems whenever readers do not fully understand what those words mean. The definitions may be close to being right, but this is not acceptable in the technical world.

It is probably a good time to point out that each of us belongs to one or more groups. We might be scientists, truck drivers, morticians, students, etc. Over time, we may move in and out of any number of such groups. Within those units, language becomes specialized and common to all in the group. Outside the group, the language is sometimes only gibberish. In the group, jargon is fine; but outside the group, use language that universally communicates your message. However, even within a group you must be precise when relating vital information—assumptions often backfire, and jargon always assumes.

When you have a choice between writing *precisely what you want to say* or something more convoluted in an attempt to sound *impressive*, try to choose the more straight-forward approach whenever you can. This will minimize the chance for confusion and maximize comprehension. It will also allow a wider audience to understand what you are trying to communicate. Note the following examples where a synonym could replace a more complex word:

**Clarity**

| | |
|---|---|
| altercation —> dispute | utilize —> use |
| obfuscate —> confuse | facile —> easy |
| innocuous —> harmless | oblique —> indirect |
| ulterior —> hidden | elucidate —> clarify |

However, you should realize that while synonyms have basically the same meaning, there often are subtle differences between them, even if only slight. Be sure to use the best word to communicate your thoughts accurately. Sometimes this will require that you choose a more "sophisticated-sounding" word over a more common synonym.

Just for fun, several humorous examples of convoluted speech are listed below. Can you figure out the meaning of each statement?

1.) Slumbering canines are best left in a recumbent position.
2.) It has been posited that a high degree of curiosity proved lethal to a feline.
3.) There is a large body of experimental evidence which clearly in-dicates that members of the genus *Mus* tend to engage in recre-ational activity while the feline is remote from the locale.
4.) Even with the most sophisticated experimental protocol, it is ex-ceedingly unlikely that you can instill in a superannuated canine the capacity to perform novel feats of legerdemain.

Here are the same statements, only simplified for clarity:

1.) Let sleeping dogs lie.
2.) Curiosity killed the cat.
3.) While the cat's away, the mice will play.
4.) You can't teach an old dog new tricks.

Engineers should avoid the use of clichés in their technical writing. Clichés, when they were first coined, conveyed a clear message. Over time, with changes in language and lifestyle, the meaning of many clichés have become unclear. Here are some familiar examples of what you should avoid:

**Clichés**

| | |
|---|---|
| a can of worms | at loggerheads |
| acid test | by the same token |
| bated breath | avoid like the plague |
| in the nick of time | |

## Redundancy

You may have a notion about what the above expressions mean, but your audience might have a different interpretation, especially if your audience includes those from another country.

There are unnecessary, redundant phrases that are also best avoided. Several examples follow, along with their simplified form:

| | | |
|---|---|---|
| red in color | → | red |
| on a monthly basis | → | monthly |
| at that point in time | → | at that time |
| due to the reasons that | → | due to |
| during the course of | → | during |
| very unusual | → | unusual |
| round in shape | → | round |

Technical writing needs to be clear and concise. Do your best to say what you need to say as briefly as possible while fully communicating all you need to communicate.

## Language Concerns

Jargon is informal language that is used by a particular group of people. It is language that is peculiar to that group and may not be readily understood by any other group or individual. It is not advisable to use this type of language in groups that are unfamiliar with it, because the level of understanding will drop markedly.

## Sexist Language: Pronouns and Gender

Since so many jobs that were traditionally held by men are now held both by men and women, use discretion with generic references to positions and roles.

### Examples

1.) Since English has no gender-neutral singular pronouns, masculine singular pronouns such as "**he**" are traditionally used, and technically correct, for use in gender-neutral contexts. However, since this may appear potentially insensitive to some, you may want to minimize the use of masculine pronouns whenever possible. Note the following example:

We need to hire an additional engineer. He will be in charge of plant operations.

(Since the engineer could be a man or a woman, the use of the pronoun "**he**" may not be preferable.) One option is to rephrase the statement.

We need to hire an additional engineer who will be in charge of plant operations.

2.) Sometimes it's possible to use <u>plural</u> "gender-neutral" pronouns such as "**they**" and "**we**." This is another way to avoid the situation.

**Each engineer must make his own decision.**
  becomes
**Engineers must make their own decisions.**

3.) On those occasions when using a plural pronoun is not an option—and often it's not—some writers simply alternate the use of "he" and "she." Though this is technically improper, this convention is growing in popularity.

4.) You might choose words that have no gender connotation:

**firefighter** instead of **fireman**

**hours of labor** instead of **man-hours**

**chairperson** instead of **chairman**

5.) Avoid extreme, convoluted efforts to remain gender-neutral which distract from the clarity of your communication.

**s/he** or **he/she**

It is preferable to write "**he or she**" instead. Or, as mentioned above, use plurals, change the wording.

# 10.6   Technical Communication

Because your engineering career most likely will involve a great deal of technical communication, it is vital that you understand the qualities of effective technical writing. In general, technical writing:

- identifies the main premise early on
- is clear, objective, and economical
- follows a specific format (i.e., memos, lab reports, abstracts)
- takes a problem-solving approach
- involves a specialized vocabulary
- often incorporates signs, symbols, formulas, graphs, and tables
- documents completed work

Let's look at some of the unique aspects of technical writing:

*Content*—Technical writing is typically based on research. Therefore, it must be factual and objective.

*Methodical Construction*—The way technical engineering material is written is specific and consistent throughout the profession. In general, paragraphs begin with a topical sentence, which is followed by supporting statements. Technical writers like to present the most essential information first and then present supporting and clarifying information. In this way there is no mystery as to the purpose of the communication.

*Clarity*—Technical communications must be clear and complete.

*Objectivity*—Technical writing should be unemotional and unbiased. It should not convey the feelings of the writer or others.

*Importance in Workplace*—Technical documents provide the information necessary for the proper functioning of an engineering-related business. Such documents supply information related to design, construction, research, and operations. Improper or incomplete technical communication can lead to design flaws, hazardous developments, waste of resources, production delays, loss of competitive advantage, and other undesirable outcomes. Such failure to effectively communicate can have an adverse effect on one's career.

**Note...** This chapter offers the briefest possible tips about a very important area of engineering and science. It is likely that in the future you will spend a great deal of your time in the preparation of and the reading of technical reports. Your skill in this area will in large part determine your advancement in your chosen profession.

# 10.7  Grammar and Technical Writing

Technical writing requires that you draw upon knowledge you attained in grade school but may have thought unimportant to your engineering career. We're talking about those grammar exercises you worked on so diligently when you were young—the ones you hoped you could forget about as you got older.

Mastery of good, basic grammar is important to successful engineers, as it is to all professionals. The use of proper grammar not only improves your writing's clarity, it also helps eliminate distractions to your readers and increases their confidence in what you have to say. It shows that you care about what you write, that you take your job and your responsibilities seriously.

In this section we will discuss several grammatical topics which prove to be a challenge for many young engineers. Perhaps you already are a strong writer. In that case, some of our discussion may be overly simplistic for you, but a brief review of some of these topics still may be helpful.

**Subjects and Verbs**

Let's start with the basics—the construction of complete sentences using subjects and verbs. This probably was one of the first areas you focused on while studying grammar in grade school. For starters, you must make sure that subjects which are singular are coupled with the singular verb form, and plural subjects with plural verb forms.

Here are some helpful grammar rules regarding subject/verb agreement:

1.) A verb must agree with its subject, not necessarily with any words that come between them.

   incorrect—The first item in each of the chapters were interesting.
   correct—The first item in each of the chapters was interesting.

2.) Subjects joined by "and" usually take a plural verb, unless "and" introduces a clause.

   *incorrect*—Lead and zinc is used in many ways in the plant.
   *correct*—Lead and zinc are used in many ways in the plant.
   *correct*—Zinc is a large part of the manufacturing process, and titanium is only used in rare instances.

3.) Singular subjects joined by "or" or "nor" take a singular verb.

   *correct*—Either the manager or the department head is required to approve the purchase.

4.) When two subjects differing in number are connected by "or" or nor," and one is plural, it should be placed second and should be in agreement with the verb in number.

   *incorrect*—Neither the analyzer nor the test tubes was contained in the initial laboratory.
   *correct*—Neither the analyzer nor the test tubes were contained in the initial laboratory.

5.) "Numberless" words, which can be both plural and singular, which are used as subjects require the singular form of a verb. Example: *equipment*.

   *incorrect*—The equipment were assembled by the students.
   *correct*—The equipment was assembled by the students.

6.) The verb agrees with the subject even though in many cases the subject will follow the verb.

7.) When using a linking verb (*is, are, was, were*, and forms of *to be*), the subject is the noun that precedes the verb, not a noun which follows the verb.

     *incorrect*—The abstract are printed in two parts.

     *correct*—The abstract is printed in two parts.

Read the following examples and make sure that you understand why the verb-subject agreement was handled as it was.

1.) The lab experiment demonstrates the principles we learned in class.
2.) I think labs are more enjoyable than lectures.
3.) Each of the labs, including Fluids, is going to be difficult.
4.) Twelve vials were used in the experiment.
5.) Most test tubes meet minimum safety standards.
6.) Plutonium and tritium have been used in the study.
7.) Sodium or magnesium can be used in the study.
8.) Nitrogen, as well as oxygen, was used in the study.
9.) Unfortunately, Jesse mixed the chemicals that were used in the previous experiment.
10.) Teripherium is both a gas and a solid.

## Activities

1.) Explain why the previous sentences are correct. Don't give rules. Instead, demonstrate an understanding of their construction.

2.) Select ten sentences from your readings and explain why they are constructed as they are.

## Paragraph Construction

Technical writing should be organized and straightforward. The construction of your paragraphs should be clear and easy to navigate, without the digressions and blind alleys you may encounter in other writing, such as fiction.

Therefore, paragraphs should possess:

1.) logical, ordered sentence arrangement
   a. progressing from general to specific information
   b. developing from facts
   c. with premises followed by supporting information
2.) only *meaningful* repetition of words
3.) parallel structure
4.) transitional words (since, however, in which, therefore, thus)

Present your ideas simply. Complexity breeds confusion and wastes time—both for the reader and the writer—and the subject matter of technical writing is often complex enough as it is, without being presented in an awkward, indirect fashion. Your purpose is to lead your audience through the text in such a way that your ideas will be communicated clearly, your arguments will be accepted fully, or your instructions will be followed with precision. If done properly, even the most challenging, complex subject matter can be readily understood.

Here are a few suggestions:

1.)    Select the proper format for the particular task at hand.
2.)    Develop an appropriate outline.
3.)    Edit, edit, and edit again!
4.)    Do your own "weeding." Get rid of things that may hinder the reader's understanding of the text.
5.)    Simplify wherever possible. (Remember, though, that this is a general rule. Sometimes the most appropriate word may not be a simple word.)

## Topic Sentences and Supporting Statements

A paragraph should address a single topic and have a single primary focus accompanied by supporting details. The typical paragraph should have roughly five sentences on average, all relating to one particular idea. Within the paragraph, the main focus is usually presented in what we call a *topic sentence*. The topic sentence can appear anywhere within the paragraph. However, the topic may be conveyed more subtly instead, as a general idea which the reader is able to discern. It is your job to make everything within a paragraph relate to one central idea. If you find yourself addressing more than one idea, that usually means you need another paragraph.

Here's an example of appropriate paragraph structure:

> Turbulent flow in a pipe transporting water was studied. Enhanced turbulence was created using a valve to suddenly reduce the pipe diameter from 1" to 1/2". The pressure in the flow entering the valve was not increased, but the pressure exiting the valve decreased by a factor of 2. It was decided that the enhanced turbulence would act in a similar manner for all valve reductions with these dimensions.

In technical writing, topic sentences tend to be presented at the beginning of a paragraph, as in the paragraph above. It is easy to identify the central idea as "turbulence created in a water pipe." Each sentence that follows the topic sentence relates to the idea of turbulence.

In general, the first paragraph will set the stage for the rest of your text. Subsequent paragraphs usually expand on the topic presented in the initial paragraph. For instance, the sample paragraph presented above could lead to additional paragraphs which discuss valve types, further inlet valve restrictions, flow speeds, and the water pipe itself. However, be sure that the paragraphs you write stay on the general topic of the paper as it relates to the paper's intended purpose. Avoid rambling into areas that are unrelated to that purpose. The scope of the topic and the writing's purpose determine the proper length of the paper.

## Active and Passive Voice

There always will be disagreement among writers about when and how to use active voice and passive voice. Here's a simple explanation of the two:

*Active voice expressly focuses on the "doer" of the action.*

Note the following examples:

**Sound waves reverberated off the walls.**

It is easy to see that the writer wants us to understand that sound waves are the main focus of the sentence, the entity engaged in the "doing."

Use of the passive voice, on the other hand, makes the object of the "doing" the subject.

**The walls were struck by the reverberating sound waves.**

Again, in this instance it seems that the writer wants us to focus on the walls as the receivers of the sound, not the sound itself. The main thing to remember is that with the active voice there is a distinct "doer" of the action. In the passive voice the subject is the "receiver" of the action.

Using the active voice helps make your writing more direct and tends to add clarity. The doers of action are easy to identify. It puts the doers of all actions at the forefront in the reader's mind. The passive voice, on the other hand, may sometimes results in confusion, since the doer of action is presented more indirectly.

## Pronouns

Webster's defines pronouns as "a class of words that function as substitutes for nouns or noun phrases and denote persons or things asked for, previously designated, or understood from the context."

Commonly used pronouns include:

**I, me, mine, ours, theirs, its, yours, you, them, they**

Be careful when you use pronouns. Though they are an essential part of good grammar, they have the potential to cause confusion for your readers if you're not careful.

**I rushed my experiment out of the lab because <u>it</u> was on fire.**

In the sentence above, was the experiment on fire, or the lab? To avoid confusion, the writer could have written "I rushed my experiment out of the lab because *the experiment* was on fire," if, in fact, it was the experiment that was burning. Try to "hear" your writing through your reader's ears as you write it. Does your use of pronouns leave any ambiguity?

Lastly, don't allow your pronouns and your antecedents (the nouns they represent) to be separated too far apart in your sentences. The farther apart they are, the more difficult it may be for your reader to recognize their relationship. Note the following examples:

**1.)**   If your lab partner brings a guest, tell <u>him</u> to leave the lab.
**(Who should leave—the guest, or your lab partner?)**

**2.)**   There was ongoing conflict between the marketing staff and the production team because <u>they</u> were consistently unreasonable.
**(Was the marketing staff unreasonable, or was it the production team?)**

**3.)**   It would be inadvisable to use their part on our equipment because <u>it</u> doesn't conform to specifications.
**(Which doesn't conform, the part or the equipment?)**

## Consistency

For your audience to have maximum confidence in your technical writing, it must be consistent. Poor, inconsistent writing can cause your readers to lose faith in your writing's technical content. It is imperative that your form, style, and verb tenses remain consistent throughout your text. This applies to capitalizing and abbreviating words, using acronyms, underlining, using bold-face type, and numbering figures. When you refer to an item, be sure to use that same name throughout your text.

**Your Secret Weapon!**

Get yourself a good writing guide such as *The Chicago Manual of Style* or *Webster's Concise Handbook for Writers*. These aids will answer your grammatical questions.

Misspelled words provide a sure way to lose the confidence of your audience. There is no excuse for not having every word in your text spelled correctly. Use the spell-check function in your word processing application, and if time allows, use the grammar-check function as well (if available). Punctuation also must be correct and consistent. Though it may seem awkward to you at times, periods and commas *always* go before a closing quotation mark (inside the end quote). Other types of punctuation, such as question marks, go after the closing quotation mark unless it is actually part of the quoted material. Use single quotation marks only for quotations within quotations. Otherwise, always be sure to use double quotation marks. Most writers today tend to use commas before the word *and* when it connects the last two items of a series (red, blue, and green), though technically this comma is optional.

How many errors can you find in the following paragraph?

> **Once preleminary working criteria has been established it is important to develop an overall goal for the "design process". Basically this should be a statement of objectives which can be evaluated as the design process evolve. Establishing such a goal for the project provides a means of evaluating. monitoring, and changing, if necesary, the focus of the process as it evolves through the various stages of design.**

The paragraph should read:

> **Once <u>preliminary</u> working <u>criteria have</u> been established<u>,</u> it is important to develop an overall goal for the "design process<u>."</u> Basically<u>, the goal</u> should be a statement of objectives which can be evaluated as the design process <u>evolves</u>. Establishing such a goal for the project provides a means of evaluating<u>,</u> monitoring, and changing, if <u>necessary</u>, the focus of the process as it evolves through the various stages of design.**

Okay, so you're not going to make that many mistakes in one paragraph, but this example gives you an idea of the types of mistakes to avoid. In the top paragraph, verb tenses are incorrect, punctuation is sloppy, and words are misspelled. Not only can such errors lead to questions about your competence, but they also make reading the material cumbersome, and can distract the reader from your point.

**Proper Flow**

Another valuable element of good writing is the use of transition words that help with flow and transition. The better your writing flows and transitions from one point to the next, the easier it will be for your reader to understand. Pronouns also serve to link paragraphs together, as long as they are used with clarity. Some common transition words and phrases include:

> *hence, first, finally, then, therefore, similarly, likewise, as a result*

Sometimes it will be natural to link paragraphs together by first setting up the subsequent paragraph. Then you may want to use a transition word to begin the paragraph that follows.

In the following statements, we'll look at examples of improper comma use. Note the potential confusion that results when you fail to use proper punctuation. Although by rereading such statements your readers often can figure out what you're trying to say, it places an unfair burden on them. The confusion you generate may well distract them from your content. As mentioned previously, it also may cause your readers to lose confidence in your technical competence. For example:

**Punctuation Errors: Commas**

> **As stated above the derivation of the unknown quantities...**

Upon first glance, your readers may assume that something is *stated above the derivation*. When a comma is placed between *above* and *the,* the meaning becomes clear: *As stated above,* ...

> **In this experiment error could have been introduced....**

Again, without a comma the words run together and might cause confusion. The reader may think you mean <u>experiment error</u>. Though the confusion may last for only a second, that second may cause the reader to lose the point of the entire paragraph.

We'll address only a few of the uses of colons in this section. Colons are used to introduce words or phrases that explain or summarize what came before the colon. Remember the rule from your younger years: colons link related thoughts, but one of those thoughts must be able to stand alone as a sentence.

**Punctuation Errors: Colons**

> **The lab lacked only one piece of equipment: a vise.**

Colons also are used to introduce lists.

> **Professor Harrigan cataloged the chemicals: tritium, parodesium, and grotect.**

We often use colons to introduce equations, but only with transitional words, as follows:

$$x + a = y + b$$

## Hyphens

Use hyphens when spelling numbers as follows.

> **twenty-four**
> **one hundred fifty-nine**     (note: "and" is not used after "hundred")

Also use hyphens in adjective compounds with a numerical first element.

> **4-in. pipe**
> **8-foot board**

But don't use hyphens when describing the length of an item.

> **4 inches of pipe**
> **a board 8 feet long**

## Sample Text Analysis

In this sample text, we'll look at how some of the elements we've discussed so far come together.

### Introduction

Whirlwind Corporation requires dynamic signal analysis techniques to aid the design of their new light aircraft gas turbine. The first phase of the project entails testing and evaluation of instrumented prototype systems for use in the development stage. Reviewing signal processing fundamentals and evaluating test equipment is of primary importance at this time, but the Development Group's ultimate objective is to provide a new industry standard {20B}.

Whirlwind's revolutionary turbine design will set several new benchmarks {22}. It will provide 30% better energy efficiency and 20% more peak power than similar products currently available {20B}. Moreover, it will require significantly less engine compartment space and allow designers to reduce aircraft weight {22}. Both undesirable vibration and extraneous noise emission also will be substantially reduced {22}.

Tests of the transient and steady-state operation of the new turbine will generate many signals. Of course, the Development Group will analyze the variation of the working fluid's thermodynamic and transport properties {20A}. The "frequency signature" of the vibrating components will also be determined {20A}. Small variations in this signal often indicate imminent failure. By measuring and modifying the natural frequencies of the turbine and its forcing functions, catastrophic failure due to resonance also can be avoided. Finally, the acoustic output of the turbine will be monitored, and sources of unwanted noise will be pinpointed and eliminated {20A}. Although the Development Group will collect data in the time domain, most of the analysis will be completed in the frequency domain {21}.

The Development Group's current investigation focuses on elementary sinusoidal and square wave signals {21}. Although these waves do not represent actual turbine test data, their analysis highlights difficulties encountered in digital signal processing. Specifically, the sampling frequency and the quantity of data collected determine the accuracy and precision of the results generated {21}. These parameters also predetermine the instrumentation's range of frequency.

---

Look at the underlined → items. They indicate a variety of flow techniques used to keep the reader moving easily through the text.

The use of the pronoun it links sentence 1 to sentence 2 in paragraph 2. →

Finally signals an end to the idea flow in the third paragraph. →

In column 2, notice this signal and the way it helps the flow of the text. →

*(Note that referenced technical material is labeled with bracketed numbers.)*

# 10.8  Formats

As an engineer, your written communication will take a variety of forms. These include brief notes to a superior or to a peer, hand-written reminders about a project, informal reports, memos, formal reports that may document months of work, and visual presentations that require significant preparation, yet little text. Much of your writing will need to conform to one standard format or another. In this section, a number of these formats are presented.

## Memos

The *memo* provides a convenient, relatively informal way to communicate the existence of a problem, propose some course of action, describe a procedure, or report the results of a test. Though informal, memos are not to be carelessly written. They must be carefully prepared, thoughtfully written, and thoroughly proofread for errors.

Memos begin by clearly presenting all relevant information right up front, as follows:

**To:**      name, job title
            department
            organization

**From:**    name, job title
            department
            organization

**Subject:** (or "Re:") issue addressed in the memo

**Date:**    date

The following may be used as well, if applicable:

**Dist:**    (Distribution) list of others receiving the memo

**Encl:**    (Enclosure) list of other items included with the memo

**Ref:**     (Reference) list of related documents

The format of the text of the memo is also simple and contains the following information:

*Foreword*—statement of the problem or important issues addressed

*Summary* (or *Abstract*)—statement of results, findings, or other pertinent information

*Discussion or Details*—technical information, discussion of the problem or issues, or support for the claims in the Summary

## Activity

Prepare an answer to the following memo. Keep in mind the appropriate format and approach for a good memo.

*Situation*: Assume that the following memo has been received in your office at Engo-Tech Consultants (an engineering consulting firm specializing in technology compatible with the request in the memo). Please read it and address its concerns.

To: Engo-Tech Consultants

From: Michelle De LaBaron, General Manager, Euro-Flags

Date: August 1, 2002

Re: Increased activity for Euro-Flags

To ensure the continued success of our operation at Euro-Flags we believe it is essential that we add a new technical emphasis to the variety of entertainment opportunities we currently offer.

Our mission includes more than simply existing as a provider of "fun and games." We also want to stimulate our patrons intellectually, and thrill them with the wonder of current technology.

With this in mind, we ask that you present us with a proposal which outlines how your firm could assist us in this endeavor. Please explain not only what your firm can do, but also how your proposed plans would appeal to our patrons.

# 10.9   Formal Reports

The principal purpose of the formal report is to present information gathered from an experiment or simulation in a document that will be useful to your employer and colleagues.

### Title

The title should be as brief as possible, clear, and appropriate. Seven or eight well chosen words is typical. The title doesn't need to do any more than introduce the report.

### Summary

Though the Summary (or Abstract) is presented early in a formal report, it should not be written until all other parts of the report have been completed. It should state, in simple, declarative sentences, *what* was attempted, *how* it was accomplished (including a discussion of any special techniques used), and what the *implications* are. That is, it should state the results and main conclusions.

This is both the shortest and most difficult section of the report to write. It is also the most important. In technical publications, the abstract often is the only portion of the report that many people will read. Therefore, it should be written to stand alone. It must communicate all the relevant ideas and results in one to two paragraphs (250-350 words). And be sure to discuss the details, which occurred in the past, with past-tense verbs.

### Table of Contents

In the Table of Contents, list each heading found throughout your report along with its page number. However, it is customary not to list the Abstract among the contents.

### Nomenclature

List and define all symbols used in the report. They should be listed alphabetically— Arabic symbols first, then Greek. Such a listing will enable your readers to find definitions of symbols used in the Analysis portion of your report with ease. Note that abbreviations should not be listed in the Nomenclature section.

The essential task of the Introduction is to orient the reader to the substance of the experiment or project and the context in which it was executed. The Introduction should state the *motivation* for the experiment and any relevant *background* information. It should introduce the material contained in the report by noting what is presented in each of the sections that follow.

**Introduction**

The Analysis should begin with the basic, general (and well-known) relationships and proceed to the specific formulas to be used in the interpretation of the data. It is usually appropriate to use an explanatory sketch. All the symbols used should appear in the listing of Nomenclature, described above. Analytical results which have been previously derived and are readily available (equations from a handbook, for example) can be quoted and appropriately referenced. Their derivation need not be repeated unless the derivation is critical to the engineer's understanding of the experiment.

**Analysis**

All relevant mathematical steps should be presented in the Analysis as well. Mathematical explanations and commentary are crucial to a good Analysis. Be careful not to force the reader to consult lab handouts or textbooks to understand the specific analysis involved in the experiment. If it is important enough to mention, it is important enough to explain.

A schematic representation (a diagram) of the experimental equipment or simulation program, including detailed views of unusual or important components, is a valuable aid in informing the reader about the experiment. The diagram can be used to document pertinent dimensions of the apparatus, and it can be used to specify the specific experimental equipment used for the study. If the procedure used in the experiments is not a familiar one, it will be necessary to include details of the techniques used. Your information should be clear enough that someone familiar with the general area of investigation should be able to reproduce your experiments with precision, based on the information given in this section. The report can be hindered by an incomplete or incorrect discussion of the experiment's procedure and equipment.

**Experimental Procedure**

Though you may be tempted to slip into use of the present tense when describing the procedure, be careful to describe it in the past tense.

Some major points to remember are:

- When first referencing a figure, be sure to put the figure on the same page (or the next page). Your reader shouldn't have to hunt through your report to find the figure.
- Make figures at least one-third of a page in size. Figures that are too small are harder to read.
- Figures and tables should have clear and complete titles. Beneath the title, a short explanation of the figure's contents is normally necessary.

The answers derived from the analysis of an experiment are presented in the Results section of the report. This section should contain short, declarative statements of the results which summarize specific data presented in graphs or tables. The purpose is to tell the reader (without discussion) your interpretation of the results. Keep in mind that the results could be interpreted in more than one way. Thus, the importance of your stated results.

**Results**

Realize that your readers may start reading your report in midstream, meaning that they may start with the Abstract and then go directly to the Results. Never assume that the report will be read from page 1 to the end, which is seldom the case.

You also should distinguish between "Figures," which include schematic drawings, photographs, graphs, etc., and "Tables," which are tabular presentations of data or computational results. Each type of data presentation should be numbered sequentially with its title: for example, "Figure 1. Schematic of the Combustion Chamber."

Though you are to present your data without discussion, do not make this presentation without explanation. A Results section containing only tables and graphs would be difficult to understand. Remember, you want to make comprehending your report as easy as possible for your readers.

## Discussion

The Discussion section starts with a brief summary statement of the results and then proceeds to a discussion of these results. The principal task in this section is to interpret the results, to note what went "as expected," what was unexpected, and what is of technical interest. The interpretation of the results in terms of the motivation for the experiment and its correlation to the current company projects should be the focus of the discussion. The discussion could involve a comparison with other, similar investigations or comparison with expected results. The strong points of the work should be brought out here along with any limitations, because if the writer does not point out the limitations of his work, someone else surely will later. It may also be legitimate to comment on possible future investigations. Discuss the specific results of the experiment, using references to the accuracy of the measurements. It is useful to note the estimated uncertainties and their effect on the calculated values. For example: "The pressure was $2.0 \pm 1.0$ kPa and the velocity was $30 \pm 10$ m/s. Note that the "information content" of this statement is much greater than in the statement: "The manometer liquid was fluctuating and the pressure could not be accurately measured." It is extremely important to provide specifics and avoid vague modifiers such as: greater than, about, and higher than.

Keep in mind that good paragraph construction will present a thesis statement or idea and then give supporting details for that statement. When new points need to be made, make sure you make them in new paragraphs. You should make an outline for the Discussion section (just as you should for all parts of the report) so it follows a logical progression which will support the conclusions presented in the next section, Conclusions. Carefully judge the information that you are providing to the readers to make sure it satisfies their specific expectations. For instance, they expect to find information presented in a pattern that presents what is known first and then what is new second. Sentences should begin with known information and then progress to new, related information.

## Conclusions

A typical way to begin the Conclusions section is to make this statement: "The following conclusions are supported by the results of this study," and then list the conclusions in one or more simple, declarative sentences using numbers to differentiate each separate conclusion. Remember that in your Conclusions section, engineers will be looking for concise statements that clearly communicate what your results indicate—not additional explanatory material or any plans for further investigation. They want the mass of data synthesized into the briefest conclusions you can present.

References are to be listed at the end of the report. In general, be sure to list: author's last name, then first and middle initials (and then list co-authors with initials first), title of source being quoted (in quotes), edition of source, volume of source (if applicable), city of publication (from source's title page), publisher, date of publication, and pages referenced. Note the variety of possible formats in the following samples:

**References**

> Walker, R.E., A.R. Stone, and M. Shandor. "Secondary Gas Injection in a Conical Rocket Nozzle," AIAA Journal, Vol. 1, No. 2, Feb. 1963, pp. 334-338.
>
> Turner, M.J., H.C. Martin, and R.C. Leible, "Further Development and Applications of Stiffness Method," Matrix Methods of Structural Analysis, 1st ed., Vol. 1, New York, Macmillan, 1964, pp. 203-206.
>
> Segre, E., ed., Experimental Nuclear Physics, 1st ed., Vol. 1, New York: Wiley, 1953, pp. 6-10.
>
> Book, E. and H. Bratman, *Using Compilers to Build Compilers*. SP-176, Aug. 1960, Systems Development Corp., Santa Monica.
>
> Soo, S.L. "Boundary Layer Motion of a Gas-Solid Suspension." Proceedings of the Symposium on Interaction between Fluids and Particles, Institute of Chemical Engineers, Vol. 1, 1962, pp. 50-63.

Always give inclusive page numbers for references from journal articles and a page or chapter number for books. Each reference in the text must be cited.

Lengthy calculations and side issues that are not really related to the main theme of the report should be placed in the Appendix. To decide whether something should go in the body of the text or in the Appendix, think about whether or not its inclusion in the main body of the report required as part of the description of the investigation. If the answer is No, the item either should be left out or put in the Appendix.

**Appendices**

Some assignments may require that you submit an Executive Summary instead of a full formal report. This type of report condenses the work you have done into the briefest document possible. It assumes a managerial perspective instead of a technical one, and will contain enough information that an individual with limited technical expertise will be able to make clear decisions based on your report. The Executive Summary presents just the simple facts and describes the key elements of your work in non-technical language—concise and straight to the point. A good Executive Summary contains the following information:

**Executive Summaries**

- the background of the situation or problem
- cost factors
- conclusions
- recommendations

If you have constructed Abstracts before, you will understand that they are similar to the Executive Summary. Both are quite brief. They focus on what was investigated, the conclusions your work has produced, and the recommended course of action. In some instances, this Summary will be all that is read before a decision is made either to proceed or to halt further action.

**Communications Checklists**

One way to make sure that you have successfully completed an assignment is to evaluate your work with a checklist. You can create custom checklists for each of the various communications formats you use. These checklists will guarantee that you have included everything that is required to inform the reader.

**Formal Report Checklist**

## Title Page
1. Title of paper
2. Course/project
3. Date due
4. Section meeting time
5. Name

## Abstract
6. Purpose of the lab/experiment
7. How the lab was performed
8. What was discovered, achieved, or concluded
9. Use of past tense
10. Reference to experiment, not paper
11. No personal pronouns (I, we)

## Nomenclature
12. Listed in alphabetical order
13. Upper case, then lower case (A a B b c G g1 g1a)
14. Arabic and Greek listed separately (Arabic first)
15. Only symbols are listed

## Table of Contents
16. All sections represented
17. Abstract and Table of Contents are not listed
18. Observations as headings; analysis, equipment, procedures, results as sub-headings
19. All columns aligned

## Introduction
20. Motivation for the experiment stated
21. Sufficient information to orient reader to the substance of experiment
22. Sufficient information to excite reader
23. Listing of sections which follow

## Observations
24. Mathematical model used to predict system behavior presented with ample explanation and lead-in

25. Equations numbered
26. Proper punctuation in equations
27. Equations properly spaced
28. Schematic of equipment used
29. Tables oriented correctly, labeled and referenced
30. Figures oriented correctly, labeled and referenced
31. Highlights of equipment used
32. Highlights of the procedure (not specific steps)
33. Data presented with clear indication of what it applies to
34. Clear indication of what data refers to
35. Trends in data stated (to be discussed in the Discussion section)
36. Clear indication of what reader should observe in the data

## Discussion
37. Complete discussion of the results
38. Connection of data and objectives is clearly stated
39. Comparison to similar experiments presented
40. Strong points of the study given
41. Weak points of the study given
42. Statements are specific
43. Logical progression which supports Conclusions that follow

## Conclusions
44. Begins with, "The following conclusions are supported by this study"
45. Conclusions are numbered
46. Conclusions are concise and highly specific
47. Vague statements are not used
48. Conclusions flow directly from Discussion

## References
49. Initials for first names
50. All necessary information included
51. References numbered in text like this: [1]

The following abstract satisfies the checklist features listed in the previous section. By making sure that the required items are included for each part of your report, you have a chance to review what you've included and what still needs to be addressed before the paper is submitted. (Note: The numbers that appear in the following abstract refer to the preceding checklist. References aren't numbered in a real abstract.)

## A Sample Abstract

### Abstract

Experiments {10} were conducted {9} to assess the appropriateness of digital signal analysis to the design, development, and testing of Whirlwind Corporation's new light aircraft gas turbine. Such analysis could be used to monitor the transient and steady state property variation of the new power plant, predict potentially catastrophic failure, and pinpoint sources of extraneous noise generation {6}.

Several elementary sinusoidal and square wave signals were generated {9} by a commercially available Waveteck function generator. These signals were then converted {9} to the frequency domain by LabView's "Spectrum Analyzer" via the Fast Fourier Transform. Various combinations of sampling frequency and sample size were investigated {9}. When deemed appropriate, these signals were {9} also low-pass filtered {7,10}.

The Discrete Fourier Transform accurately represented {9} only components less than half the sampling frequency. Higher frequency components were reflected {9} across the Nyquist frequency or its integer multiples. This aliasing was eliminated {9} by low-pass filtering, but on occasion, important signal components were discarded {9}. Whenever the input signal contained components that were not integer multiples of the frequency resolution, the magnitude of the corresponding spectrum peaks were diminished {9}. This leakage was reduced {9} by increasing the frequency resolution by increasing the sample size {8,10}. These signal analysis techniques proved {9} their utility and applicability to the new gas turbine project.

In this abstract, the author has addressed the important elements of the assignment, which was to tell the reader why the experiment was performed, how it was done, and what the principal results were.

Notice some of the methods that the writer has used to make the text flow smoothly. Line 8 begins the paragraph with "Several elementary sinusoidal and square wave signals," which prompts the reader to wonder what is to follow. Line 25 shows the writer relating "These signal analysis techniques" back to the previous sentence. Be sure to help your readers by presenting a logical, flowing report.

# 10.10  Other Types of Communication

**E-mail**

Electronic mail, or e-mail, has become a part of life in almost every office. Messages gather in computer mailboxes at a phenomenal rate, and the time required to respond to these e-mails continues to grow. It is, therefore, vitally important that we look closely at what we receive and what we send. E-mail must be read carefully and the responses we send must be well written. Just as with more formal reporting, if your communication includes spelling errors, poor punctuation, and sloppy construction, it will likely hamper your reader's comprehension. And it projects general incompetence—even if you are quite competent technically. Besides, e-mail provides an easily preserved record of your communication, so it's worth doing right.

E-mail's instant transmission and spontaneous feel can tempt us into serious communications errors. It's easy to get sloppy. Given the format, e-mail can be easily misunderstood—which is one reason why the variety of "smiley face" punctuation tags evolved. And once we have sent a message, it is out of our hands and we can do nothing to change the content or presentation. Here are a few simple rules to follow:

- Use proper grammar—appropriate sentence structure, subject/verb agreement, spelling, flow, etc.
- Carefully proofread, edit, and spellcheck your e-mail. Are there gaps in your train of thought, or opportunities for confusion, that may cause a reader to misunderstand your message?
- Have you thought about your response before you hit "Send"? Once it is sent, it can't be retrieved. With professional e-mail, it's often best to wait as long as possible to send your reply, even if you only have an hour. Let it rest, then re-read your reply before sending.

E-mail is a fantastic tool, but it has limitations. Never use it when a face-to-face dialog actually is needed. Do not use it when a formal document is required. E-mail does not carry the status of a formal report. And remember that e-mail is not quite as private as you think. A message may be downloaded and left on the screen of your recipient, in plain view of anyone entering his office. And in truth, many e-mail logs and data archives are more public than you may realize. It is vital, therefore, that you make every effort to use the tool wisely, and avoid letting it embarrass you or cause you undue grief.

**Progress Reports**

At some point in every job, you will be asked to indicate how you are progressing toward some goal. Managers, co-workers, stockholders—even you yourself—will want to know what you've accomplished. The Progress Report provides the means by which you can report your status.

The reports need to be attractive and easy to understand. Commonly, the progress report will have sections much like any other report prepared for management:

- Introduction
- Project description
- Summary of what you have done

Each section is discussed briefly below:

**Introduction**—Here you capture the interest of the readers by informing them of what you are going to do in the report. You expose readers to the scope of the work being done, the purpose of the work, and any major changes that have been required in the project.

**Project Description**—Here the reader is acquainted with the time-frame of the project and the current progress. Phases that have been completed and the time it took to complete them are presented in this section. It is also important to explain the tasks that remain to be completed.

**Summary**—This is where you draw the whole report together. Summarize the main points of action and reiterate where you are along the road to completion.

## Problem Statements

Problem solving is a critical aspect of engineering. Before you begin to work on a project you may be asked to present a Problem Statement pertaining to the project which explicitly states the problem to be investigated and outlines the course you intend to take. Correctly defining the problem is vital to the solution. If you are not on the right track from the beginning, you may well fail to obtain the appropriate solution. So here are some basics:

- Work with data that you already have collected—evaluate all reasonable avenues of pursuit in finding a solution as you define your problem.
- Consult your contacts who have knowledge of the problem.
- Investigate the problem first-hand.

Here's a silly anecdote which may help you remember to properly identify a problem.

> A man who had a chronically sore neck went to the doctor. He told the doctor of his suffering, that he was unable to turn his head without severe pain. Even visiting the doctor caused him pain. The doctor asked him what he had done to try to solve the problem. The man listed a variety of attempted remedies—none of which had worked. The doctor looked at the man, analyzed the situation, and calmly said, "When you get up in the morning, try taking the hanger out of the shirt you're going to wear before you put it on."

Thus the problem was properly identified and solved.

## Cover Letters

Cover Letters will be especially important as you attempt to land your first job in engineering. Such letters are sent along with resumes and transcripts. They "cover" your other material—thus the name. In many instances, they will be the only personal impression you will get to make on a potential employer. A great Cover Letter can open doors, but a poor one can slam them shut.

The Cover Letter functions as your introduction to a prospective employer. View the letter as an opportunity to communicate what you have to offer them. If they don't like what they read, you may never get the chance to talk to them in person.

So what should you include in a Cover Letter? In brief terms, here is what you need to address:

1.) the date, your address and phone number, and the name and address of the person to whom you are writing

2.) 1st paragraph—the reason you are writing the letter, the source of your information about the employer, and what you would like to do for the employer (the position for which you would like to apply)

3.) 2nd paragraph—a brief discussion of your resume, hitting the highlights

4.) 3rd paragraph—"current" information which may not yet be appropriately included in your resume (courses you are currently taking, research you may be involved in, activities you are about to engage in)

5.) 4th paragraph—summary paragraph in which you thank the reader for his or her consideration "in advance" (You may also include your phone number, and mention that you will call to confirm receipt of the letter.)

Remember, you want your Cover Letter to present the most positive impression possible. A good deal of time should be spent perfecting it. Cover Letters and Resumes should be printed on the same kind and color of paper. Always use a high-quality laser printer to print these items.

## Sample Cover Letter

August 10, 2002

Ms. Lydia Baron
Human Resources Director
Harrigan Corporation
120 Rollaway Road
Chicago, IL 60606

Dear Ms. Baron:

I am very interested in the summer internship program offered at Harrigan Corporation. I know from my research that Harrigan's rapidly expanding automotive parts operation has an outstanding reputation. The opportunity to gain valuable experience through working in that division would be of tremendous benefit to me in my professional development.

My interest in engineering began at an early age, and I look forward to a lifetime of continual engineering education. I am completing my junior year at Michigan State University, having taken a number of courses toward my degree in mechanical engineering. As my resume shows, I have concentrated a great deal on developing my computer skills as part of my engineering background.

This semester I am taking two courses—Advanced Mechanics and Finite Element Analysis—which have been particularly valuable in preparing me for working at Harrigan Corporation. I have studied the current work schedule of the Brighton facility and understand that these courses fit well with your plans for the new turbo charger assemblies.

A copy of my resume is enclosed for your consideration. I will contact you during the week of August 20 to see if you require any additional information. If you have any questions for me in the interim, please contact me at (517) 555-1212, or at smith@pilot.msu.edu. I look forward to discussing with you the job opportunities at Harrigan Corporation.

Sincerely,

Jason D. Smith

222 Landon Hall
Michigan State University
East Lansing, MI 48824
(517) 555-1212
smith@pilot.msu.edu

Enclosure

## Résumés

There is a wide variety of appropriate formats for Résumés, so ultimately you have to decide upon a format that you will be happy with. Regardless of the specific format, every Résumé must present your name, address and phone number; your educational background; your previous employment (if any); and extracurricular activities. The following simple format is merely one of many. Key elements are highlighted in bold type.

Make sure that you get as many people as you can to read and critique your Résumé. Listen to and evaluate all the comments you get, and then customize the Résumé to fit your needs.

(See actual sample Résumé on next page.)

## Sample
## Résumé Format

**Your Name** (centered)

**College Address**                                          **Permanent Address**

**College Phone Number**                                **Permanent Phone Number**

**Objective:**  An entry level position in (state field or position sought).

**Education:**
- your degrees
- your school
- graduation date (or anticipated date)
- grade point average and basis (e.g., 3.4/4.0)

**Work Experience**:
- your job titles (list most recent first)
- the companies for which you worked
- dates for which you worked for those companies
- responsibilities in those positions

**Activities:**
- all meaningful activities you have participated in

# Jason D. Smith

222 Landon Hall
Michigan State University
East Lansing, MI 48824
(517) 555-1212
smith@pilot.msu.edu

37 Dilgren Avenue
Port Anglican, VT 12225
(234) 555-1212

| | |
|---|---|
| **Objective:** | An internship in the manufacturing area of a mid-sized company |
| **Education:** | Bachelor of Science in Mechanical Engineering<br>Michigan State University, East Lansing, Michigan<br>Expected Date of Graduation: May 2003; GPA: 3.6/4.0 |
| **Computer Skills:** | MS Office 2000, Lotus, C++, MATLAB |
| **Employment:** | May 2002—August 2002; Production Engineer<br>Labardee and Sons, Aberdeen, Texas |

- Developed new production plans for the Labardee grinding machine
- Consulted with primary Labardee clientele

June 2001—August 2001; Machinist
Rawlins Machining, Deer Park, New York

- Produced updated schematics for all standing jobs
- Instituted changes in assembly line procedures

| | |
|---|---|
| **Honors/Activities:**<br>Team, | Ames Scholarship, Dean's List (4 semesters), MSU Swim racquetball |
| **References:** | Available upon request |

**Thank-You Letters**

After interviewing for a new job or co-operative position, you should send Thank-You Letters to the individuals who interviewed you. You shouldn't wait more than 48 hours to send these letters. They should be businesslike and concise. And remember that a strong follow-up letter may be just the thing that puts you ahead of the competition.

Here's what you should include in your Thank-You Letter:

*1st paragraph*
— Thank the interviewers for their time, and reiterate your interest in working for the company.

*2nd paragraph*
— Briefly restate your qualifications. This is the time to address any positive qualities you may have failed to mention during the interview.

*3rd paragraph*
— Close the letter with a final thank-you, and express your interest in hearing back from the interviewer. You may even want to give a specific time frame within which you will follow up with a phone call. Provide the interviewer with phone numbers where you can be reached and your e-mail address, if you have one.

*Additional advice:*
— Write or modify each Thank-You Letter separately. Don't use generic form letters. Try to personalize your summary of the interview in each letter.
— Even if you are rejected for the position, sending a Thank-You Letter may open up other doors or future opportunities with that company.

# 10.11  Relevant Readings

Here are just a few of the many technical communication books that are on the market. They all contain pretty much the same information, presented in varying formats. Having such a resource on your shelf will be of great value.

Fear, D., Technical Writing. Glenview, Illinois: Scott, Foresman and Company. 1981.
(For the writer who wants instruction in clear, concise steps.)

Eisenberg, A., Effective Technical Communication, 2nd Edition. New York: McGraw-Hill. 1992.
(General overall coverage of the major topics of technical writing; includes proposals, letters, and reports.)

Hirschorn, H., Writing for Science, Industry, and Technology. New York: D. Van Nostrand. 1980.
(Particularly helpful appendix. An approach to writing that takes the writer from the beginning of the process to the end product.)

Houp, K. and T. Pearsall, *Reporting Technical Information*, 6th Edition. New York: Macmillan. 1988.

> (The best of the bunch, both for future use and present needs. Gives lots of examples, and helps greatly in the writing process.)

Huckin, T., and L. Olsen, *English for Science and Technology*. New York: McGraw-Hill. 1991.

> (Good book for the non-native English speaker about the process of report construction.)

Mathes, J. C., and D. Stevenson, *Designing Technical Reports*. New York: Macmillan. 1991.

> (A process approach to the writing of a technical presentation—written from the information perspective, not from the "form of the report" perspective.)

*Michaelson, H., How to Write and Publish Engineering Papers and Reports. Phoenix: Oryx Press. 1990.*

> (Primarily aims at the writing of papers for publication. Does a good job of looking at quality of writing and the concern for the reader.)

Miles, J., *Technical Writing—Principles and Practices*. Chicago: SRA Research Associates. 1982.

> (A general text to help with basic problems. Includes a great deal on the process of getting started.)

Turner, M., Technical Writing. Reston, Virginia: Reston Publishing. 1984.

> (A teaching text with a good visual format. Nice section on memo writing.)

*Also:*

Beer, D. A Guide to Writing as an Engineer. New York: John Wiley & Sons. 1997.

Eisenberg, A. A Beginner's Guide to Technical Writing. New York: McGraw Hill. 1998.

Finklestein, L. Pocket Book of Technical Writing. New York: McGraw Hill. 2000.

Pfeiffer, W. Proposal Writing. New Jersey: Prentice-Hall. 2000.

Pfeiffer, W. Technical Writing . New Jersey: Prentice-Hall. 2000.

## Exercises and Activities

10.1   Prepare a five-minute presentation to be given to a group of your fellow students concerning an engineering topic of interest to you. Prepare overheads for your presentation. You may not need more than one overhead per minute of presentation.

10.2   Prepare a presentation, similar to the one above, to be given to a group of elementary students. What information did you change?

10.3   List at least five clichés that you have used or heard that could make comprehension difficult for a foreign student.

10.4   Write a memo informing your fellow students of a seminar you are presenting.

10.5    Write a memo to your supervisor requesting that your work activity be reviewed for a possible raise.

10.6    Write a formal report about a lab that you have completed, using the suggestions presented in this chapter.

10.7    Write an Executive Summary of a lab experiment that you have completed.

10.8    Create a checklist similar to the one presented in section 10.9 which you could use in conjunction with a paper you are required to write for another course. Evaluate how well your checklist improved the quality of your paper.

10.9    Write an e-mail to your local newspaper responding to an issue (deer hunting, guns, smoking, abortion, air bags, etc.) for the "Letters to the Editor" section.

10.10   Engineering is not recognized as a profession by federal agencies since it does not have any entrance requirements. In fact, a college degree is not even required. Write an e-mail to your state representative supporting or opposing this position of the federal agencies.

10.11   a) Write your résumé.

        b) Write a cover letter to the OLG Manufacturing Company seeking employment in the engineering department.

        c) Write a thank-you letter to a Dr. Jan Harris, as though she interviewed you for an engineering position.

10.12   Locate a grammar or writing textbook in your library. Carefully read through the Table of Contents. Write down any topics with which you are unfamiliar. (For instance, your knowledge of the definition and use of *verbs* may be satisfactory, but perhaps you are much less familiar with *subordinate conjunctions*.) Read through the material in the book related to any unfamiliar or forgotten parts of speech, and write a brief report detailing new information you learned.

# 11. Ethics and Engineering

## 11.1 Introduction

In addition to technical expertise and professionalism, engineers are also expected by society and by their profession to maintain high standards of ethical conduct in their professional lives. This chapter will cover the most important issues of ethics insofar as they are relevant to careers in the various engineering disciplines. A brief concluding section will also summarize the various legal guidelines and requirements with which engineers are expected to be familiar.

To begin, here is a summary of the basic areas and kinds of questions to be covered.

**Summary**

1. Since engineering ethics is a branch of ethics, we start with a clarification of the nature of ethics, and how ethical values and concerns differ from other kinds of values and concerns.
2. Next, the specific concerns of engineering ethics will be examined, and distinguished from other concerns about values which engineers may also have. Such non-ethical concerns include personal preferences, social or political values, and more technical professional values related to becoming and remaining an excellent engineer from a purely engineering or scientific point of view.
3. After the preliminary clarifications and distinctions above, we then examine the basic issues, concepts and special topics of engineering ethics. Most of these are also included in or assumed by the codes of ethics provided by various professional engineering societies.

It will be useful to examine one of these codes in more detail, namely the National Council of Examiners for Engineering and Surveying (NCEES) Model Rules of Professional Conduct. The study of this code will give engineers a good understanding of the kind of real-world requirements which professional engineering societies uphold for their members. The discussion will also be supplemented with some model answers to typical ethics questions, which provide a useful test of familiarity with and understanding of the ethical issues covered in such codes. In addition, more wide-ranging discussion and essay ethics questions are included to encourage more principled and independent thinking about ethical issues by engineers.

## 11.2  The Nature of Ethics

Generally, ethics is concerned with standards, rules or guidelines for moral or socially approved conduct such as being honest or trustworthy, or acting in the best interest of a society.

*Ethical standards exist independently of any particular group of experts.*

Not all standards or values are ethical standards, though. For example, personal preferences and values such as individual choices of food or clothing aren't important enough to qualify as ethical values or choices. Generally speaking, ethical standards apply only to conduct which could have some significant effect on the lives of people in general.

Thus for an engineer, using a substandard grade of steel in the construction of a bridge would definitely violate ethical standards because of the potential for safety hazards for people in general. But on the other hand, using an inferior brand of ketchup on one's fries does not violate any ethical standards, since at worst it would only violate personal, non-ethical standards of what tastes good to oneself.

Ethics and ethical standards should also be distinguished from matters of legality and legal standards. Roughly speaking, the distinction is that legal standards are defined in legal documents by some properly appointed legal body, and those documents and legal experts determine what the law is and who should obey it.

In ethics, by contrast, it is instead assumed that ethical standards exist independently of any particular group of experts (being accessible to all through the exercise of their own thinking capacities), and that a codification or written-down form of the standards merely describes or summarizes what those pre-existing ethical standards are, rather than (as in the case of the law) defining their very nature.

*Even legal standards must give way to ethical standards in the case of conflicts.*

For example, if there were a legal requirement that any load-bearing beam in a bridge must be able to bear five times the average real-world stress on the beam, then it is a simple matter of calculation to determine whether a beam meets the legal standard or not. However, an ethical standard for what would be an adequate safety factor in such a case is a very different matter, and it cannot be settled by calculation or appeal to a rule-book, since an ethically concerned engineer could convincingly argue that conventional or routine safety factors are really ethically inadequate and unacceptable. Such a person cannot be answered merely by appeal to some rule-book or to current legal standards for such matters.

Another vital point about ethics is that its standards are always more important than any other standards. Thus in the contrast between personal and ethical standards discussed above, if one's personal standards conflict with ethical standards, then one must suppress one's personal standards to resolve the conflict in favor of the ethical standards. For example, if one personally values finding the highest-paying job, but it turns out that the job in question involves some ethically wrong conduct, then it is one's ethical duty not to accept that job, but to take a lower-paying, more ethical job instead.

It is generally held that even legal standards must give way to ethical standards in the case of conflicts. For example, suppose that a law governing motor vehicles did not currently require a vehicle recall in the case when engineering defects were found in its construction. Nevertheless, if a good case could be made that this was an ethically unacceptable situation, then the law would have to be changed to conform to the ethical standard. So here too, other kinds of values or standards must be suppressed or adjusted in the case of a conflict with ethical standards, which are always of overriding importance.

The subject of ethics includes both theories about the nature of ethics (this part of ethics is sometimes called *meta-ethics*), and recommendations of appropriate standards or guidelines for morally right or good behavior by people (sometimes described as *normative ethics*). *Engineering ethics* mainly deals with such normative or ethical standards as to how responsible engineers should behave.

**The Subject Matter of Ethics**

Three main varieties of meta-ethical theories are ethical skepticism, relativism, and absolutism. Skeptics believe that ethical beliefs and standards are merely a matter of subjective personal opinion or biased feelings, so that there really are not any justifiable, reliable ethical standards. However, we can ignore this theory here, since professionals in any field, including engineering, are required as part of their professional obligations to adhere to high ethical standards. So ethical skeptics should seek some other line of work than engineering or the other professions.

*Ethical skeptics should seek some other line of work than engineering.*

*Ethical relativism* claims that in some way ethical standards are relative to a society or culture. In contrast, *ethical absolutism* claims that the same ethical standards apply to all societies and cultures. Applied to engineering ethics, this would be a dispute about whether or not it would be acceptable for different societies to have different engineering ethics standards for their professional engineers.

Fortunately, we can avoid this dispute here, at least at the present introductory level, in a similar manner to that in which we avoided ethical skepticism. As already noted, professionals in any field, including engineering, are required as part of their professional obligations to adhere to high ethical standards. But inevitably those high standards in practice are those approved or debated within the specific society in which an engineers beginning their own professional lives. So for present purposes we can put aside as peripheral or irrelevant the question of whether or not other societies should adopt codes of engineering ethics similar to our own.

Returning to normative ethics, recall that it involves recommendations of appropriate standards or guidelines for morally right or good behavior by people, and that engineering ethics mainly deals with such normative or ethical standards as to how responsible engineers should behave. We will be examining these standards as applied to engineering in succeeding sections. But for the present, here is a brief summary of some standard approaches to normative ethics.

**Normative Ethics**

There are two main approaches to normative ethics. The first views the subject primarily in terms of rules or principles for right conduct, developed on various different bases such as that of intuition, reason or social tradition. Fortunately for us, in spite of the disagreements as to the basis or source of normative ethical principles, there is very general agreement on what the resulting rules should be—rules such as "Always tell the truth," "Keep your promises," "Protect innocent life," and so on.

The other main approach to normative ethics focuses instead on ethical behavior viewed as whatever actions will best promote the general welfare or benefit of all people. This is often called a *"utilitarian"* approach to normative ethics, and it can be summed up in the slogan "The greatest good for the greatest number." Another common description for this approach is as a *"consequentialist"* ethic, in that it morally judges an action in terms of whether or not it would have better consequences than any other action one could perform.

## Conflicts in Ethics

Potentially at least, it seems there could be a conflict between the rule-based versus the utilitarian or consequentialist approaches to normative ethics. For example, following the principle "Always tell the truth" may apparently not have the best consequences in a particular case, while the action which does seem to have the best consequences (possibly a case of lying) may violate a moral principle.

However, fortunately, most utilitarians agree that a deeper analysis will almost always show that what their consequentialist ethic says is morally correct is substantially the same as what conforms to the ethical standards recommended by the rule-based ethical theories. So we will find that in practice both ethics in general and engineering ethics in particular contain elements from both approaches, which generally co-exist without any fundamental difficulty.

Another example of this practical convergence in spite of initial theoretical differences is found in the issue of intentions versus actions. Rule-based theories tell us that being ethical involves following a morally correct rule or principle, whatever the consequences of so doing.

*Sincere intentions or wishful thinking have no special ethical value, nor are they legitimate excuses for ethically bad behavior.*

However, the following concern could be raised: Since human beings are fallible and imperfect, the best they can do is to sincerely intend to follow a morally correct rule. So whatever happens as a result of their actions in doing so, no matter how bad it is, nevertheless they will be morally blameless because their intentions were good. (Example: a well-intentioned person tries to push a passerby out of the way of a speeding vehicle, but only succeeds in pushing him or her more directly into its path.)

On the other hand, from the utilitarian or consequentialist approach, a person's intentions are morally irrelevant; only the actual consequences of their actions are relevant to judging the morality of their actions. So as before, it initially looks as if there could be a fundamental conflict between the rule-based and utilitarian, consequence-based approaches.

Fortunately, though, here too a deeper analysis can resolve the apparent clash between these approaches. Though it is true that humans are imperfect, nevertheless we also have enough intelligence and rationality so that we can legitimately be held responsible for cases in which we actually fail to follow the ethically correct rule. Though mismatches between good intentions and good actions can occur, it is the responsibility of ethical persons to eliminate or avoid all potentially serious cases where such clashes might occur.

A standard example of this is driving while drunk. Even if a person sincerely did intend to drive well and not injure anyone while he/she was drunk, nevertheless they can be held responsible for any injuries they cause in spite of their good intentions. Since their moral duty is to always follow ethical rules such as "Cause no harm to innocent persons," they have a duty to avoid getting into situations of this kind where they know they may be unable to avoid breaking such rules. Thus they should either not drink before driving, or not drive after drinking, to ensure that they do not violate the moral rule.

Hence in practice, and on a deeper analysis, both the rule-based and consequence-based theories of ethics would agree that "good intentions are not enough" in ethics, and that ethical behavior requires actually doing what is right—sincere intentions or wishful thinking have no special ethical value, nor are they legitimate excuses for ethically bad behavior.

## Special Topics to Consider

Before concluding this section, a couple of special topics should be mentioned which do not fit easily into the standard classification of rule versus utilitarian approaches, but which are nevertheless generally regarded as being ethically important.

The first of these is the concept of *justice*. Intuitively, it is ethically wrong to treat people in unjust or unfair ways. People should be given what they deserve, and also all people should be treated equally. But it is difficult to reduce justice to a series of rules, and justice may even conflict sometimes with a utilitarian approach, since what is best for most may be unfair to minorities.

The second special topic is that of *human rights*. For example, the U.S. Declaration of Independence talks about the rights to life, liberty and the pursuit of happiness. As with justice, moral rights are not easily reducible to rules, and they may conflict with the greatest good for most people in some cases.

## Descriptive Ethics

In addition to the term "ethics" being used to refer to meta-ethics or normative ethics, sometimes it is used more descriptively (as in "the ethics of a society") to refer to the actual ethical or moral customs or guidelines that people do follow in the given society. To some extent engineering ethics is also descriptive in this way, in that it includes the actual standards which professional engineers have adhered to and continue to adhere to as registered engineers.

Nevertheless, the standards of engineering ethics should always be viewed as having normative force as well—as part of normative ethics, not just descriptive ethics. Thus they should be viewed as standards which one is required to follow as a necessary guide to one's future conduct, not merely as standards which as a matter of fact engineers have followed up to now.

### Example 11.1

A person's behavior is always ethical when one:
A) Does what is best for oneself
B) Has good intentions, no matter how things turn out
C) Does what is best for everyone
D) Does what is most profitable

**Solution:** Since ethics is concerned with standards (other than purely legal ones) for socially approved conduct and with promoting the general welfare, C is the only acceptable answer. The other three answers (doing what is best for oneself, acting with good intentions, and doing what is most profitable) may or may not involve socially approved conduct or promote the general welfare, depending on the individual case, and therefore they cannot guarantee ethical behavior.

## Example 11.2

Which of the following ensure that behavior is ethical?
  I.    Following the law
  II.   Acting in the best interest of a society
  III.  Following non-legal standards for socially approved conduct

  A)  All of the above
  B)  II and III only
  C)  None of the above
  D)  I only

**Solution:** Given that ethics is concerned with standards (other than purely legal ones) for socially approved conduct and with promoting the general welfare, the choices rate as follows: Following the law (I) cannot ensure ethical behavior, because a given law might itself be unethical (example: earlier laws legitimizing slavery.) But both II and III do ensure that behavior is ethical. Hence the correct option is B.

# 11.3  The Nature of Engineering Ethics

Much of engineering ethics is an applied or more specific form of what could be called "general ethics," that is, ethical standards which apply to any human activity or occupation. For example, ethical duties of honesty, fair dealing with other people, obeying the relevant laws of one's country or state, and so on apply in any situation. Thus some of the ethical obligations of engineers are of this general kind also found in many other occupations and activities.

*A conflict between personal standards and engineering ethical standards must be resolved in favor of supporting the relevant engineering ethics standards*

Other kinds of standards found in engineering ethics are more specific but not unique to engineers, such as the principles governing the ethical activities of any professionals in their contacts with clients or customers. These too are an important part of engineering ethics, and they show up in the large overlap in the contents of many codes of ethics derived from many different professions (engineering, law, medicine, academia, etc.).

Then finally there are some standards applying primarily to the profession of engineering, such as those dealing with the proper, ethical manner of approval of designs or plans by managers which require them to have professional engineering qualifications and expertise. These standards are what make engineering ethics distinctive relative to other kinds of applied ethics, but they are only one small part of the whole of engineering ethics, and so should not be over-emphasized.

Now to the question of how ethical standards for engineers differ from other standards of value which engineers might use. This brief discussion will parallel that given above for ethics. The aim is to get clear on which of the choices an engineer might make are relevant to engineering ethics, as opposed to other choices which are instead based on other kinds of values or standards.

As with ethics in general, purely personal choices or values are not relevant to engineering ethics. For example, an engineer who is submitting a bid to a potential client on an engineering project might personally decide to factor in a somewhat higher rate of profit on the project for him/herself than is usual in such projects.

However, as long as there is an open bidding process (where others are free to submit possibly lower bids), and as long as the personal pricing would not affect the quality of the work which would be done, then the pricing level remains a purely personal or economic decision with no ethical implications.

On the other hand, if an engineer were to decide to maximize his/her profit in another way—by submitting a low bid, but then secretly drastically compromising on the quality of materials used so as to achieve maximum personal profit—that would be a violation of engineering ethics standards, which generally require the use of high-quality materials and construction methods acceptable to all parties signing the initial project agreement. This would be a case in which a conflict between personal standards (maximum profit) and engineering ethical standards (honoring the contract and using high-quality materials and methods) must be resolved in favor of supporting the relevant engineering ethics standards in this case.

These contrasts between personal versus ethical values apply also to the related contrast of corporate or economic values versus ethical values. Thus the examples just given of legitimate versus ethically illegitimate maximizing of profits would apply just as well if the motivation were the interest of some corporation or economic group—it would be equally wrong for a corporation or engineering firm to seek profit at the risk of lower engineering quality of products or services. Thus it is the duty of an engineer to uphold engineering ethics standards even if (to take an extreme case) his/her job is at risk, such as when doing the ethically right action would conflict with the non-ethical interests of a boss or corporate structure.

*Engineering ethics codes also generally include prohibitions on unethical behavior while off the job as well.*

The relation of engineering ethics issues to legal issues applying to engineers should also be discussed. Although (as we saw previously) legal and ethical values are basically different kinds of values which should be distinguished, nevertheless it is generally true that one has an ethical obligation to obey the prevailing laws in a given jurisdiction. Also, persons voluntarily entering the engineering profession agree to abide by codes of ethics which explicitly or implicitly require compliance with existing laws, so here too good engineering ethics standards will generally be is in close harmony with existing legal codes governing the practice of engineering.

Another area of close harmony between different kinds of values in engineering ethics is found in the case of technical or scientific values. Any scientific community dealing with pure sciences or applied sciences such as engineering will have technical values concerning the proper and correct ways to construct theories, carry out calculations, make statistically satisfactory estimates where exact results are not possible, and so on. These themselves are not ethical values (good science is not the same as good ethics). But nevertheless engineers do have an ethical obligation to use good scientific methods at all times, and so here too (as with the law) there is a close correlation between practicing good science and being an ethical engineer.

One other topic should be mentioned here. Naturally, most of engineering ethics concerns an engineer's behavior while engaged in professional engineering activities. But one should be aware that engineering ethics codes also generally include prohibitions on unethical behavior while off the job as well, if those activities would affect public perceptions of one's professional integrity or status. This would include activities such as gambling which might tend to bring the profession of engineering into disrepute, deceptive or other inappropriate forms of advertising of services, or any other activities suggesting a lack of integrity or trustworthiness in an engineer.

## Example 11.3

Engineers should follow their professional code of ethics because:
- A) It helps them avoid legal problems, such as getting sued
- B) It provides a clear definition of what the public has a right to expect from responsible engineers
- C) It raises the image of the profession and hence gets engineers more pay
- D) The public will trust engineers more once they know engineers have a code of ethics

**Solution:** These choices are a little harder than those in previous questions, because even the wrong choices do have some connection with ethics. But as long as you follow the initial instruction (choose the best and most relevant answer), you shouldn't have any difficulty answering it.

Choice A: Avoiding legal problems is generally a good thing, but it's not the most relevant reason why the code should be followed (and in some cases, strictly following the code might make it more likely that you'd get sued, e.g., by a disgruntled contractor if you refuse to certify shoddy workmanship).

Choice C: Raising the image of the profession is a good thing, but not for the reason of getting more pay; that would be to act from self-interest rather than ethical motivations.

Choice D: Increased public trust is generally a good thing, but the mere knowledge that there is such a code means little unless engineers actually follow it, and for the right reasons.

Choice B: The best and most relevant answer. The code of ethics is designed to promote the public welfare, and hence the public has a right to expect that responsible engineers will follow each of its provisions, as clearly defined by the code.

## Example 11.4

Engineers should act ethically because:
- A) If they don't, they risk getting demoted or fired
- B) The boss wants them to
- C) It feels good
- D) That's the way responsible engineers behave

**Solution:** As with problem 11.1, some of the wrong answers here do have some connection with ethics, so as before the task is to pick the best and most relevant answer.

Choice A: Even though ethical behavior may usually improve one's job security, it's not the most relevant reason for being ethical.

Choice B: Doing what the boss wants is a self-interested reason for acting, not an ethical reason.

Choice C: Even though, hopefully, being ethical will make engineers feel good, it's not the main reason why they should be ethical.

Choice D: The right answer. There's a very close connection between acting in a responsible manner and acting ethically.

## Example 11.5

The first and foremost obligation of registered professional engineers is to:
   A)  The public welfare
   B)  Their employer
   C)  The government
   D)  The engineering profession

**Solution:** As before, the best and most relevant answer should be chosen.

Choice B:  Engineers do have ethical obligations to their employer, but it's not their foremost or primary obligation.

Choice C: As with the employer, engineers do have some obligations to the government, but again, not the first or primary one.

Choice D:  Obligations to the engineering profession are probably even more important than those to employers or the government, but still they are not the foremost obligation.

Choice A: Correct. The foremost, primary obligation of engineers is to the public welfare.

## Example 11.6

Registered professional engineers should undertake services for clients only when:
   A)  They really need the fees
   B)  Their own bid is the lowest one
   C)  They are fully technically competent to carry out the services
   D)  Carrying out the services wouldn't involve excessive time or effort

**Solution:** Choice A: Personal financial need is a matter of self-interest, not of ethics.

Choice B: The competitive status of a bid is a matter of economics, not ethics.

Choice D: The amount of time involved in carrying out services is also a matter of economics and self-interest, rather than of ethics.

Choice C: Correct. Ethical engineers will not undertake to provide services for clients unless they are fully technically competent to carry them out.

# 11.4 The Issues and Topics of Engineering Ethics

Now we will examine the basic issues, concepts, and special topics of engineering ethics. Recall that most of these are also included in or assumed by the codes of ethics provided by various professional engineering societies. Hence as noted in the Introduction, we will use the most relevant one—the NCEES Model Rules of Professional Conduct—as the basis for the discussion. Its study will provide a good understanding of the typical actual ethical requirements imposed by professional engineering societies on their members.

**The Preamble**

The Preamble (the Introductory section) to the Rules is important in that it describes their purpose, which is to safeguard life, health and property, to promote the public welfare, and to maintain a high standard of integrity and practice among engineers. Where it is helpful, we will slightly paraphrase the statements taken from the code so as to clarify their meaning.

This part of the code makes explicit its connection with general ethical concerns which all ethical persons should adhere to. We will examine each of the items covered in turn in this statement of the purpose of the code.

*Engineers will be honest and trustworthy, and maintain high standards of professional conduct and scientific expertise in their work.*

**1.) First, normative ethical theories universally agree that it is ethically wrong to cause harm to people.** This is why the Preamble mentions the purpose of "safeguarding" life, health and property. Thus ethical persons in general, including engineers, should avoid doing anything which would damage or adversely affect other people. More positively, one should also take measures which will safeguard or preserve people from possible future harm.

For engineers, the more positive measures would include such things as building devices with extra "fail-safe" features included which make harmful consequences of their use as unlikely as possible. Overall, then, ethical persons take great care not to cause harm to others, and they take whatever extra steps are necessary to minimize risks of potential harm to others as well.

**2.) The second purpose of the code is to "...promote the public welfare."** Here too, standard normative ethical theories fully support this rule. It states a duty or obligation not simply to act in a harmless or safe way (as previously discussed), but in addition to take active steps so that one's professional activities will result in definite benefits and improved conditions for the general public.

For example, as an engineer planning a new highway this rule would require one not only to plan and build it in a safe manner (as previously discussed), but also to do such things as choose the shortest feasible route between its endpoints, or to choose that route which would permit the most efficient road-construction techniques, hence maximizing the utility of the highway to the general public and minimizing its cost to them as well.

**3.) The third and last purpose of the code, according to the Preamble, is "...to maintain a high standard of integrity and practice among engineers."** Thus the code also has the purpose of ensuring that engineers will continue to be honest and trustworthy, and to maintain high standards of professional conduct and scientific expertise in their work.

Note that this Preamble makes it explicit that the reason why engineers should adhere to the code isn't simply because it is ethical to do so, or because their professional organization tells them to do it. Instead, it emphasizes that there are vital practical benefits to society which can only be achieved by engineers committing themselves to rigorously follow the code at all times.

Other issues covered in the Preamble are as follows:

**Other Issues**

> *"Engineering registration is a privilege and not a right. This privilege demands that engineers responsibly represent themselves before the public in a truthful and objective manner."*

A privilege is a socially earned license to do something which is granted by society only under certain conditions. For example, a driving license requires the passing of a driving test and other requirements. Similarly, engineering registration must be earned by recipients (they do not automatically have a right to that status) and the granting of it requires them in return to adhere to ethical standards such as the stated one.

> *"Engineers must compete fairly with others and avoid all conflicts of interest while faithfully serving the legitimate needs and interests of their employers and clients."*

This statement sums up a range of ethical requirements which are more fully covered in the body of the code, so the various issues which it raises will be discussed along with those requirements below.

The first group of rules in the code address **the engineer's obligation to society**. As before, we will slightly paraphrase the statements taken from the code so as to clarify their meaning:

**The Engineer's Obligation to Society**

> *1.)* While performing services, the engineer's foremost responsibility is to the public welfare.

This rule is also featured in a related form in the Preamble, as a duty to promote the public welfare. The idea here of responsibility to the public welfare includes the idea of safeguarding the public from harm, as more fully spelled out in the next rule:

> *2.)* Engineers shall approve only those designs that safeguard the life, health, welfare and property of the public while conforming to accepted engineering standards.

These two rules together imply a much broader context of responsibility for engineers than those arising from any one task or project. Designs and materials that seem perfectly adequate and ethically acceptable within the bounds of a given project may nevertheless be unacceptable because of wider issues about the public interest.

For example, until recently a refrigeration engineer could have specified Freon (a chlorinated fluorocarbon or CFC product) as the prime refrigerating agent for use in a product, and defended it as an efficient, inexpensive refrigerant with no risks to the purchaser of the appliance. However, we now know that there are significant risks to

*Designs and materials may be unacceptable because of wider issues about the public interest.*

the public at large from such chemicals because of the long-term damage to the environment they cause when they leak out, perhaps many years after the useful life of the product is over. Rules 1 and 2 tell engineers that they must always keep such wider, possibly longer-term issues in mind on every project they work on.

> *3.)* If an engineer's professional judgment is overruled resulting in danger to the life, health, welfare or property of the public, the engineer shall notify his/her employer or client and any authority that may be appropriate.

An important rule, which may place the engineer in a difficult position if his/her employer or client is among those contributing to the problem. But the engineer's duty in such cases is clear: "...any authority that may be appropriate" must be notified, even if the employer/client tries to prevent it. (Cases of this kind are popularly referred to as "whistle blowing.")

*The duty to be forthcoming about all pertinent or relevant information in reports.*

> *4.)* Engineers shall be objective and truthful in professional reports, statements, or testimonies and provide all pertinent supporting information relating to such reports, statements, or testimonies.

> *5.)* Engineers shall not express a professional opinion publicly unless it is based upon knowledge of the facts and a competent evaluation of the subject matter.

Rules 4 and 5 together implement the general ethical requirement that one ought to tell the truth in the specific context of an engineer's professional duties. Note that the duty as mentioned in Rule 4 is not simply to be truthful in what one says, but also to be forthcoming about all pertinent or relevant information in reports, etc.

Rule 4 also mentions being "objective," which adds the element of being unbiased and basing one's beliefs and reports only on objective, verifiable matters of fact or theory.

Rule 5 enlarges on the idea of being objective in one's reports—others should be able to rely upon one's professional opinion, and they can do this only if one knows all of the relevant facts and is completely competent to evaluate the matter being dealt with.

> *6.)* Engineers shall not express a professional opinion on subject matters for which they are motivated or paid, unless they explicitly identify the parties on whose behalf they are expressing the opinion, and reveal the interest the parties have in the matters.

Rule 6 expresses what is sometimes called the duty of full disclosure. Even if one honestly seeks to be truthful and objective (as in Rules 4 and 5), still doubts might be raised about one's motivation or objectivity unless one reveals on whose behalf one is expressing an opinion, and the interests that such persons have in the case. This rule is also related to the issue of conflicts of interest (see Rules 6 and 8 in the second section, below).

> *7.)* Engineers shall not associate in business ventures with nor permit their names or their firms' names to be used by any person or firm which is engaging in dishonest, fraudulent, or illegal business practice.

This might be called the "clean hands" rule (shake hands only with those whose hands are as ethically clean as your own). It isn't sufficient to be completely ethical in one's own (or one's own company's or firm's) practices; one must also ensure that others do not profit from one's own good name if their own activities are unethical in some way. This Rule 7 is clearly related to Rule 1 concerning the public welfare—one must promote this in one's external dealings just as much as in one's own activities.

8.) Engineers who have knowledge of a possible violation of any of the rules listed in this and the following two parts shall provide pertinent information and assist the state board in reaching a final determination of the possible violation.

Rule 8 generalizes Rule 3 (a duty of disclosure when one's professional judgment is overruled) to a duty of disclosure in the case of any of these rules, when one has knowledge of possible violations of them. In terms of the public welfare, it is very important that each profession regulates itself in this way, so as to minimize or eliminate future infringements of its rules. Strict adherence to this rule also will also lead to wider appreciation and respect for the profession of engineering because of its willingness to "clean its own house" in this way.

The second group of rules in the code address **the engineer's obligation to employers and clients.**

### The Engineer's Obligation to Employers and Clients

1.) Engineers shall not undertake technical assignments for which they are not qualified by education or experience.

2.) Engineers shall approve or seal only those plans or designs that deal with subjects in which they are competent and which have been prepared under their direct control and supervision.

Rules 1 and 2 require an engineer to be professionally competent, both in undertaking technical assignments and in approving plans or designs. Rule 2 in fact requires a double kind of knowledge, both technical competence in the matters to be approved and that one has had direct control and supervision over their preparation. Only thus can one be sure that one's approval is legitimate and warranted.

*A professional engineer must be trustworthy as well as honest, and being trusted not to reveal confidential information is an important kind of trust.*

3.) Engineers may coordinate an entire project provided that each design component is signed or sealed by the engineer responsible for that design component.

Rule 3 in effect invokes Rule 2. As long as each component of a project is satisfactorily approved as per Rule 2, then it is permissible for an engineer to coordinate an entire project. This also underlines the importance of Rule 2, as project managers have to rely heavily on the validity of the approvals for each prior part of a project.

4.) Engineers shall not reveal professional information without the prior consent of the employer or client except as authorized or required by law.

This confidentiality requirement is the other side of the coin of duties to tell the truth (as in Rules 4 and 5 of the first Obligations to Society section). Just as one must not lie or misinform, so also must one restrict to whom one reveals professionally relevant information. A professional engineer must be trustworthy as well as honest, and being trusted not to reveal confidential information is an important kind of trust.

Confidentiality is a central factor in assuring employers and clients that one's professional services for them are indeed for them alone, and that they can rely upon one's discretion in not revealing to others any private information without their full consent. This rule is also related to Rules 5 through 8 below, in that any revealing of information to others would probably create conflicts of interest or other "serving more than one master" problems of those kinds.

> 5.) Engineers shall not solicit or accept valuable considerations, financial or otherwise, directly or indirectly, from contractors, their agents, or other parties while performing work for employers or clients.

This is the first of four rules dealing with problems of "conflicts of interest." These are cases where one has some primary professional interest or group of interests—generally, to carry out some project for an employer or client—but where other factors might enter into the picture which would activate other, non-professional interests one also has which would then conflict with the professional interests.

In the case of Rule 5, the concern is that soliciting or accepting such things as gifts, hospitality, or suggestions of possible future job offers for oneself would activate non-job-related, personal interests of yours (for additional pay, career advancement, etc.) which would then be in conflict with one's primary professional interests and duties concerning a current project.

Special attention should be paid to this and the other conflict-of-interest rules, and they should be strictly observed. People are sometimes tempted to think that breaking these rules is ethically harmless, on the grounds that if one has a strong enough character, then one will not actually be professionally influenced in a detrimental way by gifts, etc., and hence that accepting such inducements cannot do any harm.

*Even the appearance of a conflict of interest can create serious ethical problems.*

However, it is important to realize that even the *appearance* of a conflict of interest (however careful one is to avoid actually undermining one's professional interests) can create serious ethical problems. One basic concern about this is the potential loss of trust it could cause in an employer or client. Just as clients need to know that the engineer will keep their information confidential (as in Rule 4), so also they need to know that the engineer is single-mindedly working with only their interests at heart.

Any doubts raised because of the appearance of a conflict of interest could be very damaging to the client/engineer professional relationship. This is why it is necessary for the professional engineer to avoid doing anything which would create conflicts of interest or even the appearance of them.

Further rules are necessary to deal with conflicts of interest, because unfortunately in some situations the appearance or possibility of conflicts of interest may be virtually unavoidable, no matter how ethically careful everyone is. However, fortunately at the same time there is a powerful method available for minimizing any ethically bad effects of such situations. This is the method of full disclosure of potential conflicts to all interested parties, and it is addressed in the following two rules:

> 6.) Engineers shall disclose to their employers or clients potential conflicts of interest or any other circumstances that could influence or appear to influence their professional judgment or the quality of their service.

*7.)* An engineer shall not accept financial or other compensation from more than one party for services rendered on one project unless the details are fully disclosed and agreed to by all parties concerned.

Both these rules address issues of full disclosure, or keeping all the relevant parties fully informed as to areas of potential conflict or potentially undue external influences. The basic idea behind full disclosure is that it can maintain trust and confidence between all parties in several important ways, as detailed below.

First, if engineer A informs other party B about a potential conflict or influence, then A has been honest with B about that matter, hence maintaining or reinforcing B's trust in A. Furthermore, if B is not further concerned about the matter once he/she knows about it, then the potential problem (namely, the apparent conflict or potentially bad influence on A's professional conduct) has been completely defused.

Suppose on the other hand that B is initially concerned about the issue even after it was honestly revealed to B by A. Even so, the problem is already lessened: at least A has honestly revealed the area of concern. Things would be much worse if B later discovered the problem for him/herself, in a case when A had not fully disclosed it—that would be very destructive of trust between A and B.

Furthermore, now that B knows about the area of concern, and knows that A is fully cooperating with him/her in disclosing the potential problem, both of them can proceed to work out mutually acceptable ways of minimizing or disposing of the problem to their joint satisfaction. Thus even if the full disclosure does lead to an initial problem which needs to be resolved, nevertheless it is one which does not break down the trust between A and B. In fact it may even reinforce trust, in that A's willingness to fully disclose a potential conflict/influence and negotiate with B about it is good evidence for B of A's professional honesty and sincerity.

*When a governmental body is involved, there is another interested party which is not directly represented in negotiations, namely the electorate.*

*8.)* To avoid conflicts of interest, engineers shall not solicit or accept a professional contract from a governmental body on which a principal or officer of their firm serves as a member. An engineer who is a principal or employee of a private firm and who serves as a member of a governmental body shall not participate in decisions relating to the professional services solicited or provided by the firm to the governmental body.

Rule 8 deals with a special case of potential conflicts of interest, namely, when one of the interested parties is a governmental body. In such a case, a somewhat stricter rule is required than for the more usual cases involving only non-governmental agencies.

In non-governmental cases, it is ethically sufficient to fully disclose potential conflicts of interests to all parties, and then to negotiate with the other parties as to how to deal with the potential conflicts. For example, if one's engineering firm has an official who was also on the board of directors of a bank, it would be ethically acceptable to accept a professional contract from that bank, as long as all parties were fully informed about the official's joint appointment prior to the agreement, and also as long as they could come to agree that the joint appointment was not an impediment to their signing a contract.

Returning to the governmental case, the reason why the stricter Rule 8 is required when one of the parties is a governmental body is as follows: In the case of agreements among private persons or businesses, they themselves are the only parties having a legitimate interest in the negotiations, and hence whatever they freely decide among themselves (of course assuming that no other ethical rules or laws are being broken) is acceptable.

On the other hand, in the case when a governmental body is involved, there is another interested party which is not directly represented in negotiations, namely the electorate or citizens of the jurisdiction covered by that governmental body. The governmental body must act only in ways which fully respect the interests and concerns of the electorate. In such a case, it is impossible to ensure that full disclosure to all of the citizens of the electorate of the potential conflicts of interest would be made in a case such as that envisaged in Rule 8, and hence there is a need for the stricter rule which completely prohibits conflicts of interest of the kinds defined in Rule 8. Only then can public trust both in the engineering profession and in governmental bodies be preserved.

## An Engineer's Obligations to Other Engineers

The third group of rules in the code primarily address **an engineer's obligations to other engineers**. These rules can be considered as more precise specifications of various rules already introduced, when applied to the specific context of obligations to other engineering professionals. The first rule also covers obligations to potential employers (whether they are engineers or not) when one is seeking employment.

> *1.)* Engineers shall not misrepresent or permit misrepresentation of their or any of their associates' academic or professional qualifications. They shall not misrepresent their level of responsibility nor the complexity of prior assignments. Pertinent facts relating to employers, employees, associates, joint ventures or past accomplishment shall not be misrepresented when soliciting employment or business.

This Rule 1 is an application or more specific form of Rules 4 and 5 of the first section (concerning an engineer's obligation to society), which require objectivity and truthfulness in all professional reports, statements and opinions.

*This duty not to misrepresent applies also to issues of prior levels of responsibility and to the complexity of previous assignments.*

Rule 1 requires one not to misrepresent one's own qualifications, or those of associates, which is a very important kind of truthfulness and objectivity. The rule further specifies that this duty not to misrepresent applies also to issues of prior levels of responsibility and to the complexity of previous assignments as well. The rule concludes with a statement about any and all pertinent facts in one's previous history, and it requires that they too should not be misrepresented.

It might be thought that this Rule 1 is unnecessary, since it simply applies some general ethical principles already in the code to special cases and circumstances centering around issues of employment and qualifications. However, as with issues of conflicts of interest, there are some special temptations in these areas which are best addressed by explicitly spelling out what is ethically required for dealing with such situations.

A typical temptation in this area might go something like this: Even a well-meaning, otherwise generally honest engineer might be tempted to make his/her qualifications seem more impressive to a potential employer. One might rationalize that one really would be the best person to do the required job, and that therefore one is doing the employer a favor by making one's qualifications seem more impressive and thereby inducing them to hire oneself, rather than some less-qualified person whose paper qualifications might misleadingly look as good as one's own.

Clearly there could be many variants on this manner of thinking. But all would involve rationalization (the inventing of dubious reasons for what one wants to believe anyway) and wishful thinking, rather than the objectivity and true rationality and truthfulness which are required of professional engineers. So Rule 1 serves a useful function in explicitly stating requirements which some might otherwise be tempted to ignore or overlook.

> *2.)* Engineers shall not directly or indirectly give, solicit, or receive any gift or commission, or other valuable consideration, in order to obtain work, and shall not make a contribution to any political body with the intent of influencing the award of a contract by a governmental body.

Rule 2 continues the prohibitions against conflicts of interest which were found in Rules 5 through 8 in the previous section (rules covering the engineer's obligations to employers and clients). Those previous rules mainly covered cases where an engineer was already employed, while Rule 2 here specifically applies to attempts to obtain future work, including the award of a contract by a governmental body.

This Rule 2 also emphasizes that it is just as wrong to attempt to unduly influence someone else (a potential employer, for instance) as it would be to allow others to unduly influence oneself. Thus the rule underlines that it is just as ethically unacceptable to try to cause conflicts of interest in others as it is to allow oneself to be enmeshed in improper conflicts of interest.

It should be noted that the second part of Rule 2, "... and shall not make a contribution to any political body with the intent of influencing the award of a contract by a governmental body," specifically mentions the intent of the person making the contribution. Contributions are not prohibited, only contributions with the wrong intent.

This part of the rule could be difficult to apply or enforce in practice, because it may be very hard to establish what an engineer's actual intent was in making a political contribution. Also, the freedom to make political contributions to organizations of one's own choice is generally viewed as an ethical right which should be limited as little as possible. So this part of the rule very much depends on and appeals to the ethical conscience of the individual engineer, who must judge his/her own intentions in such cases and avoid such contributions when their own intent would be self-interested in the manner prohibited by the rule.

*It is ethically unacceptable to try to cause conflicts of interest in others as it is to allow oneself to be enmeshed in improper conflicts of interest.*

Note also that the first part of Rule 2, "Engineers shall not directly or indirectly give, solicit, or receive any gift or commission, or other valuable consideration, in order to obtain work ...," also mentions the reason or intention behind giving or receiving gifts, etc., in the phrase "in order to obtain work." However, in practice it is much easier to judge when gift-giving is ethically unacceptable than when political contributions are unacceptable, since there are more behavioral and social tests for suspicious inducements to obtain work than there are for suspicious political support. So it is much easier to police and regulate infringements of this first part of Rule 2 than it is for the second part.

> *3.)* Engineers shall not attempt to injure, maliciously or falsely, directly or indirectly, the professional reputations, prospects, practice or employment of other engineers, nor indiscriminately criticize the work of other engineers.

Rule 3 as it stands is somewhat unclear, so some discussion is required to bring out its ethically legitimate core.

First, how should the subordinate phrases "maliciously or falsely" and "directly or indirectly" be interpreted? On one possible interpretation, Rule 3 says outright that engineers should never attempt to injure in any way or for any reason the professional reputations of other engineers. On this interpretation those phrases just give examples of possible modes of injury which are prohibited, leaving unmentioned any other possible modes of injury which are nevertheless also assumed to be prohibited.

However, another interpretation is possible, according to which it is only certain kinds of injury which are prohibited by Rule 3, namely those spelled out by those same phrases interpreted so that the "directly or indirectly" part modifies the "maliciously or falsely" part. On this interpretation, Rule 3 prohibits only malicious or false attempts (whether carried out directly or indirectly) to injure the reputations of other engineers.

This second interpretation would ethically permit attempts to injure the reputations of other engineers, as long as the attempts were carried out in a non-malicious and honest, truthful way (and presumably with the public welfare in mind as well).

Some support for this second interpretation can be derived from the final clause of Rule 3, "nor indiscriminately criticize the work of other engineers." Clearly, in this case it is not all criticism of the work of other engineers which is being prohibited, but only indiscriminate criticism. Thus the last section of Rule 3 outlaws criticism which is over-emotional, biased, not well reasoned or factually inaccurate, and so on, but it does not prohibit well-reasoned, careful, accurate, factually-based criticisms of other engineers.

*If criticism of other engineers ever becomes necessary, it should be done in a very cautious and objective manner, and with all due respect for the professional status of the person being criticized.*

Thus in support of the second interpretation, it might be said that there too, it is not all attempts to injure reputations, which are being prohibited, but only those which are malicious, false, indiscriminate or otherwise highly ethically questionable in the methods they employ.

However, there is still something to be said in favor of the first interpretation as well (which, it will be recalled, involves a blanket condemnation of all attempts to injure reputations). Those defending it might do so as follows: It is arguable on general ethical grounds that any attempt to injure someone's reputation must be viewed as going too far and therefore becoming unethical. Even if one is convinced that someone else's work is shoddy, dishonest, and so on, at most (it could be argued) one has a duty to point out the problems and shortcomings in their work. It is a big leap from criticizing an engineer's actions, on the one hand, to condemning the engineer in a way designed to injure his/her reputation, on the other hand.

Thus (on this line of thinking) one should criticize an engineer only out of a disinterested desire to let the truth be known by all, not with the aim of injuring someone's reputation—it is for others to judge whether the truth of what one has pointed out will diminish the reputation and so on of the engineer in question.

Fortunately, it is not necessary to definitively decide between these different interpretations of Rule 3. What is important is to become sensitive to the ethical issues involved in each interpretation. And for the purposes of conforming to Rule 3, both sides can agree that if criticism of other engineers ever becomes necessary, it should be done in a very cautious and objective manner, and with all due respect for the professional status of the person being criticized.

## Example 11.7

With respect to the Moral Rules of Professional Conduct for engineers:
- A) The rules are a bad thing because they encourage engineers to spy on and betray their colleagues
- B) The rules are a useful legal defense in court, when engineers can demonstrate that they obeyed the rules
- C) The rules enhance the image of the profession and hence its economic benefits to its members
- D) The rules are important in providing a summary of what the public has a right to expect from responsible engineers

**Solution:** Remember here that the best/most relevant answer is to be chosen, because each answer has some truth to it, but only one has the most truth.

Choice A: It is true that the rules require those who have knowledge of violations of the rules to report such cases to the relevant State Board. And this could involve one in collecting more information on the possible infringements, and hence exposing those involved to the scrutiny of the State Board. However, the generally negative connotations of "spying" and "betraying colleagues" do not make the rules bad—the activities in question are a necessary part of responsible reporting of possible violations to the authorities, and hence are morally fully justified.

Choice B: It is true that proof in court that one has followed the rules may be a useful legal defense. However, this is only a secondary, indirect effect of the ethical value of the rules themselves, and so B does not provide the best answer.

Choice C: Again, it is true that the adoption of the rules by the engineering profession will enhance its image and economic benefits. But like B, this is only a secondary and derivative effect of the rules.

Choice D: The right answer. Since the basic function of the rules is to provide a guide for ethical conduct for engineers, the rules also provide a useful summary for the public of what they have a right to expect from responsible engineers.

## Example 11.8

The Model Rules of Professional Conduct require registered engineers to conform to all but one of the following rules—which rule is not required?
- A) Do not charge excessive fees
- B) Do not compete unfairly with others
- C) Perform services only in the areas of their competence
- D) Avoid conflicts of interest

**Solution:** There may be one or two problems of this kind in the exam, which can be solved by memorizing the rules or by checking them directly to see which are or are not included. However, it is better to acquire a good understanding, the basic ethical concerns behind the rules, in which case the right answer here will be clear immediately.

Rule A is not required because fees are a matter of free negotiation between engineers and clients. Hence a fee which might seem excessive to some may be acceptable to others because of an interest in extra quality, or unusually quick delivery time, and so on. The other three rules are required.

## Example 11.9

You are a quality control engineer, supervising the completion of a product whose specification includes using only U.S.-made parts. However, at a very late stage you notice that one of your sub-contractors has supplied you with a part having foreign-made bolts in it—but these aren't very noticeable, and would function identically to U.S.-made bolts. Your customer urgently needs delivery of the finished product—what should you do?

A) Say nothing and deliver the product with the foreign bolts included, hoping this fact won't be noticed by the customer

B) Find (or, if necessary, invent) some roughly equivalent violation of the contract or specifications for which the customer (rather than your company) is responsible—then tell them you'll ignore their violation if they ignore your company's violation

C) Tell the customer about the problem, and let them decide what they wish you to do next

D) Put all your efforts into finding legal loopholes in the original specifications, or in the way they were negotiated, to avoid your company's appearing to have violated the specifications

**Solution:** Choice A: This is wrong because it is dishonest—it violates the requirement of being objective and truthful in reports, etc.

Choice B: This is wrong because "two wrongs don't make a right." Negotiations with clients should always be done in an ethically acceptable manner.

Choice D: This would violate at least the spirit of the initial agreement. The ethical requirement of being objective and truthful includes an obligation not to distort the intent of any agreements with clients.

Choice C: The correct answer. Being honest with a client or customer about any production difficulties allows them to decide what is in their best interest given the new disclosures, and provides a basis for further good-faith negotiations between the parties.

## Example 11.10

You are the engineer of record on a building project which is behind schedule and urgently needed by the clients. Your boss wants you to certify some roofing construction as properly completed even though you know some questionable installation techniques were used. Should you:

A) Certify it, and negotiate a raise from your boss as your price for doing so

B) Refuse to certify it

C) Tell the clients about the problem, saying that you'll certify it if they want you to

D) Certify it, but keep a close watch on the project in future in case any problems develop with it

**Solution:** There are some temptations and half-right elements in some of these, but they must be resisted as not being completely ethical.

Choice A: Even if one informs one's boss of the problems, and negotiates his/her consent to your certifying it, nevertheless it is always wrong to certify work which does not measure up to the highest professional standards.

Choice C: Wrong for the same reason as A. Even if one honestly reveals the problems to the clients, and gets their consent, nevertheless it is the professional duty of an engineer not to certify dubious work.

Choice D: The initial certification would be wrong, no matter how carefully one monitors future progress in the hope of minimizing any future problems.

Choice B: The right choice. Whether or not one's boss is happy with this, it is one's professional duty to refuse certification in such cases, even if as a result one is reassigned or fired.

## Example 11.11

You are an engineer and a manager at an aerospace company with an important government contract supplying parts for a space shuttle. As an engineer, you know that a projected launch would face unknown risks, because the equipment for which you are responsible would be operating outside its tested range of behaviors. However, since you are also a manager you know how important it is to your company that the launch be carried out promptly. Should you:

A) Allow your judgment as a manager to override your judgment as an engineer, and so permit the launching

B) Toss a coin to decide, since one's engineering and managerial roles are equally important, so neither should take precedence over the other

C) Abstain from voting in any group decision in the matter, since as both a manager and an engineer one has a conflict of interest in this case

D) Allow your judgment as an engineer to override your judgment as a manager, and so do not permit the launching

**Solution:** Note that a real-life case similar to this problem occurred with the Challenger space-shuttle disaster.

Choice A: Wrong, because engineers have special professional duties and ethical commitments which go beyond those of corporate managers.

Choice B: Wrong for same reason as A; ethically, engineering responsibilities are more important than managerial responsibilities.

Choice C: Wrong because one's duties as a professional engineer require appropriate action even if other factors may seem to point in the opposite direction or toward abstention.

Choice D: The correct choice. Whatever other duties an engineer has, his/her professional engineering responsibilities must always be given first priority.

## Example 11.12

Your company buys large quantities of parts from various suppliers in a very competitive market sector. As a professional engineer you often get to make critical decisions on which supplier should be used for which parts. A new supplier is very eager to get your company's business. Not only that, but you find they are very eager to provide you personally with many benefits—free meals at high-class restaurants and free vacation weekends for (supposedly) business meetings and demonstrations, and other more confidential things such as expensive gifts that arrive through the mail, club memberships and so on. What should you do?

    A) Do not accept any of the gifts that go beyond legitimate business entertaining, even if your company would allow you to accept such gifts

    B) Report all the gifts, etc., to your company, and let them decide whether or not you should accept them

    C) Accept the gifts without telling your company, because you know that your professional judgment about the supplier will not be biased by the gifts

    D) Tell other potential suppliers about the gifts, and ask them to provide you personally with similar benefits so you won't be biased in favor of any particular supplier

**Solution:** Choice B: Wrong, because even if one's company finds such gifts awkward, it is still one's duty as a professional engineer not to become involved in such conflicts of interest.

Choice C: Also wrong. It doesn't matter whether you believe you can remain unbiased, because you still have the conflict, and the possible perception by others that you might be biased by it also remains an ethical problem.

Choice D: Wrong. Remember, two wrongs don't make a right, and the same principle applies no matter how many wrong actions are involved.

Choice A: The right answer. As with the reasoning on choice B, it makes no difference whether others (having less demanding ethical standards than engineers) would find such things acceptable or not. You must not accept any gifts the acceptance of which would involve you in conflicts of interest.

# 11.5 Engineering Ethics and Legal Issues

This section will provide an introduction to the various ways in which the law regulates and affects the profession of engineering. The impact of laws and regulations on various engineering ethics issues will also be discussed.

Engineers, like any citizens, are of course expected to obey the general legal rules and regulations of the societies in which they live. But here we will concentrate on those laws and legal concepts which have special relevance to engineers and the engineering professions. Every engineer needs to have a good basic grasp of these legal matters.

First is a very important area of the law which is central not only to engineering but to many professions and businesses as well. It concerns the regulation of the basic transactions between professionals and their clients or customers.

**Contract Law**

A *contract* is a mutual agreement between two or more parties to engage in a transaction which provides benefits to each of them. Here is a breakdown of this concept into its elements, each of which is required for a valid contract to exist. For simplicity we will concentrate on two-party contracts.

    1) Mutual consent. Each of the parties to a contract must consent to the contract and agree to be bound by its terms.

2) Offer and acceptance. There must be an offer of some kind by one of the parties (e.g., to manufacture some equipment or carry out some services), and acceptance of that offer by the other party.

3) Consideration. Each party must provide something of value to the other party.

The first two conditions, mutual consent and offer/acceptance, are needed because merely discussing a possible contract in provisional terms establishes no legally binding agreement. A contract exists only when each party has established that the other consents to it, and that one is offering something which the other has decided to accept. In other words, a legally enforceable agreement requires a definite promise by each party to do something specific; vague talk about possible actions by each is not sufficient.

The issue of consideration is also important, but in more complex ways. The idea behind it is that there cannot be a legally enforceable contract unless each side stands to benefit in some way from the contract, so that after the contract is fulfilled, each has received some benefit which they did not have prior to being fulfilled.

Without such evidence of definite benefits for each party, it would often be impossible to decide whether or not each party had fulfilled their side of the agreement. For example, if an engineer were to hire a relative to do some work, and paid them for doing it, but did not specify in any way exactly what the work would be, then the courts would likely rule that no valid contract existed and so the payments to his/her relative were illegal or invalid.

Another kind of example where lack of consideration raises legal and ethical questions is as follows: If an engineering firm provided certain services to a client, but never asks for nor receives any payment for them, there would be strong suspicion that the services provided were some kind of bribe or illegal payback for other hidden services or benefits rather than part of a valid contract between the parties.

Some cases involving the issue of consideration are less clear-cut. For example, if an engineer decides to undertake a difficult and complex project, yet does so at a very low rate of pay far below standard prices for such work, could such a contract later be held to be invalid because of the very inadequate reward provided to the engineer? Generally the courts have held that it is up to the parties to a contract themselves to decide what constitutes adequate consideration or reward, so the law will not provide protection to engineers against any possibly self-destructive intentions or other errors of judgment in which they may become involved.

On the other hand, issues about specific amounts of consideration or reward are often central in settling contractual disputes between parties; see the section on breach of contracts below.

It should be noted that a contract does not have to be in writing to be valid. For example, an engineer could make a contract over the phone with a client. But a written contract is always desirable for clarity, and it provides documentation should any questions about the terms and conditions of the contract arise later. So engineers are well advised to prepare specific documentation on any agreements they make as soon as possible after the agreements are finalized.

*Without evidence of definite benefits for each party, it would often be impossible to decide whether or not each party had fulfilled their side of the agreement.*

*A contract does not have to be in writing to be valid.*

## Breach of Contract

One of the main values of having a legally enforceable system of contracts is that in the case of any disputes between the parties, the contract provides an independent check or test of the validity of the claims of each party to the dispute. For example, if a client orders some parts from an engineering firm, but later is dissatisfied with them and wants replacements or a refund of payments, then the issue can be resolved by checking in the contract to see whether the parts conform to what is specified for them in the contract. If the engineering firm has carried out its part of the contract by supplying parts as specified, then legally the client has no case against it. Each party to a contract must decide ahead of time whether they really want what they are agreeing to in a contract, and therefore they cannot blame the other party to the contract even if they are unhappy about the outcome of the agreement later.

*In order for an actual breach of contract to occur, there must be some actual violation of the terms of the contract.*

In order for an actual breach of contract to occur, there must be some actual violation of the terms of the contract by one or both of the parties involved. For example, if specified items were not supplied, or were supplied but were of substandard quality, or if items were not supplied until long after a deadline had expired, then the other party could claim that a breach of contract had occurred and that they are entitled to compensation or termination of the contract.

Generally speaking, those suing for breach of contract will try to obtain sufficient damages or compensation to recover the value that they would have obtained under the contract had it not been violated. In other words, the other party is being required to provide an equivalent value to that which they had previously offered under the provisions of the contract.

One further distinction should be noted here: that between a material and an immaterial breach of a contract. Material breaches concern vital elements of the contract, the presence of which would justify termination of a contract and a suit for damages. But there may be less severe deviations from a contract (such as a contractor not cleaning up a site at the end of a day's work, or being a few days late in delivering a product) which would be handled differently. Usually immaterial breaches will require the offending party only to make reparations, or to accept a reduced fee for their work.

*We should try to "read between the lines" in terms of the intent of those documents as understood by those who formulated them.*

Often the distinction between a material versus immaterial breach of a contract will depend on issues about consideration (which is, you will recall, the value being received by each party to the contract). For example, suppose engineering firm A fails to supply client B with certain parts X as specified in a contract, but supplies substitute parts Y instead.

If the contract specified parts of X, then there has certainly been a breach of the contract. But whether it is a material or immaterial breach will depend on whether the substitute parts Y are similar enough in value to parts X so that client B should be willing to accept them as a substitute for parts X. Intuitively, if B is just as well off (or nearly so) with the parts Y as he would have been with parts X, then the breach of the contract would only be an immaterial one. However, if parts Y are substantially worse than parts X, there would be a material breach of the contract, requiring stronger legal action.

This material/immaterial distinction has ethical implications as well. Generally, any party to a contract who is unable to fulfill the exact provisions of the contract is under an ethical as well as a legal imperative to do everything possible to provide at least an equivalent value to the other party to a contract.

This is an important distinction. It comes about because laws or contracts typically are not completely specific in their terms, and so are open to varying degrees and types of interpretation in problematic cases.

How should such interpretations, or attempts to "read between the lines," be carried out?  A standard view is that we should try to do so in terms of the intent or desired interpretations of those documents as understood by those who formulated them.

In this way ethical standards enter as an important element in understanding the "spirit" of laws and contracts. Because typically, lawmakers or contracting parties wrote their rules or agreements with general ethical standards in mind as to what would constitute the best and most ethically acceptable laws or contracts, given the purposes they were trying to achieve. Thus basic issues of engineering ethics are relevant even in cases when legal documents are being dealt with, insofar as such documents leave any room at all for disputes about their proper interpretation.

**The Letter versus the Spirit of the Law**

# Exercises

**Introductory Note**

As with any form of ethics, engineering ethics demands more than an impersonal, spectator point of view from us. It is important to consider ethical problems not simply as some other person's problem—to be handled with impartial advice or citing of ethical rules—but instead as a vital problem for oneself which must be resolved in terms that you personally can live with.

To encourage this perspective, many of the cases are presented in the second person, i.e., as problems for you rather than some abstract person. In answering them it is important to maintain this perspective, and honestly explain how you would actually deal with each situation yourself.

11.1 Your company urgently needs a new manager in another division, and you have been invited to apply for the position. However, one of the job requirements is competence in the use of some sophisticated CAD (computer aided design) software. Currently you do not have that competence, but you are enrolled in a CAD course, so eventually you will have that competence (but not before the deadline for filling the new job).

The application form for the job requires you to state whether or not you have the necessary CAD software competence. Is it ethically permissible to claim that you have (in order to get the job for which otherwise you are well qualified), even though strictly speaking this is not true?

Would your answer be any different if your current boss informally (off the record) advised you to misrepresent your qualifications in this way, giving as his/her reason that you are overall the best qualified person for the new managerial position?

11.2 You are a supervisor for a complex design project, and are aware of the NCEES code rule according to which engineers may coordinate an entire project provided that each design component is signed or sealed by the engineer responsible for that design component.

You are very overworked, and so have got into the habit of, in effect, rubber-stamping the design decisions of the subordinate engineers working on your project. You do no testing or checking of your own on their projects, and routinely approve their plans (after they have signed or sealed them) with only the most superficial read-through of their work. In your mind this policy is justifiable because the NCEES rule says nothing about the supervisory responsibilities of someone in your position. Besides which, you trust the engineers to do a good job, and so see no need to check up on them.

Is the supervisor's position here ethically justifiable?

11.3   Newly hired as a production engineer, you find a potential problem on the shop floor: workers are routinely ignoring some of the government-mandated safety regulations governing the presses and stamping machines.

The workers override safety features such as guards designed to make it impossible to insert a hand or arm into a machine. Or they rig up "convenience" controls so they can operate a machine while close to it, instead of using approved safe switches, etc., which requires more movement or operational steps. Their reason (or excuse) is that if the safety features were strictly followed then production would be very difficult, tiring and inefficient. They feel that their shortcuts still provide adequately safe operation with improved efficiency and worker satisfaction.

Should you immediately insist on full compliance with all the safety regulations, or do the workers have enough of a case so that you would be tempted to ignore the safety violations?  And if you were tempted to ignore the violations, how would you justify doing so to your boss?

Also, how much weight should you give to the workers' clear preference for not following the regulations: ethically, can safety standards be relaxed if those to whom they apply want them to be relaxed?

11.4   You and an engineer colleague work closely on designing and implementing procedures for the proper disposal of various waste materials in an industrial plant. He is responsible for liquid wastes, which are discharged into local rivers.

During ongoing discussions with your colleague, you notice that he is habitually allowing levels of some toxic liquid waste chemicals, which are slightly higher than levels permitted by the law for those chemicals. You tell him that you have noticed this, but he replies that, since the levels are only slightly above the legal limits, any ethical or safety issues are trivial in this case, and not worth the trouble and expense to correct them.

Do you agree with your colleague? If not, should you attempt to get him to correct the excess levels, or is this none of your business since it is he rather than you who is responsible for liquid wastes?

If he refuses to correct the problem, should you report this to your boss or higher management? And if no one in your company will do anything about the problem, should you be prepared to go over their heads and report the problem directly to government inspectors or regulators?  Or should one do that only in a case where a much more serious risk to public health and safety is involved?

11.5    Your automotive company is expanding into off-road and all-terrain vehicles (ATVs). As a design engineer you are considering two alternative design concepts for a new vehicle, one having three wheels and another the four.

Engineering research has shown that three-wheeled vehicles are considerably less stable and safe overall than four-wheeled ones. Nevertheless, both designs would satisfy the existing safety standards for the sale of each class of vehicles, so either design could be chosen without any legal problems for the company.

However, there is another factor to consider. Market research has shown that a three-wheeled version would be easier to design and produce, and would sell many more vehicles at higher profit margins than would a four-wheeled version.

Is it ethically acceptable for you as an engineer to recommend that your company should produce the three-wheeled version, in spite of its greater potential safety risks? Or do engineers have an ethical obligation always to recommend the safest possible design for any new product?

11.6    Your company has for some time supplied prefabricated wall sections, which you designed, to construction companies. Suddenly one day a new idea occurs to you about how these might be fabricated more cheaply using composites of recycled waste materials.

Pilot runs for the new fabrication technique are very successful, so it is decided to entirely switch over to the new technique on all future production runs for the prefabricated sections. But there are managerial debates about how, or even whether, to inform the customers about the fabrication changes.

The supply contracts were written with specifications in functional terms, so that load-bearing capacities and longevity, etc., of the wall sections were specified, but no specific materials or fabrication techniques were identified in the contracts. Thus it would be possible to make the changeover without any violation of the ongoing contracts with customers.

On the other hand, since there is a significant cost saving in the new fabrication method, does your company have an ethical obligation to inform one's customers of this, and perhaps even to renegotiate supply at a reduced cost, so that one's customers also share in the benefits of the new technique?

More specifically, do you have any special duty, as a professional engineer and designer of the new technique, to be an advocate in one's company for the position that customers should be fully informed of the new technique and the associated cost savings?

11.7    Your company manufactures security systems. Up to now these have raised few ethical problems, since your products were confined to traditional forms of security, using armed guards, locks, reinforced alloys which are hard to cut or drill, and similar methods.

However, as a design engineer you realize that with modern technology much more comprehensive security packages could be provided to your customers. These could also include extensive video and audio surveillance equipment, along with biometric monitoring devices of employees or other personnel seeking entry to secure areas which would make use of highly personal data such as a person's fingerprints, or retinal or voice patterns.

But there is a potential problem to be considered. A literature search reveals that there are many ethical concerns about the collection and use of such personal data. For example, these high-tech forms of surveillance could easily become a form of spying, carried out without the knowledge of employees and violating their privacy. Or the data collected for security reasons could easily be sold or otherwise used outside legitimate workplace contexts by unscrupulous customers of your surveillance systems.

Your boss wants you to include as much of this advanced technology as possible in future systems, because customers like these new features and are willing to pay well for them.

However, you are concerned about the ethical issues involved in making these new technologies available. As an engineer, do you have any ethical responsibility to not include any such ethically questionable technologies in products which you design and sell, or to include them only in forms which are difficult to misuse? Or is the misuse of such technologies an ethical problem only for the customers who are buying your equipment, rather than it being your ethical responsibility as an engineer?

11.8   Can you think of any issue which should be of ethical concern to engineers but is not adequately addressed by current codes of engineering ethics?

If so, should it be added to the current rules, or is it something which cannot adequately be summed up in an ethical rule?

11.9   Can you think of any rule included in current codes of engineering ethics which is unnecessary, too restrictive or even ethically wrong? If so, should it simply be removed from the current rules, or is it something which serves a useful function even though it is not ethically required?

If you don't think there are any superfluous rules, instead pick one which you think is one of the least ethically important, and explain why you think it is not important.

11.10   As an engineering expert on a state planning board, you have to decide which traffic safety projects (involving installation of traffic lights, road-widening, etc.) should be funded by the state. Previously these matters were decided politically: each region received roughly equal funding, which tended to maximize voter satisfaction and ensure re-election for the politicians. But now non-elected engineering experts such as yourself will decide these matters.

On what grounds should decisions such as these be made? What would be the soundest and most ethically justifiable way for you to allocate the funds?

In particular, would a utilitarian approach (the greatest good for the greatest number) be the best way to allocate funds? But there is a potential problem with this approach. Since the greatest number of people live in cities and suburbs and very few in rural areas, it is likely that scarce funds would always tend to be allocated to cities with this approach, with almost none going to sparsely-populated areas. This policy may tend to save the greatest number of lives or minimize accidents overall, but is it not very unfair to those living in or traveling through rural areas?

But on the other hand, how could you justify spending scarce resources in rural areas when you know that this would benefit far fewer people than if you spent the money in cities instead?

# 12. Engineering Work Experience

## 12.1  A Job and Experience

This is the job-seeker's first, natural question. It was used in a national publicity campaign by the National Commission for Cooperative Education a number of years ago. It's one that all engineering students have considered or will consider at some point during their college career. Career-related experience, commonly called "experiential learning," has become an important supplement to a student's engineering education. Throughout the years, and despite the changing times of up-and-down job markets, numerous employers of engineering graduates, and engineering graduate schools have continued to stress the fact that they seek to fill their annual openings by selecting only from those candidates who have done outstanding things during their college years to distinguish themselves from their peers. In most cases, those students who have participated in some form of experiential learning have gained the types of skills and competencies that prospective employers and graduate schools are seeking.

*"How do you get experience without a job, and how do you get a job without experience?"*

A college education has evolved over the past century to be quite a different experience from that of our predecessors. For most engineering students during the first half of the 20th century, college was completed in four years. Since studying engineering has always been a rigorous experience, most students chose to devote their time and efforts to experiences in the classroom, library, or laboratory. Those who chose to work while going to school were viewed by the conventional student population as "non-traditional," representing the "working class" who were forced to work for pay solely as a means of affording their schooling. However, in the latter part of the 20th Century the costs of a college education soared. It is now the rule, rather than the exception, that most students are forced to work for pay in order to keep up with steadily increasing tuition, room and board, and associated expenses such as books, computers, and transportation. (Again, some countries fully subsidize college education—with one of the benefits being that students can concentrate fully on their studies, with less need for extracurricular employment during school.)

For many students, the idea of combining their academic studies with a career-related employment experience is something they had not previously considered. The thought of stretching their college education to five or even six years can seem to be an unnecessary burden. To their parents who grew up with the notion that a college education is a four-year plan with a direct path into the world of employment, lengthening the time spent in school is also an uncomfortable concept.

Consider the change in employment trends in recent years. The current generation of college students was born, for the most part, in the late 1970s and early 1980s. During the relatively short span of their lifetime, this group has witnessed firsthand several major shifts within corporate America. During the years when many of these students were born, a major recession was experienced in some of the largest manufacturing sectors. Especially hard hit was the automobile industry and the surrounding Midwest region of the country. Dubbed the "Rust-Belt," this area saw a significant reduction in the work force, especially in the traditional "blue-collar" jobs. Jobs that had traditionally been viewed as secure for a worker's lifetime were now eroding and in many cases disappearing. During this period, the employment of professionals remained fairly stable, and new engineering graduates were having only minor problems gaining employment at graduation time.

*Most students have to get a job to help keep up with steadily increasing tuition, room and board, and associated expenses.*

The period 1983-1986 saw a revival in the U.S. economy and this was reflected by an increase in hiring of new engineers, with salaries well above the rate of inflation.

The period 1988-1994 was one of major restructuring within corporate America. Many major firms instituted a series of layoffs and experienced periods of downsizing. This period was notable for how the "white-collar" work force was affected. Professionals either were being laid off or provided with incentives to retire early. The net effect was a significant restructuring, reflected by one of the weakest labor markets ever for engineers.

From 1994 to the present, the economy has rebounded vigorously. Increases in productivity, the general vitality of the global marketplace, mergers and acquisitions, and a lower birth rate which has produced one of the lowest supplies of graduating engineers in decades all have contributed to one of the strongest job markets in many years. The current group of graduating engineers are receiving multiple job offers— some with extra incentive packages including tuition reimbursement, extra benefits, and signing bonuses.

Note that the potential long-term economic impact of the catastrophic terrorist attacks of September 11, 2001, are yet to be fully realized.

Truly, the current generation of college students has witnessed a roller-coaster type of economy. Many realize that the rules of the game have changed and that lifelong job security with a single employer is a thing of the past.

However, one fact remains clear from employers in good times and bad: the need for college graduates with experience is a constant. Many employers expect today's college graduates to have significant work experience. Why? Because engineering graduates with career-related work experience require less training and can produce results more quickly than the typical recent college graduate. To an employer, this is sound economic policy.

There are many forms of experiential learning available to engineering students. This chapter will discuss several of the most popular options, including on- and off-campus jobs, summer work, volunteer experiences, academic internships, research assistantships, and perhaps the most popular and beneficial program for students, cooperative education. Each of these has its own advantages and disadvantages, depending on a student's particular situation and background. These issues will be examined as part of this chapter.

## 12.2  Summer Jobs and On- and Off-Campus Work Experiences

Most engineering students arrive at college with some type of work experience. Some have mowed lawns, done baby-sitting, or worked in fast-food franchises, while others may have had the opportunity to actually work in a business or industry that was involved in some type of engineering activity. Whatever the experience, it can be helpful in preparing you for your future. Career-related work experience is clearly preferred by engineering employers, but one has to start someplace in developing a work history.

> *Career-related work experience is clearly preferred by engineering employers.*

There is value in working in the residence hall cafeteria, or delivering pizzas, or helping out in one of the campus offices. These experiences begin to build a foundation which your classroom learning and other career related experiences will supplement. On- and off-campus jobs and summer work experiences may not seem highly relevant, but they can help develop important skills. Through such experiences, any students have strengthened their communication skills, learned to work as part of a team, and developed problem-solving abilities and an organizational style that can be applied to engineering courses and future employment positions. Jobs that you hold as a student will influence your perceptions about the things you like to do, the conditions and environments in which you like to do them, and the types of people and things with which you like to work. These jobs can be useful in helping you identify your personal strengths and weaknesses, interests, personal priorities, and the degree of challenge you are willing to accept in a future position.

"Real world" experience can come in many forms. While career-related experience is important, do not underestimate the importance of learning some basic "on the job" skills. If you are seeking to develop some basic job skills or merely trying to earn some extra money to assist with your college expenses, you should not be afraid to explore some of the many employment options that are currently available on and off campus during the school year and in the summer.

## 12.3  Volunteer or Community Service Experiences

Another popular form of experiential learning is found in the many opportunities associated with volunteer or community service experiences. While these programs are available to engineering students of all class levels, they can be particularly beneficial to freshman or sophomore students who are trying to gain some practical experience. Generally, volunteer or community service projects can be short- or long-term arrangements. Typically, these positions are with nonprofit organizations such as human service groups, educational institutions at all levels, a variety of local, state and federal government agencies, or small businesses. These groups are particularly interested in engineering students for their strong backgrounds in math, science, and computer skills, which can be applied to a variety of projects and activities.

> *They can be particularly beneficial to freshman or sophomore students who are trying to gain some practical experience.*

Even though these positions usually are not paid, students can gain significant experience and develop important skills. Students have found that working with different groups can enhance their communication abilities, as they are often interacting with people much different from those in their collegiate environment. Through the various projects and activities, students can develop organizational skills, initiative, and the desire for independent learning that can be useful in many settings including the classroom and other employment situations. Students involved in volunteer and community service experiences find these opportunities helpful with career decisions, while building confidence with the satisfaction of helping others and providing needed services.

For those students interested in developing a basic set of skills which can be useful throughout their professional engineering career, volunteer and community service work is worth exploring. For more information, contact your campus placement center or a particular firm or agency in your area. A faculty member or academic advisor may be able to provide assistance in this very valuable endeavor.

## 12.4 Supervised Independent Study or Research Assistantship

The *Supervised Independent Study* or *Research Assistantship* is a form of experiential learning which is designed primarily for the advanced undergraduate engineering student who is considering graduate school and/or a career in research and development. Generally, a supervised independent study or research assistantship is a planned program of study or research that involves careful advanced planning between the instructor and student, with the goals, the scope of the project, and evaluation method specified in writing. Some of these opportunities are paid, while others involve the award of credit, and possibly both.

For students considering graduate school or a career in research and development, these opportunities can provide experience working in an environment which can be quite different from other engineering functions. Graduate school and research work involve an application of technical skills at a very theoretical level. Students should enjoy the challenges of advanced level problem solving which will require a very in-depth use of their engineering, math and science background.

To learn more about supervised independent study programs or research assistantships, talk with some of your professors whom you enjoy working with and with whom you share similar engineering interests.

## 12.5 Internships

What is an internship? Of all forms of experiential learning, some think that this is the most difficult to describe, as it can have different meanings, uses, and applications depending on a variety of circumstances. Generally, *internships* can be either paid or unpaid work experience that is arranged for a set period of time. This usually occurs during the summer so there is the least disruption to a student's academic schedule.

For most students and employers, an internship is a one-time arrangement with no obligations by either party for future employment. Objectives and practices will vary greatly depending on the employer and the student's class level in school. Some internships involve observing practicing engineers and professionals to see what working in the field is "really like." The intern's actual work often involves menial tasks designed to support the needs of the office. However, other internships can be structured as a "capstone experiences" which permits students to apply the principles and theory taught on campus to some highly technical and challenging real-world engineering problems.

At most engineering schools, internships are treated as an informal arrangement between the student and the employer. Therefore, this activity is usually under the general supervision of an experienced professional in the field in a job situation which places a high degree of responsibility on the student for success or achievement of desired outcomes. While many campus placement centers provide facilities and assist with arrangements for internship interviews, there generally is not a formal evaluation of the internship job description by any engineering faculty member. As a result, it is often difficult for students to understand how, and whether, this internship will be related to their engineering field of study. Since the typical engineering internship does not involve faculty input, there is usually not any formal evaluation or monitoring of the experience by any college officials. For these reasons, college credit is usually not awarded for these experiences.

*For most students and employers, an internship is a one-time arrangement with no obligations by either party for future employment.*

For many students, however, an internship can be a very valuable part of their college education. Under the best of circumstances, the internship will be structured by the employer so it will be directly related to the student's field of study. It is advantageous for the student to try to obtain, in advance, a description of the duties and responsibilities of the position. The student may wish to review this material with an academic advisor or faculty member to gain an opinion on the relevance to the student's selected field. These individuals may be able to make suggestions and comments on ways that the student can maximize the benefits of the experience. The most productive internships are obtained by those which involve upper-level under-graduates who have completed a significant amount of engineering course work. It is often easier for employers to match these students and their qualifications to specific projects and activities which will take advantage of the student's academic experiences. Students gain some practical real-world experience which will provide them with a better perspective of the engineering field that has been selected. For the employer, internships provide a mechanism to supplement their workforce in order to complete many short-term engineering projects. It also enables the employer to evaluate students in a variety of work-related situations. Many employers will use internship programs to screen candidates for possible full-time employment after graduation.

*Many employers will use internship programs to screen candidates for possible full-time employment after graduation.*

For students with restrictions related to time constraints, curriculum flexibility, scheduling difficulties, location or geographical preferences, the one-time internship program can serve as a valuable supplement to their engineering education. While internships do not provide the depth of experience found in other forms of experiential learning, they can help students gain an appreciation for the engineering profession and perhaps provide additional opportunities for full-time employment after graduation.

# 12.6 Cooperative Education

At most engineering schools, cooperative education programs are viewed and promoted as the preferred form of experiential learning.

> *"Engineering cooperative education (co-op) programs integrate theory and practice by combining academic study with work experiences related to the student's academic program. Employing organizations are invited to participate as partners in the learning process and should provide experiences that are an extension of and complement to, classroom learning."*

This background statement, endorsed by the Cooperative Education Division of the American Society of Engineering Education, describes the basic foundation of *cooperative education* and some of the critical components of a quality engineering co-op program. Perhaps the most important part of any cooperative education program is its linkage to a college's engineering programs. Co-op is considered to be an academic program, and in fact at many colleges of engineering it is administered centrally, by the college itself.

*As students mature academically and professionally they become stronger students and valuable employees.*

As an academic program, co-op is structured so student assignments are directly related to their major field of study. Typically, participating employers, who have been previously approved by the program administrators will forward detailed job descriptions, including salary information of available positions, to co-op faculty and staff. These descriptions are reviewed to evaluate their correlation to the academic content and objectives of their educational program. Employers also provide a set of qualifications for the candidates they are seeking appropriate matching of students with the positions they have available.

The integration of school and work is another important component of cooperative education programs. This is typically achieved by alternating several periods of increasingly complex co-op assignments with periods of study at school, almost always with the same employer. This program is called the *"alternating-term co-op"* and is the most common form of co-op scheduling. For example, a student might be at the XYZ Company during the fall semester, back at school for the spring semester, then return to XYZ for the summer semester, back to school for the fall, etc. This type of schedule allows students to apply classroom learning to real-world problems in an engineering setting. Upon returning to school, students are then able to relate the learning in the classroom to the experience gained on the job. After another semester of schoolwork, students return to the employers with an even greater knowledge of engineering and scientific principles, which can be used for even more complex assignments in the workplace. As students mature academically and professionally they become stronger students and valuable employees. It is truly a winning combination for the employers, the students and the schools.

Another form of cooperative education scheduling which is popular at some colleges and universities is called "parallel co-op." Under this arrangement, students usually enroll for a partial load of class work while engaging in co-op assignments for 20 to 25 hours per week. This type of co-op program allows students to continue to make progress toward their degrees while gaining valuable work experience. The disadvantage of this arrangement is that students may encounter difficulties if the demands of the co-op position begins to interfere with the time needed for school assignments, or if the time devoted to schoolwork begins to hamper the performance on the co-op assignments. Great care and judgment should be exercised when students and employers consider undertaking this co-op option.

Under another arrangement, co-op assignments are scheduled consecutively, during back-to-back semesters. This option is particularly useful for co-op projects that require a longer commitment than a single semester, or where the activity is seasonal in nature. For example, many civil engineering students are employed by highway construction firms for assignments that typically last for a summer and fall semester. A student involved in this type of co-op may work two consecutive semesters, followed by a semester or two in which they attend school and then a return to the employer for one or two additional co-op assignments. This pattern can provide many of the same benefits of the alternating plan if it is scheduled in a way which ensures that the academic course work is still a major component of the overall plan.

Some co-op schedules are actually a combination of these different patterns. Due to the sequential scheduling of a typical engineering curriculum, it can be difficult to adhere to a true alternating co-op program. Therefore, some students and employers will develop a series of assignments that may begin as an alternating plan for the first two or three rotations, but which may then require a back-to-back schedule or a parallel semester to complete the full schedule plan due to a student's particular academic schedule.

Cooperative education is rarely a summer-only program. Interrupting work for two semesters every year makes it difficult to maintain continuity in the integration of academics and the employment experience. In addition, most participating employers have needs which require them to have co-op students available for these positions throughout the year. In fact, most co-op employers plan their hiring so that while one student is away at school, another student is on-site working on a given co-op assignment.

In order for both an employer and a student to gain maximum benefit from the co-op experience, most employers and schools require that a student make a commitment for three to four semesters. This guarantees stability for the employers while ensuring the student at least a full year of job-related experience. Most engineering schools typically will recognize the student's participation in co-op after completion of the equivalent of one year of work experience by placing an official designation on the student's transcript and/or diploma.

## ...Advantages for Students

Cooperative education programs provide students with an advantage for employment or graduate school consideration. Many employers seek to fill permanent positions with students who have had significant work experience. These employers use the cooperative education program to evaluate candidates on the job. For many students this usually leads to an offer of permanent employment upon graduation. Students are under no obligation to accept full-time employment from their co-op employer, but studies have demonstrated that many will accept these offers. Students who choose to work for their co-op employer often begin with higher seniority and benefits, including larger starting salaries, increased vacation time, and other fringe benefits. Those students who elect to interview with other firms usually find that their

**Advantages of Cooperative Education Programs**

co-op experience places them far ahead of their peers, as they are sought more often for interviews and plant trips, and receive more offers, and earn higher starting salaries than their classmates.

Students who have participated in co-op programs usually comment on how they have been able to develop their technical skills. The nature of co-op assignments allows students to use the skills gained in the classroom and apply them to actual engineering problems. In the classroom, students' assignments and laboratory experiments are graded based on how close they come to achieving the correct answer. As they quickly learn on the job, the real-life assignments and projects often are not directed toward a specific answer, but a variety of potential solutions. This is where the student's technical background is developed and applied. Much of the actual technical learning occurs when students are forced to apply various principles to these unique situations. The learning that is achieved as part of the co-op experience provides students with new knowledge that often will make them better prepared for classroom challenges. The opportunity to "learn by doing" has been an important component of human development throughout time.

*Students often begin with higher seniority and benefits, including larger starting salaries, increased vacation time, and other fringe benefits.*

Many students who participate in cooperative education programs also find that the experiences can help solidify career decisions. Many engineering students are not fully aware of the spectrum of options and opportunities available to them. By engaging in the daily responsibilities of engineers, interacting with both the technical and non-technical individuals that work with engineers, and applying their academic background to their positions, students begin to gain a better understanding of the engineering profession. As a result, many find that their selection of a particular discipline has been confirmed, reinforced and even strengthened. Others find that they have been exposed to new fields and technologies of which they had not previously been aware. Some change majors to reflect their new interests, while others modify their course schedule to include new courses to incorporate some of the exciting new areas to which they have been exposed.

*Co-op experience provides students with new knowledge that often will make them better prepared for classroom challenges.*

All cooperative education assignments are, by definition, paid positions. Therefore, students find that the salaries they earn as co-op participants can greatly assist them in meeting the high costs of education. Recent studies have demonstrated that average salaries paid to engineering co-op students are in the neighborhood of $2,400 per month. For students who are working between four and seven months per year, co-ops can provide significant resources to meet their education expenses. It should be noted that average salaries can vary greatly depending on such factors as the employer, geographical location (pay is often higher in high-cost-of-living areas), the students' majors, class levels, and the amount of experience they have had.

## ...Advantages for Employers:

Many students often wonder why employers become involved in cooperative education programs. Some of the most common reasons given by participating employers are:

- The cost of recruiting co-op students averages sixteen times less than recruiting college graduates.
- Almost 50% of co-op students eventually accept permanent positions with their co-op employers. The retention of college graduates after five years of employment is 30% greater for co-op graduates.

- Typically, co-op students receive lower salaries and fewer fringe benefits than permanent employees. Total wages average 40% less for co-op students. In addition, employers are not required to pay unemployment compensation taxes on wages of co-op students if they are enrolled in a qualified program.
- The percentage of minority members hired is twice as high among co-op students as among other college graduates, thus assisting co-op employers in meeting EEO objectives.
- Co-op programs provide an opportunity to evaluate employees prior to offering them full-time employment.
- The co-op graduate's work performance is often superior to that of a college graduate without co-op experience. Co-op students are typically more flexible, and easily adapt to a professional environment.
- Regular staff members are freed up from more rudimentary aspects of their jobs to focus on more complex or profitable assignments.
- Co-op programs often supply students who have fresh ideas and approaches., and who bring state-of-the-art technical knowledge to their work assignments.
- Co-op graduates are capable of being promoted sooner (and farther) than other graduates.
- Co-op programs build positive relationships between businesses and schools, which in turn helps employers with their recruiting.

## ...Advantages for Schools:

There also are many benefits to educational institutions which offer cooperative education programs to their engineering students.

- Cooperative work experiences provide for an extension of classroom experience, thus integrating theory and practice.
- Cooperative education keeps faculty members better informed of current trends in business and industry.
- Co-op programs build positive relationships between schools and businesses, and provide faculty members with access to knowledgeable people working in their fields.
- Co-op programs enhance the institution's reputation and attract students interested in the co-op plan to their school.
- Cooperative education provides schools with additional business and industry training facilities that would be unaffordable otherwise.
- Cooperative education lowers placement costs for graduates.

Co-op students have found that many professional development skills so critical for success in today's workplace can be enhanced through their co-op experiences.

**More
Benefits of Co-op**

1) **Written and oral communication skills:** The ability to communicate ideas, both in written form and orally, is a skill that is critical. Most co-op assignments require students to document their findings and report them to others. This may take the form of an e-mail to a supervisor, a technical memo, a presentation to a group of engineers and technicians, or even a formal presentation to a group of directors or other executives. While communications and presentations may not be a favorite activity for many

engineering students, co-ops provide students the opportunity to improve and develop these critical skills. For those who find this communication a liability, cooperative education assignments can help turn it into an asset.

2) **Networking:** Most co-op students find the opportunity to develop new contacts and work with people from a variety of backgrounds a great asset to their short- and long-term career objectives. Co-op students come in contact with many individuals who are able to provide advice, lend support with projects, and help get things done more efficiently. These same people can recommend new assignments to co-op students, and maybe even write an important letter of recommendation.

3) **Self-discipline:** Most co-op students quickly discover that the transition from school to work requires that they develop an organizational style which may be superior to that used in college. They find that within the co-op structure they are given a great amount of responsibility and freedom to succeed (or fail) with their projects. Time management, punctuality, adequate preparation, organization, and specific protocol and procedures are learned on the job. As they acquire and develop these skills, they find that they are able to apply them back at school, which in turn makes them more productive and successful students.

4) **Interactions with a variety of people and groups:** Co-op students deal with a wide range of individuals as they complete their tasks. Typical individuals may include engineers, technicians, scientists, production workers, labor union representatives, clerical staff, managers, directors, and perhaps even CEOs. Each of these will contribute to the students' learning as they experience the successes, challenges, frustrations, and friendships that will evolve from these interactions. These individuals can help the co-op student evolve from a student to a successful engineer.

5) **Supervisory and management experience:** Many cooperative education assignments require that the student take responsibility for other individuals or groups in order to accomplish their tasks. These opportunities to manage the work of others can provide invaluable experience to students, especially those who aspire to eventual management or leadership positions.

**A Note of Caution**

As discussed above, cooperative education programs can provide many positive benefits to students, employers, and schools. However, a co-op program may not be the appropriate form of experiential education for every engineering student. There are many aspects to the co-op structure that may not be consistent with a particular student's overall goals and objectives.

The biggest drawback to co-op participation is the intermingling of work and study. During each of the semesters that students are off campus working on their co-op assignments, they are not making progress toward their degree requirements. This means that typical co-op students add an extra year to the duration of their academic programs. For many, this means that co-op participation will extend their engineering degree program to five or even six years. The overwhelming majority of co-op students, schools, and employers feel that this sacrifice is minor compared to the outstanding benefits received from this type of education. However, students and parents should carefully weigh this factor.

There are other factors which also should be considered. The continual relocation between school and the work is a minor inconvenience, but one which takes time and effort—especially if the two sites are a substantial distance apart.

The engineering curricula also can present a difficulty due to the nature of the sequential offering of certain classes and their prerequisites. Students should work out a long-term schedule with their academic adviser and co-op coordinator so that any class scheduling problems can be handled.

To find out more about engineering cooperative education programs at your school, contact your co-op office or placement center, or talk with your academic advisor or a faculty member. If your school doesn't have a co-op program, you may want to write or talk with a corporation of interest to you. You may be able to work out a co-op assignment with the help of an advisor, or faculty member, or the campus career center.

## 12.7  Which Is Best for You?

Many forms of experiential learning have been discussed in this chapter. Each of these opportunities has advantages and disadvantages associated with it. Only you can determine which program is best suited to your particular situation. Begin by asking yourself these types of questions:

- Is obtaining career-related work experience a high priority in my educational planning?
- Am I willing to sacrifice personal convenience to gain the best possible experience?
- Will I be flexible in considering all available work opportunities, or do I have special personal circumstances (class schedules, time constraints, geographic or location restrictions) which limit my choices?
- Do I have the drive and commitment necessary to succeed in a co-op program?

Your answers to such questions will be important as you move forward in your engineering career. Work with your academic advisor, a professor, or your campus placement office to get as much information as possible to make an informed choice in determining which is right for you. The resources are available to you—it is up to you to take advantage of them.

# Exercises and Activities

12.1  Develop a list of your skills and abilities. Determine which could be improved with work experience and which will be developed in the classroom.

12.2  Interview one of your professors who is involved with a research project. Discuss the advantages and disadvantages of a career in research.

12.3  Talk with other students who have done volunteer and community service projects. List the benefits they have gained which have helped them in their engineering classes.

12.4  Visit your engineering co-op office or campus career planning or placement center and read the recruiting brochures of some of the employers in which you have interest. Make a list of the types of qualifications these firms are seeking for full-time and co-op or intern positions.

12.5  Attend a co-op orientation session. List some of the advantages and disadvantages as they relate to your personal priorities.

12.6  Attend your campus career/job fair. Talk with at least three recruiters from different firms. Write a report comparing their views of cooperative education experiences, internships, and other forms of experiential learning.

12.7  Visit with some upper-class engineering students who have been co-ops or interns. Write an essay that compares and contrasts their experiences. Which students do you feel gained the most from their experience, and why?

12.8  Visit the campus placement center. Gather information concerning the starting salaries and benefits of those students who have participated in cooperative education compared to those who have done internships or some other form of experiential learning.

12.9  Visit the engineering co-op office and read some of the job descriptions of current openings. Select one of these descriptions and write an essay discussing the benefits you feel would be gained from this position.

12.10  Meet with your academic advisor and develop a long-range course schedule. Modify this plan so it shows how a co-op and an internship would affect this schedule. Write a short paper discussing the advantages and disadvantages of extending your education to include extra time for work experience.

12.11  Develop a list of your personal career priorities. How can this list be enhanced by some form of experiential learning?

12.12 Develop a list of your personal and educational abilities that you think need to be strengthened. Write an essay that discusses how work experience could improve these skills for you.

12.13 Interview an engineering professor or graduate student who had been a co-op student as an undergraduate. Discuss how the co-op experience influenced them to pursue an advanced degree in engineering.

12.14 Visit <http://epics.ecn.purdue.edu> and review the volunteer student team projects that have been developed at this university. Prepare a short paper that discusses the benefits gained by these students and how these experiences can be an advantage in academic studies.

12.15 Attend a campus career fair. Talk with at least three recruiters about their opinions concerning career-related work experience. Develop a report that discusses the reasons that employers prefer to hire students for permanent positions who have had significant work experience as part of their undergraduate education.

12.16 Talk with an advanced engineering student who changed their engineering field of study as a result of a co-op or intern work experience. What factors in their experience caused them to change their major? Do they view this work experience as a disadvantage or as an advantage to their overall career planning?

# Index